The International Library of Environmental, Agricultural and Food Ethics

Volume 24

Series editors

Michiel Korthals, Wageningen, The Netherlands
Paul B. Thompson, Michigan, USA

The ethics of food and agriculture is confronted with enormous challenges. Scientific developments in the food sciences promise to be dramatic; the concept of life sciences, that comprises the integral connection between the biological sciences, the medical sciences and the agricultural sciences, got a broad start with the genetic revolution. In the mean time, society, i.e., consumers, producers, farmers, policymakers, etc, raised lots of intriguing questions about the implications and presuppositions of this revolution, taking into account not only scientific developments, but societal as well. If so many things with respect to food and our food diet will change, will our food still be safe? Will it be produced under animal friendly conditions of husbandry and what will our definition of animal welfare be under these conditions? Will food production be sustainable and environmentally healthy? Will production consider the interest of the worst off and the small farmers? How will globalisation and liberalization of markets influence local and regional food production and consumption patterns? How will all these developments influence the rural areas and what values and policies are ethically sound? All these questions raise fundamental and broad ethical issues and require enormous ethical theorizing to be approached fruitfully. Ethical reflection on criteria of animal welfare, sustainability, liveability of the rural areas, biotechnology, policies and all the interconnections is inevitable.

Library of Environmental, Agricultural and Food Ethics contributes to a sound, pluralistic and argumentative food and agricultural ethics. It brings together the most important and relevant voices in the field; by providing a platform for theoretical and practical contributors with respect to research and education on all levels.

More information about this series at http://www.springer.com/series/6215

Ian Werkheiser · Zachary Piso
Editors

Food Justice in US and Global Contexts

Bringing Theory and Practice Together

 Springer

Editors
Ian Werkheiser
Department of Philosophy
University of Texas Rio Grande Valley
Edinburg, TX
USA

Zachary Piso
Department of Philosophy
Michigan State University
Michigan, MI
USA

ISSN 1570-3010   ISSN 2215-1737   (electronic)
The International Library of Environmental, Agricultural and Food Ethics
ISBN 978-3-319-57173-7   ISBN 978-3-319-57174-4   (eBook)
DOI 10.1007/978-3-319-57174-4

Library of Congress Control Number: 2017938125

Printed on acid-free paper

This Springer imprint is published by Springer Nature
The registered company is Springer International Publishing AG
The registered company address is: Gewerbestrasse 11, 6330 Cham, Switzerland

Acknowledgements

This volume is informed by many conversations over several years at the Workshop on Food Justice and Peace: Bringing Theory and Practice Together at Michigan State University. Thus, the editors wish to thank all the conference participants, whose comments improved the chapters and volume as a whole. We would also like to thank the other organizers of the conference, Samantha Noll, Julia Gibson, and Kenneth Kwabena Edusei; the farmers and food justice practitioners who hosted site visits for the workshop; the food providers who fed us all during the events; the many programs at Michigan State University who supported the workshops; and the Shalom Center for Peace and Justice who supported the first annual workshop in 2013. The editors would also like to thank Kyle Whyte and Paul Thompson, whose examples and guidance have informed our own approaches to engaged scholarship.

Contents

Contributors

Tawny Alvarado Michigan Public School Teacher, Michigan, USA

Melanie Bowman University of Minnesota, Minneapolis, USA

Avi Brisman Eastern Kentucky University, Richmond, USA

Marisela A. Chávez University of Kansas, Lawrence, KS, USA

Mallory Cochrane Our Table Cooperative, Sherwood, USA

Person Cole Michigan Public School Teacher, Michigan, USA

Elisabeth (Elle) Mari University of California, San Diego, USA

Shane Epting University of Nevada, Las Vegas, USA

Rebecca L. Farnum King's College London, London, England

Lisa Heldke Gustavus Adolphus College, St. Peter, USA

Emily A. Holmes Christian Brothers University, Memphis, USA

Matt Jason Michigan Public School Teacher, Michigan, USA

Elissa Johnson Syracuse University, Syracuse, USA

Lisa Oliver King Our Kitchen Table, Grand Rapids, MI, USA

Amber Leasure-Earnhardt Center for Agriculture and Food Systems, Vermont Law School, Royalton, VT, USA

David J. Leichter Department of Philosophy and Cultural Studies, Marian University, Fond du Lac, WI, USA

Seven Mattes Department of Anthropology, Michigan State University, East Lansing, MI, USA

Jonathan McConnell Philosophy and Literature Program, Purdue University, West Lafayette, IN, USA

R.W. Mittendorf Chaffey College, Rancho Cucamonga, CA, USA

Samantha E. Noll The School of Politics, Philosophy, and Public Affairs, Washington State University, Pullman, WA, USA

Christopher Peterson Loch Holland Farm, Saulsbury, TN, USA

Anne Portman University of Georgia, Athens, GA, USA

William D. Schanbacher University of South Florida, Tampa Bay, FL, USA

Nigel South University of Essex, Colchester, UK

Carrie A. Scrufari Center for Agriculture and Food Systems, Vermont Law School, Royalton, VT, USA

Sarah Riggs Stapleton Department of Education Studies, University of Oregon, Eugene, OR, USA

Maya Terro FoodBlessed, Beirut, Lebanon

Paul B. Thompson W. K. Kellogg Chair in Agricultural Food, and Community Ethics, Michigan State University, East Lansing, MI, USA

Laurie Thorp Michigan State University, East Lansing, MI, USA

Joseph A. Tuminello III Department of Philosophy and Religion, University of North Texas, Denton, TX, USA

Rebecca Valentine Center for Agriculture and Food Systems, Vermont Law School, Royalton, VT, USA

Melissa Washburn Michigan Public School Teacher, Michigan, USA

Justine MacKesson Williams Department of Anthropology, University of North Carolina at Chapel Hill, Chapel Hill, NC, USA

Introduction: Bringing Theory and Practice Together for Food Justice

Food Justice

Food justice is an increasingly common concern among activists, policymakers, academics, and the general public. Achieving more just food systems in terms of distribution, participation, recognition of people's identities and backgrounds, and a host of other concerns beside is a vexed, "wicked" problem (Whyte and Thompson 2012). It is made all the more difficult given that those goals must be pursued while also increasing the ability of those systems to feed the world. The overarching thesis of this book is that making more just food systems requires work that bridges theory and practice. The chapters in this book are offered partly as examples of what such engaged scholarship and theoretically informed practice can look like, and partly as attempts to wrestle with and work through recalcitrant problems that exist at these intersections. Thus, these chapters are signs of hope for the success of public philosophy and other forms of engaged scholarship to address wicked problems, but they are also signs that there is much work still to do.

One sign that food justice requires new approaches and engaged scholarship is the ways in which it presents unique challenges and questions (albeit ones with implications for and that overlap with other justice concerns). For one thing, food is something everyone engages with every day. When we eat we engage with food literally viscerally, connecting with our emotions, tastes, memories, ethical commitments, personal identities, family identities, relationships with our environment and society, and a host of other important parts of who we are. Moreover, almost all of us engage with food in other ways, as we produce, procure, prepare, and dispose of food in various ways as well. This intimate relationship we all have with food makes it an ideal boundary object (Star and Geisemer 1989), bringing together a wide range of discourses to the same table. As Gottlieb and Joshi argue, food justice "…resonates with many groups and can be invoked to expand the support base for bringing about community change *and* a different kind of food system." (Gottleib and Joshi 2010, p. 5, emphasis in original). It also brings up issues not germane to any other discourse, including the importance of taste to motivating justice, and the justice implications of eating and being eaten. Perhaps most importantly, we have very powerful feelings around food, a fact which can effectively motivate a justice movement but also lead to serious missteps. As Thompson (2015) argues:

> The emotional underpinning of food behavior may encourage mental leaps and even lapses of logic when food norms are threatened or questioned. Social psychologists have argued that food is a domain where "magical thinking" predominates. Metaphor becomes metonymy, and cognitive categories that may have their basis in social habits appear to take on deep metaphysical significance (p. 37).

This makes food both a powerful tool and a dangerous one. Only the cooperation among a wide variety of careful thinkers, grounded in real-world situations, is likely to build a successful movement for food justice. Encouraging this interplay and showing ways in which it is already happening is a central goal of this book.

Another sign that food justice is a suite of wicked problems which do not fit neatly into the pre-established boundaries between disciplines, or between scholarship and practice, is the difficulty in defining exactly what is meant by "food justice." As a result of this difficulty, many definitions among activists end up listing goals and values important to the group providing the definition. Consider:

> "Food justice is the right of communities everywhere to produce, process, distribute, access, and eat good food regardless of race, class, gender, ethnicity, citizenship, ability, religion, or community. [It] Includes: freedom from exploitation; ensures the rights of workers to fair labor practices; values-based: respect, empathy, pluralism, valuing knowledge; Racial Justice: dismantling of racism and white privilege; [and] gender equity" (Institute for Agriculture and Trade Policy 2012).

and

> "Food Justice is communities exercising their right to grow, sell, and eat healthy food. Healthy food is fresh, nutritious, affordable, culturally-appropriate, and grown locally with care for the well-being of the land, workers, and animals. People practicing food justice leads to a strong local food system, self-reliant communities, and a healthy environment" (Just Food).

The first definition highlights injustices based on identity, structural injustices, and exploitation, paying particular attention to race and gender injustices in the food system (among others). The second definition highlights potential injustices for non-human animals, the environment, and communities as well as individuals. Both definitions unite their lists of highlighted groups and injustices around some rights to food, but neither definition gives much reason to think their lists are exhaustive or that there is a rule for determining who should be on the lists. These evolving definitions suggest it is important for scholars to engage with current and emerging aspects of food justice on the ground, lest they miss emerging issues and aspects of the food justice movement. At the same time, scholars can hopefully put these emerging issues into a framework which provides some perspective on what might initially seem disparate, and which locates underlying causes and leverage points practitioners can use.[1]

Engaged Scholarship

The examples and discussions in this book of the interplay between theory and practice can be seen as part of the growing conversation on the importance and difficulty of engaged scholarship. For academics, engaging with stakeholders can be necessary to fully understand the complex systems in which our society is enmeshed. For activists, policymakers, and professionals, engaging with scholarly resources can be vital to effectively deal with the wicked problems our society faces.[2] An illustration of this growing recognition of the importance of engaged scholarship can be found in the new Public Philosophy Network, and its Journal. Christopher Long, one of the creators of the Public Philosophy Journal along with Mark Fisher, describes part of the mission of the Journal by saying "The PPJ is informed by a commitment to the idea that philosophically trained thinkers have an important role to play in addressing issues that concern all members of the general public" (Long 2013). At the same time, Long does not see this "important role" to be unidirectional education by scholars toward the general public. Rather, it is a

> Collaborative activity in which philosophers engage dialogically with activists, professionals, scientists, policy-makers, and affected parties whose work and lives are bound up with issues of public concern. Public philosophy is thus not limited to questions concerning the practical applicability of theoretical problems, rather it is informed by the recognition that all theoretical problems are ultimately rooted in questions of wide public interest (ibid.).

In such a model of engaged scholarship, it is clear that all parties involved can benefit. Similar conversations about the need for transdisciplinary collaboration are occurring in other disciplines as well (indeed, philosophy lags behind in many respects). To pick one example, various versions of participatory research have existed in social sciences for decades, and since at least the 1990s these approaches to research have taken seriously the expert knowledge and analytical expertise of local stakeholders (Chambers 1994; Kemmis and McTaggart 2000).

---

[1]Many of the authors in this volume offer such frameworks of food justice, with differing emphases and approaches. Thus as editors deciding what to include in this book, we took a wide view to allow for this fertile dissensus, and defined "food justice" as the discourses among various groups inside and outside academia concerned with those problems found in food systems which can be described as injustices rather than mere accidents, or even harms. This definition encompasses the issues alluded to in the above definitions. It also leaves the door open for emerging concerns that may become relevant to food justice without deciding a boundary in advance, other than that the boundary be explicitly focused on injustice.

[2]For a relatively early call for engaged scholarship of this kind, see (Boyer 1990).

At the same time as there have been more calls for engaged scholarship, there has also been a growing awareness that such projects are difficult to execute properly. It is all too easy for scholars to stay trapped in disciplinary frameworks, and those discourses may be insufficient to handle aspects of problems which have been traditionally underrepresented (Van de Ven 2007). Furthermore, the training or experience both theorists and practitioners have received often do not prepare them to engage with one another productively, and their home institutions may well not reward them for their efforts (Cantor 2006; Colbeck and Wharton-Michael 2006). This is perhaps particularly true in the US, where channels for academics to connect with the problems of the day (and do so in a way that supports their professional success) are noticeably underdeveloped. Negotiating the epistemic authority and power differentials can be a fraught process (Healy 2001), and it is not obvious that participatory research must necessarily increase social justice. Indeed, it may have the opposite effect on some communities. The potential of distortions as decisions are made about issues like who counts as a relevant stakeholder, or what counts as meaningful participation (García-López and Arizpe 2010) is high, and the consequences can be serious.

Food Justice Bringing Theory and Practice Together

Given the importance of engaged scholarship and the promise it holds for making more just food systems, but bearing in mind the potential problems and pitfalls, there is a need for models of successful projects in public scholarship and for frank discussions of recalcitrant problems. This volume presents many such examples and discussions, from a host of different disciplines and perspectives. The contributors to this volume were participants in a workshop which itself was one such model of public, engaged scholarship and bridge-building. For several years, Michigan State University hosted the Workshop On Food Justice and Peace: Bringing Theory and Practice Together. The workshop was intended as a transdisciplinary space to forge connections between theories and between theory and practice, with all papers written to be accessible to a public audience. Papers were informed by disciplines such as philosophy, anthropology, economics, gender and sexuality studies, geography, history, literary criticism, philosophy, sociology, and the human dimensions of agricultural and environmental sciences. Further, the Food Justice Workshops brought together leading scholars on food justice from across the country, in particular representatives of many food justice projects that were ongoing in Michigan. The participants visited local successful projects with those representatives in order to learn from them and have those experiences inform the conversations happening at the conference. The conference participants, both scholars and practitioners, brought considerable skill and poise to the thorny problems of bridging theory and practice, as is evident in their work for this book. The chapters in this volume are written by authors with a variety of backgrounds and experiences, but all are illustrations of engaged approaches to highly complex, wicked problems, in this case those problems arising in food justice.

One central commitment of this book is that cooperation and conversation between theorists and practitioners is necessary to the work of both. Hence we understand the chapters of the volume existing between theory and practice, in that liminal space where theorists grapple with concrete problems and practitioners reflect on tensions they encounter on the ground. For theorists, these problems reflect the experiences to which their theories are accountable–no matter how elegant a theory of justice might seem in the abstract, it must be able to respond to community objections that the theory is leaving out something crucial. For prac-titioners, food justice initiatives often confront challenging conceptual and theo-retical questions—for instance, balancing nutritional guidelines with recognition of a community's foodways may require reflection and deliberation on what really matters to us.

Our view is that these challenges between theory and practice arise commonly but unpredictably. Such challenges do not respect the boundaries of traditional academic disciplines, and will often require interdisciplinary approaches, which figure prominently in the chapters to come. Chapters draw on theories and methods from agricultural science, aesthetics, anthropology, critical race theory, religious studies, and others. Furthermore, these challenges often complicate the boundaries between the academy and the public, requiring trandisciplinary engagements that draw on innovative and participatory methodologies to create understandings that navigate diverse experiences. As was mentioned earlier, part of the project of this volume is to model this sort of boundary work, to create space to share ideas that do not otherwise find a natural home in the current landscape of scholarship (Klein 2015). Another part of this project is critical. It is far too easy for theorists to insulate themselves from the wicked problems that would require reconstruction of systematic theories, but if theories cannot account for recalcitrant experiences, we need better theories (Lake et al. 2015). Meanwhile, practitioners often face time and resource constraints that do not allow for reflection and deliberation, and food justice organizations must often dedicate themselves to specific pursuits while losing sight of the intersections between these pursuits and other food justice goals. Practice in these cases must integrate the values at stake across complex food systems, and learning how to integrate often takes the form of drawing from case studies of particular problems and issues. While no volume could provide a com-prehensive set of pertinent case studies (for example, most though certainly not all of the papers come from out of experiences in the US), the chapters to come cover a wide range of issues that speak to a diverse range of experiences. They investigate problems from the commodification of indigenous food knowledge to institutional barriers to gleaning to the ethics of hunting. In so doing they illustrate the wide-ranging applicability of public, engaged scholarship.

Four Contexts of Food Justice

The volume is organized into four sections, or contexts, that represent intersections between justice and particular dimensions of our food systems. The four contexts are food justice in a global context, food justice and the built environment, food justice and governance, and food justice and animal lives. The sections are themselves set up to encourage bridges between theory and practice. Each section opens with a vignette written by a practitioner followed by an introduction that provides a roadmap to the section's context. These introductions begin the work of bringing practice and theory together, connecting concrete cases like those featured in the vignette with the four chapters that follow. Each section's introduction is written by scholars with many successful experiences engaging in transdisciplinary projects. This is helpful both to situate the theoretical discourses in the rest of the section, and to show the ways in which engaging with reports from practitioners can ground theoretic discussions and deepen our understanding of the practical example.

Many of the traditional debates on food justice fit naturally into this cartography. Questions of food security versus food sovereignty, as well as local food versus global food trade, have been the focus of food justice scholars and practitioners grappling with food justice in an international context. How we deal with food deserts, and how we create urban food systems, figure into conversations of food justice and the built environment. Whether we should eat animals, and if so how we should raise them, has incited ethical theories and activism since long before the food justice movement. It is the section on food justice and governance that offers a new label for a constellation of conversation that arise between the theory and practice of food justice, but have so far evaded categorization. What ties the chapters of this section together is a common challenge—how can we radically reimagine and reconstruct our food systems against a background of social practices and institutions that had evolved alongside the previous regime of food production, distribution, and consumption? Here the uniqueness of food as a justice problem is clear—our lives revolve around food in very material ways, and changes in how we grow, harvest, and eat create tensions in the infrastructure of industrialized societies. These tensions should be taken seriously, and there may be times where we yield on our food justice projects because they require changes to our institutions and our governance that we cannot justify. Other times the import of food justice critiques is wider than initially appreciated; the seriousness of injustices in such cases does justify radical changes in social institutions seemingly far removed from food issues, such as intellectual property or classroom design.

Together, we believe these four sections provide four important orientations to the multidimensional justice challenges that confront food systems in theory and in practice. While we leave it to the section introductions to preface each chapter, it is worth noting that many of the chapters show how these orientations intersect in particular cases. A chapter on the human–whale bond in Japan speaks not only of food justice in an international context but also of food justice and animal lives; a chapter on teacher-led food justice projects speaks on food justice and the built environment as well as food justice and governance. The contextual nature of food justice projects requires that we learn to reason by analogy from particular cases, and the chapters in this volume grapple with an array of theoretical and practical challenges. These chapters provide models of how to orient ourselves to particular types of real-world challenges, to be able to anticipate obstacles and opportunities for better theory and practice, and to do justice to the lived experience of human and non-human actors throughout a system. While the works in this book are most relevant to audiences interested in food justice, it is hoped that these examples will inform, support, and inspire the liminal work of scholarly and policy-oriented readers dealing with any wicked problem.

In the conclusion to this volume, we reflect on the nature of and the process to produce this volume and the workshops that preceded it. As we discuss, it is possible and desirable to create food justice projects like these in the spaces between fields. Doing so requires conscientious work to build communities which can jointly address problems while also supporting individuals as they fulfill their responsibilities to the disciplinary communities they are coming from. This process can be difficult; these grounded projects have the opportunity to address wicked problems which otherwise can prove intractable. As a result, examples of such projects and discussions of success conditions are important in pushing food justice discourses forward.

Ian Werkheiser
Zachary Piso

References

Boyer, Ernest. 1990. *Scholarship reconsidered: Priorities of the professoriate*. Princeton, NJ: Carnegie Foundation for the Advancement of Teaching.

Cantor, Nancy. 2006. Taking Public Scholarship Seriously Office of the Chancellor. Paper 34. surface.syr.edu/chancellor/34.

Chambers, Robert. 1994. The origins and practice of participatory rural appraisal. *World Development* 4 (7): 953–969

Colbeck, Carol, and Patty Wharton-Michael. 2006. Individual and organizational influences on faculty members' engagement in public scholarship. In *A laboratory for public scholarship and democracy*, ed. R. Eberly and J. Cohen, 17–26. San Francisco: Jossey-Bass.

García-López, Gustavo A., and Arizpe, Nancy. 2010. "Participatory processes in the soy conflicts in Paraguay and Argentina" Ecological Economics 70 (2): 196–206.

Gottlieb, Robert, and Anupama Joshi. 2010. *Food Justice*. Cambridge, MA: MIT Press.

Healy, Karen. 2001. Participatory action research and social work: A critical appraisal. *International Social Work* 44 (1): 93–105

Institute for Agriculture and Trade Policy. 2012. Draft principles of food justice. iatp.org.

Just Food. n.d. What is food justice? justfood.org.

Kemmis, S., and McTaggart, R. 2000. Participatory action research. In *Handbook of qualitative research*, ed. N.K. Denzin and Y.S. Lincoln, 2nd ed., 567–607. Thousand Oaks, CA: Sage.

Klein, Julie Thompson. 2015. From multidisciplinarity to boundary work. In *Remapping performance: Common ground, uncommon partners*, ed. Jan Cohen-Cruz, 89–94. New York: Palgrave Macmillan.

Lake, Danielle L., Lisa Sisson, and Lara Jaskiewicz. 2015. Local food innovation in a world of wicked problems: The pitfalls and the potential. *Journal of Agriculture, Food Systems, and Community Development* 5 (3): 13–26.

Long, Christopher. 2013. What is public philosophy? cplong.org.

Star, Susan L., and James Geisemer. 1989. Institutional ecology, 'translations' and boundary objects: Amateurs and professionals on Berkeley's museum of vertebrate zoology. *Social Studies of Science.* 19: 387–420.

Thompson, Paul B. 2015. *From field to fork: Food ethics for everyone.* Oxford, UK: Oxford University Press.

Van de Ven, A.H. 2007. Engaged scholarship: A guide for organizational and social research. Oxford: Oxford Unversity Press.

Whyte, Kyle, and Paul Thompson. 2012. Ideas for how to take wicked problems seriously. *Journal of Agricultural and Environmental Ethics* 25 (4): 441–445.

Part I
Food Justice in a Global Context

Chapter 1
A Vignette from FoodBlessed

Rebecca L. Farnum and Maya Terro

Abstract In this vignette, Rebecca Farnum and Maya Terro discuss a local food waste project in a poverty stricken neighborhoods in Lebanon, which must take into account the Syrian refugee crisis as part of its mission.

Keywords Food justice · Food waste · Transnational movements · Refugees

There are any number of lenses through which food justice can be viewed: Self-determination and governance, workers' rights, environmental sustainability, health, animal rights, transportation. So many issues impact and are impacted by food systems. For Maya Terro, a Lebanese activist interested in corporate social responsibility, sustainability, and development, the lens is waste. 30% of all the food produced globally is wasted and ends up being thrown away each year. This food waste is equal to three times the amount of food needed to feed the 890 million persons suffering from malnutrition. In Lebanon alone, millions of tons of good food are thrown away every year.

Perplexed by the paradox of so many people going hungry surrounded by so much wasted food, Maya became a food activist. In 2012, she co-founded Food Blessed, a local hunger relief initiative that works with businesses and civil society to reduce the number of people going hungry in Lebanon through mindfulness about food use. FoodBlessed is a community-based and volunteer-driven nonprofit organization that links donors with food surplus get that excess to food insecure beneficiaries while simultaneously addressing and spreading awareness on food waste and food rescue. We give restaurants and catering agencies practical tips for cutting down on their food waste generation and individuals the opportunity to serve food to those in need through volunteering. In fewer than four years, we have

R.L. Farnum (✉)
King's College London, London, England
e-mail: rebecca.farnum@kcl.ac.uk

M. Terro
FoodBlessed, Beirut, Lebanon
e-mail: maya.terro@foodblessed.org

© Springer International Publishing AG 2017
I. Werkheiser and Z. Piso (eds.), *Food Justice in US and Global Contexts*,
The International Library of Environmental, Agricultural and Food Ethics 24,
DOI 10.1007/978-3-319-57174-4_1

3

successfully distributed more than 270,000 meals that were prepared by community cookouts, from surplus food rescued from events, and/or from surplus vegetables and fruits salvaged from dumpsters and local markets. Some 700 "Hunger Heroes" have found community and food education while volunteering with us.

FoodBlessed is fundamentally a local, community-based organization. We are bottom-up, volunteer-driven, self-funded. This is important to us: we believe that community-based organizations are more effective in addressing food justice than larger charitable organizations. Local groups understand on-the-ground nuances better. We insure sustainability by directly involving our donor and beneficiary communities at all stages of our work—planning, implementing, and evaluating programming.

And our mission is primarily local. We work in Beka'a and Akkar, neighborhoods with some of the highest poverty rates in Lebanon, to directly alleviate hunger. But we also know that our work—and the forces that make our work necessary—are global. We do not work in a vacuum. FoodBlessed will not succeed in its mission of zero food waste and zero food insecurity without global food justice. Local food justice is not possible in global systems of injustice and unsustainability. As local activists, then, we must engage with and address global issues.

In Lebanon, food justice has been fundamentally affected by global events. Most obviously, the impacts of the Syrian war have put additional pressures on the environmental and economic resources of the region. Lebanon continues to host the highest percentage of Syrian refugees in the world. Today, 1 in 5 people living in Lebanon is Syrian. Thousands of people are barely surviving with resources like jobs, food, and shelter growing increasingly scarce under pressure. Around 300,000 Lebanese citizens are unable to meet their most basic needs, including food. The poor are heavily concentrated among the unemployed and unskilled workers, especially in agriculture and construction sectors. And most of the country's growing population of 1.6 million Syrian refugees have little access to resources and few employment opportunities.

Lebanon's need for food—and food aid—is increasing. This is happening at the same time that global food prices are also on the rise. For us at FoodBlessed, this reinforces the need to minimize food waste and maximize the resources we have.

We are also global in our community. FoodBlessed is proud to engage in informal people-to-people diplomacy. Food has an amazing power to bring people together to a common table. At FoodBlessed, individuals from different countries, religions, ethnicities, and backgrounds come together. Eating, working, and talking together helps to reduce stereotypes and brings people together across cultural, national, and religious tensions. This is true between Syrian refugee and local Lebanese populations, across the many ethnic and religious divides of Lebanese societies, and via the organization's numerous international volunteers. International Hunger Heroes and non-Lebanese beneficiaries learn how to cook local dishes and are taught Lebanese gestures, greetings, and sayings. And, of course, this is a two-way street. International visitors have also shared their own favorite dishes from home with the FoodBlessed community and taught bits of their own language. Through this exchange, our work enhances intercultural understanding locally and internationally.

Right now, there is enough food in the world to make sure that every single person on the planet gets the food they need to lead a healthy life. As a global community, we are food blessed. Hunger is not an environmental reality. It is a societal choice. Social and political issues—poverty, gender inequality, and racism, to name a few—create barriers to food access and security. Around the world, food activists campaign for the human right to food and against large corporations threatening local food production and distribution systems. We work toward an understanding of hunger and nutrition that includes attention to health, culture, and sustainability, not merely basic calorie counts. We consider the impact of our eating on our neighbors, human and non-human alike. And we seek to understand how our local food issues intersect with wider global systems of access, power, and control.

Chapter 2
Introduction to Food Justice in a Global Context

William D. Schanbacher

Abstract This introduction argues that food justice concerns are always global in nature no matter how local they also are. William Shanbacher uses the lens of food sovereignty to understand how food justice can connect local causes into a global transnational movement while incorporating underemphasized issues like poverty, gender, and race.

Keywords Food justice · Food sovereignty · Globalization · Transnational movements · Refugees

2.1 Food Justice and Food Sovereignty

Right now, there is enough food in the world to make sure that every single person on the planet gets the food they need to lead a healthy life. As a global community, we are food blessed. Hunger is not an environmental reality. It is a societal choice. Social and political issues—poverty, gender inequality, and racism, to name a few —create barriers to food access and security. Around the world, food activists campaign for the human right to food and against large corporations threatening local food production and distribution systems. We work toward an understanding of hunger and nutrition that includes attention to health, culture, and sustainability, not merely basic calorie counts. We consider the impact of our eating on our neighbors, human and non-human alike. And we seek to understand how our local food issues intersect with wider global systems of access, power, and control.

Food justice is an ideal that invites us to think about health, culture, and sustainability, the tensions between global and local actions, and the impact of our food practices. Food justice theory and practice ask that we think as widely as possible,

W.D. Schanbacher (✉)
University of South Florida, Tampa Bay, FL, USA
e-mail: william48@usf.edu

© Springer International Publishing AG 2017
I. Werkheiser and Z. Piso (eds.), *Food Justice in US and Global Contexts*,
The International Library of Environmental, Agricultural and Food Ethics 24,
DOI 10.1007/978-3-319-57174-4_2

asking the questions that arise for activists like Maya Terro in the preceding vignette. When we reflect on the basic requirements of food justice, and speak about poverty, hunger and food justice, we must recognize the basic reality that if we feed each other, then we can dialogue with each other, and we can ask these important questions about food justice. Our food system is a failure. In fact, it has exacerbated all of the problems we face today. For instance, in the US food waste is estimated at between 30 and 40% of the food supply (USDA, n.d.). Globally, we waste our money on wars and politics, not food. This is not pontificating, this is a reality that we can fix, a reality that is at the core of food justice.

When we speak of food justice we need to think about it categorically differently. Perhaps we should think about it in terms of food sovereignty. Food sovereignty was defined in one of its original iterations:

> Food sovereignty is the right of peoples to healthy and culturally appropriate food produced through ecologically sound and sustainable methods, and their right to define their own food and agriculture systems. It puts the aspirations and needs of those who produce, distribute and consume food at the heart of food systems and policies rather than the demands of markets and corporations. (Declaration of Nyéléni 2007)

In the United States we also see the vision of food sovereignty taking hold. As the US Food Sovereignty Alliance states

> We support movement away from the dominant, corporate-controlled food system, which is shaped by systems of power and oppression. Our solutions must dismantle systemic food injustice rooted in race, class, and gender oppression. We respect people and other forms of life over profits. We honor everyone's work in the food system, including unpaid, underpaid, and devalued labor. We work to honor our human commonalities and restore traditional ways of growing, preparing, sharing, and eating food as a community (US Food Policy Alliance 2016).

In these two visions of food sovereignty we see connections being made between the US and global understandings of the term. As the introductory vignette of this section of the volume points out, there are many dimensions that need to be discussed. Perhaps we will move forward with these chapters and think of these questions. My research is in a sense an advocacy for food sovereignty, but it also opens up questions to multiple perspectives. As an ongoing movement, even my perspectives change. In these two definitions, one international, one specific to food sovereignty in the United States, we see that the term food sovereignty itself is a contested and dynamic term. As Annette Desmarais notes, food sovereignty is "an 'idea,' 'concept,' and 'framework'"…a 'mobilizing tactic,' a 'political project,' a 'campaign,' and a 'movement' (Desmarais 2014).

Food justice as defined through the lens of food sovereignty takes on a new tone. When seen as food sovereignty, food justice is a dynamic term that cannot be completely encapsulated by words and it cannot be a rigid term as "justice" historically has been understood. In my research with food sovereignty, and in my conversations with advocates, activists, and scholars, it is clear that we may not all agree on what justice requires. *But that might be the point, and the challenge.* Food sovereignty has become a concept and a movement that has transcended many disciplines and perspectives. Yet, as those who have worked with food sovereignty know, this is part of the challenge. We debate, we discuss, and we disagree. Within the movement, there are often conflicting interests, from peasant farmers to small-scale farmers to larger-scale farmers, who sympathize with the perspectives and arguments of food sovereignty.

For now, I think the most pressing issues include: How can race be more incorporated into the conversation? How can gender issues be included? How does the economic system fit into all of this? These are challenging questions to answer, but need to be included, and many of the chapters in this volume offer new and important potential answers. The perspectives in these chapters bring to bear experiences from the United States and abroad that must be a part of the conversation. Although the US has a vibrant and growing food sovereignty movement, food sovereignty is a global concept and movement, which brings it more in-line with traditional philosophical and legal notions of the concept of justice. Food sovereignty urges us to think on a global level even through our local food and agricultural decisions and policies.

Too frequently, food justice efforts become isolated. It is easy to become so discouraged by the injustice of large corporations and global power structures that we disengage. But the work we do as food justice activists is both inherently local and inextricably global. Food justice work is *always* within a global context, whether we explicitly acknowledge it or not. The refugee crisis, climate change, and hunger are major problems for FoodBlessed, for Lebanon, and for the world. Global forces make our work both harder and more important. But they can also provide solutions and tools for action, as we learn from each other, join in solidarity, and share resources. Together, we can tackle local issues of food injustice for the global advancement of universal justice. This volume hopes to contribute to this conversation. The chapters provide expertise in a diverse milieu of subjects and provide us will more resources for our continued effort.

David Leichter's chapter, "Remembering Food, Doing Justice: From Edible Memory to the Culinary Imaginary," introduces the ways in which memory and the participation of eating others' foods can function as an act of "mnemonic resistance" to (neoliberal) capitalist forms of domination and oppression. The act of eating others' foods can be an act of solidarity in which the "consumer" experiences, albeit indirectly, the history and identity of the Other. One of the inherent characteristics of capitalist modes of production and consumption is its tendency to make us forget our pasts. Food becomes simply a commodity necessary for day-to-day life; it loses its history and its cultural value.

Leichter attends to the ways that eating others' foods we might gain a deeper recognition of how food has shaped our own identity, but also ways in which our consumption patterns shape and re-shape the identity of others. Herein lies the risk of participating in the consumption of others' foods. We want to join in solidarity with others by sharing in their cultural and culinary past, i.e. become part of that past, but also avoid (or at the very least recognize ways we might be) appropriating or "colonizing" that past. We can share in this past, but we must not dominate it.

While recognizing the omnipresent risk of appropriating others' identities in the process of sharing food and food histories, Leichter's essay compels us to join in solidarity by learning about others' pasts, their cultures, their traditions, customs, experiences and histories. This form of knowledge formation is quite different from a capitalist form of knowledge in which knowledge formation inherently involves a form of forgetting our past.

Justine Williams' chapter, "Building Community Capacity for Food and Agricultural Justice: Lessons from the Cuban Permaculture Movement," provides an entry into the world of permaculture via a case study of the Fundación Antonio Núñez Jiménez de la Naturaleza y el Hombre (FANJ) in Cuba. Using food sovereignty and food justice as organizing principles, Willliams argues that FANJ, and permaculture in general, is an example of a positive alternative agricultural movement that can been bring people together and "bring new 'worlds' into material existence." Through emphasis on community autonomy and control over agriculture, organizations like FANJ work toward bringing about a healthier and more sustainable global food system. Focusing on cooperation, energy efficiency, resource management, and respect for natural ecological cycles, among other things, permaculture builds upon many of the existing practices of food sovereignty, but also, as one activist puts it, provides a "life philosophy."

Williams couples the concepts of "community capacity" and "skillful disclosure" as alternative frames through which to build more substantive food sovereignty and food justice movements. She responds to criticisms by food sovereignty organizations that domestic policies may at least occasionally serve to limit food sovereignty/justice goals. She does this by coupling the concept of "community capacity" (a community's commitment, resources, and skills to address local problems and opportunities) with the concept of "skillful disclosure" (new spaces, or "new worlds" created when people gather together to discuss policies and practices that can build stronger communities). Williams argues that with community capacities and skillful disclosures, local organizations have the potential to find ways to address any state-level limiting factors on pursuing food sovereignty and food justice.

Marisela Chavez's chapter, "It's Not Just About Us: Food as a Mechanism for Environmental and Social Justice in Mato Grosso, Brazil" provides a fascinating ethnographic analysis of an organization (The Landless Workers Movement, or MST) that has embraced food sovereignty as a natural extension of, and a means to, achieving full food justice. Chavez's work in Mato Grosso nicely illustrates the potential of food sovereignty's focus on food producers and food production as central to the attainment of social justice. Concrete steps are taken that actually produce food along more sustainable, and more economically sound in the long-term, agroecological lines. This food concretely improves the food sovereignty in the area, and also provides a praxis for activists in MST to engage in outreach (while selling their wares), advance political development through conversation with their fellow farmers, and pass these values and knowledge to future generations more successfully.

Chavez says something quite important when she says that different people she worked with and interviewed had different emphases and concepts of exactly what food sovereignty stood for—primarily for environmentally sound and sustainable food production, or primarily as a vehicle to social justice, or primarily as a way to support food producers' individual autonomy. As Chavez rightly points out, these are real differences but for all of them: "It's about justice—for people and the environment. They believe that looking at agriculture through a different lens helps connect people to the understanding that it's not just about people respecting the environment, but also about people having a different relationship with each other." This is the promise of food justice in general and food sovereignty in particular—to bring together people working in different places on different particular injustices primarily to build solidarity networks of aiding one another.

Seven Mattes' chapter, "Save the Whale? Ecological Memory and the Human-Whale Bond in Japan's Small Coastal Villages," presents an example of the larger issue of the ways in which culture, traditions, history, and geography shape our responses to questions of food justice. On the one hand, environmentalists have appropriated the whale as a sort of "poster-species" symbolizing the devastation of nature and animal life. The whale's beauty, grandeur and peaceful demeanor make it an ideal animal to evoke emotions and provoke protest against its hunting. On the other hand, nations such as Japan have highlighted that the global anti-whaling movement is based less on scientific evidence and concern about the future of the whale and more on the "ideological"/emotional sensibilities of anti-whaling campaigns.

Mattes argues that many "western" perspectives fail to recognize the cultural, religious, and historical value of the whale and whale-hunting for these Japanese fishing communities. The success of environmental anti-whaling campaigns comes at a cost to local communities that depend on the whale for sustenance and regard it as a symbol of their culture. In extremis, the success of anti-whaling campaigns could result in the loss of culture(s) tied to the whale. Mattes' chapter raises important questions pertaining to possible conflicts of interest among food justice/sovereignty organizations who take seriously less obvious but still very real harms, such as harms to oceanic ecosystems, harms to non-human animals, and harms to communities and cultures as a whole.

As one reads through these chapters, one cannot help but notice many themes that resonate with the suggestion that we look at food justice in a global context *as* food sovereignty—the idea that food sovereignty activists make justice demands which must be addressed by the wider society, and also that if food justice is to be achieved, food sovereignty must be at the center of any discussion of what a just food system must look like or how to get there. This volume, and these chapters in particular, will provide an important context for this debate around the role of food sovereignty, as well as new tools for working out an answer to that debate.

References

Declaration of Nyéléni. 2007. Accessed September 9, 2016. https://nyeleni.org/spip.php?
    article290.
Desmarais, Annette Aurélie. 2014. Food sovereignty: Some initial thoughts and questions for
    research. In *The global food system: Issues and solutions*, ed. William D. Schanbacher, 1–12.
    Santa Barbara: Praeger.
U.S. Food Sovereignty Alliance. Vision and operating principles. U.S. Food Sovereignty Alliance.
    Accessed September 9, 2016. http://usfoodsovereigntyalliance.org/visions-and-operating-
    principles/.

Chapter 3
Edible Justice: Between Food Justice and the Culinary Imaginary

David J. Leichter

Abstract This paper examines how eating illuminates the operations of memory, on one hand, and how practices of memory sheds light on the ways that our material food practices are imbued with meaning. In order to do so, I develop some ideas from Jennifer Jordan's recent book *Edible Memory*. In that work, she identifies the ways that eating food is a deeply personal experience, but one that shapes and is shaped by our social and material world. Edible memory, in other words, describes processes through which historical meaning is transmitted and how it is received and incorporated into one's life. While developing Jordan's analysis of edible memory, I also want to trouble it by suggesting another dimension of memory that remains under-theorized in her account. In particular, while edible memory can be seen in a variety of mnemonic practices surrounding food, Jordan's account underplays the relationality of memory and how memory operates to co-constitute identities. As such, her account needs to be further supplemented with a description of the ways that forgetting and appropriating culinary traditions can marginalize others' identities. By providing an account of this, I show how edible memory is implicated in, and framed by, relations of power, and how it can nevertheless operate as a site of mnemonic resistance that can challenge oppressive culinary practices.

Keywords Memory · Food · Heirlooms · Justice

Accounts about the relationship between food, identity, and memory appear with increasing regularity these days. *The Washington Post*, for example, recently published a story about reviving the cuisine of Appalachia. While noting common misperceptions about their heritage, the region is a rich, vibrant foodway that combines a number of different culinary traditions, cooking techniques, and ingredients. Not coincidentally, the rediscovery of Appalachian foodways has come

D.J. Leichter (✉)
Department of Philosophy and Cultural Studies, Marian University, Fond du Lac, WI, USA
e-mail: djleichter24@marianuniversity.edu

© Springer International Publishing AG 2017   13
I. Werkheiser and Z. Piso (eds.), *Food Justice in US and Global Contexts*,
The International Library of Environmental, Agricultural and Food Ethics 24,
DOI 10.1007/978-3-319-57174-4_3

with its own commodification—this time in the form of cookbooks, restaurants, and write-ups in newspapers of note (Black 2016). In a different context, the chef, restaurateur, and author Dan Barber recounted eating matzo at a Passover Seder, and expressed no small surprise when it tasted *good*. Returning to the farm where the grain was produced, he concluded that maintaining vigilance over the grains for these matzot, improved farming for the rest of the year (Barber 2016). Finally, the popular NPR podcast "The Sporkful" aired a series of episodes called "Other People's Food," which explored whether it is culturally appropriate for a chef to become renown for cooking a cuisine that was not ethnically his or her own. While Rick Bayless, the Oklahoman chef who has made his name preparing traditional Mexican cuisine sees little problem with his adoption of these foods, others see it as another instance of Western cultural appropriation (The Sporkful 2016).

These anecdotes—and many others like them—offer an opportunity to explore how food, identity, and memory can mutually reinforce one another and how they can be used to destabilize or resist oppressive or unjust social structures. In so doing, these stories illustrate the need for a deeper understanding of how memories are formed, preserved, and transmitted over time, on the one hand, and the ways that sharing memory between people occurs in time, in which the significance of the past, present, and future is actively constructed and negotiated. By identifying the mnemonic practices involved in eating, we can see how identity can be both formed and distorted in culturally specific contexts.

In what follows, I examine the role that remembering and forgetting food plays in forming identity. The intimate connection between food practices and identity, I argue, has significant implications for working out issues related to politics and memory. In particular, I examine how eating illuminates the operations of memory, on one hand, and how practices of memory sheds light on the ways that our material food practices are imbued with meaning. In order to do so, I develop some ideas from Jennifer Jordan's recent book *Edible Memory*. In that work, she identifies the ways that eating food is a deeply personal experience, but one that shapes and is shaped by our social and material world. Edible memory, in other words, describes processes through which historical meaning is transmitted and how it is received and incorporated into one's life.

While sympathetic to Jordan's account of edible memory, I also want to trouble it by suggesting another dimension of memory that remains under-theorized in her account. In particular, while edible memory can be seen in a variety of mnemonic practices surrounding food, Jordan's account underplays the relationality of memory and how memory operates in the co-constitution of identities. Memory can be a way to renew bonds of affiliation or challenge current configurations of social life. As such, her account needs to be further supplemented with a description of the ways that forgetting and appropriating culinary traditions can marginalize others' identities. By providing an account of this, I show how edible memory is implicated in, and framed by, relations of power, and how it can nevertheless operate as a site of mnemonic resistance that can challenge and resist dominant or oppressive culinary practices.

3.1 What Is Edible Memory?

While the connection between food and identity has been long documented, the role that remembering and forgetting food play in constituting identity has only recently become the subject of analysis. Jordan's recognition that memory is a central experience in many people's relationship to food offers a needed corrective to the often-underappreciated social dimensions of memory. While memory is often understood as an individual, cognitive act, Jordan's account of edible memory reveals the ways that eating is what might be called a "mnemonic practice"—a social framework or practice that preserves, transmits, and alters the ways that we make sense of the past (Olick and Robbins 1998). For Jordan, the term "edible memory" designates both "the experience of eating…food that carries a cultural story along with its genetic code," and also the deeply emotional and personal meanings that we attach to food in personal and familial settings (Jordan 2015, 35). However, edible memory does not just evoke one's personal memories about the past, but it also can motivate people to act—to buy certain foods, plant vegetables, tell stories, and eat together. Edible memory thus refers to the ways that we infuse food with a connection to the past that is both deeply personal but also always already social.

While edible memory covers a range of foods and experiences, Jordan identifies heirloom foods as a particularly illuminating example because they exemplify how the past is bound up with what we eat. Heirloom foods, such as apples, melons, carrots, beans, and, most dramatically, tomatoes, illustrate the ways memory can be used to mobilize a particular set of practices, habits, and attitudes about the past that can be used to transform current eating habits. A fruit or vegetable becomes an heirloom when it meets two conditions: (1) when it is passed down from one generation to the next and (2) when there is a break with the past (Jordan 2015, 23). An heirloom is, of course, something inherited, and foods have always been an heirloom insofar as food is shared among family members, techniques for growing and preserving foods are pass from generation to generation, and tastes for particular foods are cultivated with others (Jordan 2015, 22). However, heirloom foods are not saved or passed down in the same ways that other heirlooms are. While other heirlooms more or less remain the same object when passed down, fruits and vegetables do not have that luxury. Food rots, of course, and the heirloom tomato or antique apple that was picked and eaten is obviously not the one that growers preserve for the future. Rather, the genetic code of the seed, along with the cultivation of taste and the specific practices of growing and harvesting are preserved when the heirloom is passed down (Jordan 2015, 23–24). As a result, preserving an heirloom paradoxically requires it to be consumed.

Nevertheless, the ways that foods become heirlooms requires that they are no longer consumed in the same way as other foods, or even be consumed in the same ways as when they were in the past before they were designated as heirlooms. As Jordan puts it, "the concept of an heirloom becomes possible only in the context of the loss of actual heirloom varieties, of increased urbanization and industrialization as fewer people grow their own food" (22). A tomato, for example, becomes an heirloom when there is a temporal distance between the time when the specific varietal was considered an everyday object and the present context in which that varietal is less common. Because they are unlike the vegetables found at the grocery store, heirlooms evoke a distant past. In other words, a cultural transformation has resulted from a technological and agricultural break that puts temporal and practical distance between present generations and past foods. This transformation is important precisely because underlying it is a transformation of our relationship to the world and the land.

The attempt to revitalize the foodways of Appalachia neatly captures these elements of edible memory. Local chefs, farmers, seed savers, food activists, and gardeners in this region are returning to previously neglected and forgotten foods— such as creasy greens, goosefoot, and many heirloom beans—in order to foster the recognition and protection of the region as one of the most biodiverse places in North America. Farmers and home gardeners, for example, have cultivated 1412 distinctly named heirloom foods in the region, including more than 350 varieties of apples, 464 peas, and thirty-one different kinds of corn (Veteto 2011). The mountains, the short growing season, the immigration into the region, the Native American resistance to that immigration, and the history of coalmining have contributed to the varieties of foods grown and have influenced what has grown in the region and what people have traditionally eaten. Cultivating and consuming heirlooms shapes the identities of the individuals in the region by establishing a relationship to a shared past. Gardening, cooking, and preserving often allows people to imagine themselves as part of a culinary tradition, which in turn enables them to envision a future where people can still eat these unique foods. The seed savers who plant, cultivate, and tend their gardens, for example, often see themselves as "stewards" and preservers of both their own family's gardens and memories and shared stories and genetic codes in the plants themselves (Jordan 2015, 17).

3.2 Forgetting Foods

To see how edible memory can be a tool for resistance, we need to first address it attempts to resist and what it aims to recover. Dan Barber's visit to the fields where the spelt for Passover matzo is grown shows at that farming practices and technologies have transformed our relationship to the food and to the past, and that better food can come by returning to older practices of farming. If agricultural transformations change what we eat, then the reincorporation and modification of older farming practices for the present, and the foods that it produces, not only retrieves the flavors of the past but can also renew bonds between members of communities.

The transformation of the United States farming practices from privately owned farms to larger, corporate farming in the twentieth-century has been well documented (Pollan 2008). At the beginning of the 1900s, agriculture was a labor-intensive job that took place on a large number of small, diversified farms in rural areas where a majority of the population lived. In the 1930s and 1940s, agriculture began to transform from a wide array crops based on practices of open pollination toward a narrower range of hybrid crops. This turn occurred alongside a substantial increase in the scale of farming, the widespread application of synthetic fertilizers, and the growth of agribusinesses. Currently, the agricultural sector is concentrated on a small number of specialized farms where less than a quarter of the United States population lives (Dimitri et al. 2005). As a result of this trans-formation, United States agriculture has become increasingly efficient and has contributed to the overall growth of the United States economy.

However, the efficiency of these large farms comes at a steep price. Many plants —everything from varieties of wheat and spelt to varieties of tomatoes and apples —do not carry genetic traits that are well suited to large-scale and industrial agri-culture. Rather, the process of transforming our food into something that could be produced in mass quantities requires intervention at the genetic level so that the food can be transportable, durable, and uniform (Jordan 2015, 19). Large-scale, industrialized, and standardized techniques of farming supplants the older, local, and traditional varieties of fruits and vegetables, which in turn produces a break between current generations and past foods that fundamentally changes the ways that people and groups consume, remember, and forget their food. The disruption between the past traditions and current farming practices is, at least in part, the result of new techniques of farming and a reconfiguration of the economy. The transformation of food at the genetic level, in other words, transforms our rela-tionship with the land.

As food production became standardized, industrialized, and large-scale, the older, more localized, traditional varieties of plants and animals fell out of everyday use. Jordan notes, for example, that in West Germany from 1957 to 1974, 14,000 hectacres of fruit orchards were cleared to make way for new roads, power lines, and large-scale housing developments (108). Similarly, according to some figures, sixty percent of England's orchards have disappeared since the 1950s (Morris 2009). Closer to home, in 1872 there were more than 1100 different kinds of apples that originated in the United States; currently one apple variety, the Red Delicious, comprises forty-one percent of the entire American apple crop, and eleven varieties produce ninety percent of all apples sold in chain grocery stores (Nabhan 2010). The loss of these varieties is also a loss of traditional knowledge about how to grow, graft, select, and cook apples and which apples are suited for particular kinds of recipes.

The loss of biodiversity is one form of forgetting. Importantly, this form of forgetting is not a byproduct of inactivity; it is not a mere failure to remember. Rather, the orchestration and construction of a particular set of economic policies, as well as the dissemination of social norms designed to facilitate thinking and acting in terms of costs, benefits, and consequences, effectively becomes a kind implicit referendum on what to remember and what to forget. That is, the ways that these places and the food that grows there—or is no longer grown there—contribute to our identity is a continual process of formation and reformation. The transformation of landscape transforms individuals' relationship to it, and so the erasure and loss of foodways through the implementation of certain economic policies is not just a transformation of the landscape; it is an erasure of possibilities, varieties, and insurances against future failure and an erasure of cultural heritages, identities, and people's autonomy.

If biodiversity is not integrated into the culinary lives of large numbers of people, there will be fewer opportunities to reminisce about other foods. One farmer Jordan interviewed remarked, "people don't reminisce about great red Batavian lettuce they ate when they were kids they way they recall tantalizing tomatoes and mouth-watering melons" (Jordan 2015, 121–122). The lack of emotional resonance and narratives recounted about foods such as beets, turnips, carrots, or celery means that fewer heirloom varietals of these foods will be grown. The lack of stories behind these foods, even as some remain labeled as 'heirlooms,' prevents them from standing out in the culinary and cultural landscape. In fact, as Jordan notes, the designation of heirloom often functions to pick out a much narrower range of food than fits the standard definition—varieties that are open pollinated, developed before World War II, and possess some kind of meaning and story—and often are used to target relatively affluent, white audiences (Jordan 2015, 128). The concept of heirloom produce, in short, has specific historical, geographical boundaries.

If Dan Barber's visit to the farms where the spelt for matzo is grown indicates how returning to less efficient and older practices of farming, it does so in order to maintain a particular cultural identity and set of culinary practices. More than this, however, his story indicates that there are two further points to draw from the recognition of the transformation of the places where our food is grown. The first is that the places where food is grown and produced is constructed out of social relations, which include trading connections, political and economic policy, thoughts of home, and unequal relationships of power (Massey 1995). That is, the identity of the place is not "internal" to it, as though there were specific characteristics, such as the *terroir* or the unique taste of food from a specific location, that render the place unique. Rather, the "uniqueness" or identity of a place arises out of its relationship to other places. The second point is that there is an experience of disruption between the past of these places and their present or future (Massey 1995). The break between the present and the past, in other words, often motivates an artificial construction of the way the place "actually existed" in the past. In this case, it is that the past revealed in the experience of edible memory connects us to the *real* character of the past. While the capitalist system of food production aims at uniformity, the experience of the heirloom is often a self-conscious attempt to maintain a particular vision of the past. However, before addressing the limitations of the heirloom, it will be necessary to see how edible memory in general operates to transmit a past.

3.3 From Edible Memory to the Culinary Imaginary

Because food materially embodies the past and makes older meanings reverberate in the present, the role that it plays in transmitting memory cannot be understated. By eating, we are "exposed in very visceral ways to 'memories' that are not [our] own, [and] we come to be familiar with them and to feel [we] have some sense of the experience [ourselves]." In other words, eating food allows us to come to listen to and respond to a story about someone else's past, a past that is not our own (Jordan 2015, 36–37). Eating, in other words, allows us to consume other's memories. As a result, food operates at the intersection between two temporal planes: it transmits meaning *over* time from one generation to the next, and it situates meaning *in* time, with others who share the present with us.

A culinary tradition is not a static system of shared rules or essential traits that is passed down unchanged rather, as with any tradition, as Paul Ricoeur argues, any tradition is a process of structured improvisation, the interplay of both sedimentation of traditions and innovation (Ricoeur 1988). This means that a culinary tradition also consists in the interplay of new creative interpretations of food, which themselves were once creative interpretations. There is no culinary tradition or edible memory, in other words, without creativity because this tradition is a record of past innovations with food. Similarly, there is no creativity without memory, because such creativity takes place in the context of traditions. To continue to transmit an effective meaning, a culinary tradition must transform existing categories of food; it cannot exist outside of those categories. This suggests that edible memory is not a simple recovery of lost or forgotten foodways, but is instead needs to be understood as a site of a creative response to the past within the terms of the needs of the present.

The interplay between the sedimentation of traditions and the creative innovation that reconfigures these traditions suggests that edible memory is not a static phenomenon. To make sense of the interplay between sedimentation and innovation through which a culinary tradition is transmitted, edible memory needs to be supplemented further by two important ideas: Pierre Bourdieu's conception of the *habitus*, on the one hand, and Benedict Anderson's conception of an "imagined community," on the other. The *habitus*, for Bourdieu, is the bodily site where meaning is transmitted. As the site of the non-cognitive and non-verbal language of the body, the *habitus* refers to "systems of durable, transposable dispositions," trained capacities, and patterned behaviors of thinking that can be both passed on to others but nevertheless remain a remarkably coherent paradigm for conveying attitudes and values (Bourdieu 1989, 72). These dispositions are not necessarily the product of formal training, but are instead the result of social interactions. For example, my desire for a particular kind of coffee in the morning is the result of a number of different elements, including my memories of sipping coffee in the morning at the breakfast table with my parents, my desire to present myself as having an appreciation of its taste, my appropriation of social and cultural meanings of coffee drinking, as well as my chemical addiction to caffeine. The *habitus*, in other words, predisposes me to generate new forms of action that reflect the ways that I have been socialized to drink coffee. Broadly speaking, then, the internalization of these dispositions and the integration of experiences of previous generations into our own way of being in the world forms our tastes and expectations.

If the *habitus* identifies the site where broader social norms of taste become deposited and sedimented in individuals, Benedict Anderson's conception of the "imagined community" extends memory beyond the individual and even the immediate community. It is a "deep, horizontal comradeship" between individuals who nevertheless do not know each other. This vision of a shared group is "imaged" and defined not by its reference to any actually existing group or aggregate of people, but instead by the style in which it is imagined (Anderson 1991, 6). To return to the example above, drinking fair-trade, single-origin coffee at the coffee shop where I write implicates me in an imagined community of socially conscious, middle class academics. Imagined communities are thus constructed not only through the actual bonds that tie people together but also through a common set of ideals that establish affiliative bonds, which are those bonds between people who may not ever know one another.

Affiliative relationships are constituted in and through the public domain of symbols, images, and narratives. Eating an apple, for example, not only expresses one's taste for fruit, but because the apple also conveys a meaning of being American—such as "mom's apple pie" and Johnny Appleseed mythically planting orchards throughout the Midwest—it establishes culinary bonds between people and creates a regional cuisine that form preferences for certain tastes and foods. Eating, then, conveys a sense of one's tastes, both in the sense of one's preferences and one's ability to distinguish oneself culturally, and it helps to set out the conditions in which such inheritances can be brought into productive tension with others. Our embeddedness in our families' food habits and practices operates as a site where nonverbal and noncognitive acts of meaning-transferal occur, such that the experience of previous generations' is integrated into the present.

Because edible memory is the process by which certain foods are invested with memory, it also affirms particular values associated with that meaning. The practices and mnemonic modes through which people recall the heritage, provenance, and *terroir* of heirlooms is one way that people are situated differentially and in relation to one another. The category of heirloom food arises because it is situated against a group of food that is designated as being "conventionally grown." This suggests that that the ways that people remember food and impart meaning to them situates them vis-à-vis other ways of remembering food. In other words, the significance of heirlooms arises because of the ways that this designation situates some groups' food and food memories as being different from others. "Heirloom," thus is not just a term that designates a particular kind of tomato, and distinguishes it from other tomatoes. The term also invests the experience of the tomato with a particular normative meaning that separates the one group from another. As a result, designating a vegetable as an "heirloom" is not a neutral act, but is instead an act that has significant social and political consequences.

The "imagined community" and *habitus* that structure the meaning of heirlooms is part a broader set of "alternative food practices." To eat heirlooms and participate in their cultivation marks one off from "conventional" growers and the dominant ways that people produce and procure food. Alternative food practices advocate more ecologically sound and socially just farming methods and are often concerned with issues surrounding marketing, distribution (Slocum 2007, 522). The primary aim of such practices is to ensure that healthier, culturally appropriate food options exist for all citizens across the United States. Some food activists and consumers maintain and support local farmers through farmers' markets, restaurants, and agricultural policy change. Others focus on nutrition education and obesity prevention. Environmental groups, a third group, advocate for organic, free-range, and antibiotic-free food. Finally, some organizations are structured around issues of social justice for migrant laborers and food insecurity (Slocum 2007, 522). Broadly speaking, the imagined community that arises through heirloom cultivation is an environmentally concerned, locally oriented, and socially responsible one. In so doing, these practices cultivate a deep and long connection to place, and arise when we are able to know where our food comes from. In so doing, the cultivation of heirlooms helps to support the cultivation of important connections between individuals, groups, and their land.

Sharing memories—whether it takes the form of "where were you when…?" or "remember when…?"—is one of the most significant ways that we share time with others. By exchanging memories we learn how to remember and learn how to form relationships with others. Similarly, by eating food, we learn how to reform our senses of self as we reflect on our own pasts as understood by others and we learn to understand others' past. In other words, it offers an opportunity to engage different perspectives on its meaning and opens up a space in which the meaning of our own past can be unsettled and revised. In so doing, eating food with others, as well as eating others' food, can realign our experiences and give new meaning in the present. By engaging with older food practices, we can learn not only how to inhabit the past, but we can also learn how their practices of growing and cultivating food can offer viable alternatives to the current food system.

The function of edible memory as laid out thus far shapes individual and collective identities by providing a set of imagined communities that tie people together and by offering them a set of stylized habits and disposition that motivate action. As a result, by remembering an imagined community, one that has a past that extends beyond one's personal memory, is part of the process of acquiring a social identity (Zerubavel 2004, 3). Eating particular foods, in other words, is how members of a group orient themselves in the world and situate themselves in time as part of an ongoing intergenerational community. By being part of an historical, culinary community we become responsible for the history of that community, one that extends beyond the boundaries of what individuals themselves can remember. However, because remembering a past means to choose which parts of the past to remember and forget, we need to consider how food has been used to construct particular identities and how it has been used to marginalize other groups.

3.4 The Limitations of Edible Memories and Culinary Imaginaries

While memories of heirlooms and other foods contribute to how we think of ourselves, the discourse surrounding heirlooms and other antique varietals often lack critical normative resources that could help distinguish the responsible incorporation of past foods from the potentially harmful appropriation of other people's food heritages. Indeed, what distinguishes which foods should be remembered from which should be forgotten? More importantly, how can sharing edible memories across pasts that are linked by a common, though potentially toxic, history help to restore relationships and what forms should this sharing of food and memory take?

In fact, the understanding of the way that food can be used to retrieve a past detailed thus far is both deeply internalist and profoundly essentialist. It takes specific features or characteristics of food and makes a problematic generalization about the food or the group that eats the food. As a result, this notion of memory fails to capture the ways that places are interconnected and, more importantly, the ways that they presume a particular identity of the place that is recovered through memory.

Rick Bayless's comments about cooking and eating Mexican food offer a striking example of a conception of food and place that is both internalist and essentialist. As he recounts his memory of eating his first chayote, he encourages listeners to imaginatively invest themselves in the sensations that he describes. The flavors that he describes are characteristic of the cuisine of Mexico. However, his recounting also prevents listeners from raising questions about the relative privilege he had in moving across international borders, and the cultural and economic conditions that make Mexican food appear to be at once "exotic" and strangely familiar to the United States citizen's palate. Bayless's account is neither hopelessly naïve nor is it strikingly sophisticated. What marks it out is that it appears to be a "normal" experience with a new food. When he went to Mexico for the first time, he says that he felt "at home," able to move about freely, sampling and tasting the new (to him) foods. His shared memory suggests that these foods were abundant and available to everyone. In the process of sharing these foods and these memories, Bayless's edible memory is an attempt to get people to invest meaning in his past in order to create connections to others, both past and present.

Bayless offers an essentialist and internalist understanding of cuisine. He conceives himself as a sort of culinary ambassador who brings back "real" or "authentic" Mexican food to make and sell in the United States. By indicating that there is nothing offensive about a white person cooking a cuisine that is not ethnically his own, he ignores the processes of intercultural exchange that often influence a region's cuisine and affect the relationships between Mexicans and Americans. Rather than merely retrieve the past, edible memory needs to be understood relationally—that is, as occurring between people or groups. An imagined community, in other words, arises as a result of relations of differences between communities and cuisines. In order to see the relational dimensions of memory, there needs to be a further understanding of the ways that the identity of a place or of a particular food tradition is always in a process of being formed. Identities are neither irrevocably destroyed with the importation of new foods, nor are they necessarily preserved when eating old foods. Rather, the edible memories that come with new foods require us to attend to the ways that we become responsible for a past that we do not personally remember but one that is nevertheless our own. As a result, there needs to be a deeper appreciation for understanding how cooking others' cuisines can undermine and displace their memories. In fact, it may be that designating a food as a "traditional food" or an "heirloom food" removes them from the social and cultural contexts in which they were first produced and eaten. As Sue Campbell notes, the process of sharing memories often neutralizes the significance of the past by rendering them essential characteristics of that identity (Campbell 2014, 80). Using a food to evoke a memory and getting others to "discover the wonders of Mexican cooking," Bayless mixes in his own recollections of going to Mexico a regional cuisine to form an idealized, ahistorical conception about "authentic Mexican cuisine" that "real" Mexicans eat.

Heirlooms often operate as part of an imagined past to configure a culinary identity. In so doing, they identify particular features of food that are then used to characterize or define a particular group. Furthermore, rather than historically situating this identity, these foods often appear to be from an almost timeless, ahistorical past. Nostalgia for food covers the contexts in which those foods were first produced and eaten with a veneer that shields the memory from being fully contested (Campbell 2014, 80). But more than this, the kinds of foods that receive the designation of "heirloom" are often American and European foods. It is rare to see a tomatillo, yam, eggplant, or Asian or African food designated as an heirloom (Jordan 2015, 128). This decontextualization and dehistoricization of food enables producers and consumers to appropriate foods without considering the moral order or collective narratives of the group whose identities are bound together with this food and that have played an important role in constituting the meaning of these foods. The normative meaning of heirloom foods—with its connotations of authenticity, natural, nostalgic, and organic—is the background, and unremarked upon, justification for saving seeds, breeding cultivars, and even choosing the tomato at the grocery store or farmers' market.

To label a food as an heirloom, then, is not an apolitical act. Designating food as an heirloom can neutralize the past and thus close off spaces for challenging and contesting the values that it promotes. The process of decontextualizing our food by imbuing it with a nostalgic halo can also work to reinforce the whiteness of alternative food practices. Julie Guthman, for one, notes that the alternative food movement, which includes the ways that we think about the pasts of our foods, has adopted a number of discourses that are predicated on a whitened cultural history. The discourses of "organic," "natural" and "pure," in various contexts have also often been used to produce and perpetuate racial difference. Additionally, the desire to tend the land, a phrase often used in those discourses that tend to value heirlooms, neglects both the indigenous people who have been prevented from cultivating their own land and the racialized history of agrarian and labor relationships. Guthman further notes that the "messianic disposition"—the intention of "doing good" on the behalf of others—that inheres in many alternative food practices has the markings of a colonialist disposition. Finally, the discourse of returning to the land neglects understanding the often defensive and xenophobic reasons why many want to return to the rural life (Guthman 2008, 435–436). Ultimately, the discourse in such movements is one that is often framed around the improvement of others while also eliding the historical developments that have produced these material and cultural distinctions in the first place.

In short, the discourses surrounding edible memory and the promise of returning to heirlooms or other cultures' foods are not entirely innocent of race. As a result, we need to take care in understanding the ways that particular food traditions operate and the "imagined communities" that our own culinary imaginaries construct. By engaging those worlds of edible memories in which we are not at ease, we can nevertheless engage their values, give weight to their fragile cultural practices, support identities, and undertake new relationships that can forge further bonds of solidarity.

3.5   Towards Responsibility with Edible Memory

Despite Guthman's trenchant observations about the racialized discourse of alternative food practices, generally, there may be ways that edible memory can be used to resist and challenge dominant culinary imaginaries. While edible memories often are used to constitute a stable identity and ensure group cohesion around shared experiences and shared foods, they do so by problematically generalizing a particular trait. As a result, it is important to understand how sharing memory can destabilize group identities and challenge the ways that we tend to think about the past. Because edible memory recounts the ways that material objects and practices are invested with meaning, the alternative food practices cannot be written off. Rather, the culinary connections that arise through edible memory can highlight the stakes involved in eating. As Elspeth Probyn suggests, "eating, its connections to the land and its histories, may highlight the (im)possibilities of coexistence" (Probyn, as quoted in Slocum 2007, 528).

Edible memory, on this revised interpretation, is a process of negotiating our identities and forming relationships through mnemonic practices surrounding food. Edible memory is relational insofar as it recognizes the ways that people participate in the formation of the significance of one another's pasts. As such, it encourages people to invest in one another's pasts by affirming the value of remembered foods. At the same time, the tension between tradition and innovation in the experience of edible memory implies that identities can be distorted as a result of foreign influence. Being responsible with edible memory requires paying attention to the ways that our memories allow us to form connections with others. Heirlooms, for example, are often eaten by people who have no past experience with that food. More than this, however, if one group eats others' foods for a long enough time, "[their memories] may eventually become your own (or your children's)" (Jordan 2015, 37–38). In order to ensure that this intertwining of memories is done justly, we need to look more closely at those processes whereby we make ourselves responsible for what we have not ourselves done.

Eating food, as Jordan claims, is one way that we not only connect with our own past—as when we return to the foods of our childhood in times of stress—but eating food is also a way that we are presented with a story about someone else's past. The recently opened National Museum of African-American History and Culture in Washington, D.C. offers an opportunity to consider the careful ways that food can open us to another's past and give visitors the opportunity to see themselves implicated in the past. Cooks at the museum cafeteria, Sweet Home Café, have developed a menu that tells the story of the African diaspora in the United States. Rather than taking a monolithic perspective on the kinds of foods introduced, the café instead identifies how different people living in different regions transformed their own cuisine and, conversely, how their cuisine influenced the region's foodways. By highlighting the various contributions made by African-Americans to "American" cuisine, the menu calls attention to the ways that various African cuisines were brought to the Americas via the slave trade and the ways that these foods have also become part of the broader culinary landscape. By emphasizing the historical situation, as well as the dialectical movement between sedimentation and innovation, of such dishes such as hoppin' John, fried chicken, shrimp and grits, smoked and barbequed meats, collards, po' boys, and black-eyed peas, the café not only gives visitors a respite from the intense story the museum tells about African-American history, but it also offers an opportunity for people to recognize and recover a deeper sense of their identities. As a result, eating food here becomes and important way to understand the legacy of the slave trade and of the Great Migration.

The Sweet Home Café offers an instructive lesson about the ways that edible memory highlights the relationality of social identies by bringing visitors into an emotional relation with others (Severson 2016). Memory is not just a cognitive claim about what we know about the past, but also an emotional and ethical relationship that ties us to the past. In remembering the past, we also find ourselves implicated in it. Hoppin' John, for example, could not exist in the form it does would not have been possible without the slave trade. Eating it in the museum, then, enables visitors to call attention to the ways that food positions individuals and groups in relation to one another and offers an opportunity to reconfigure that relationship to be more just. Eating food at the Café exposes one to a past that is not one's own. However, taking responsibility for that memory—whether it is an heirloom tomato, a piece of cornbread, or a mole sauce—brings us into emotional relation with past, present, and future. We find ourselves in emotionally invested relations with others when we take responsibility for them. Taking responsibility for who we are and for others is the core of the commitment involved in edible memory.

Importantly, this suggests that the places where our food grows and the food that we do eat need to be considered historically and not just spatially. As Doreen Massey puts it, "not only does that particular articulation of social relations which we are at the moment name as that place have a history, but also any claim to establish the identity of the place depends upon presenting a particular reading of that history" (Massey 1995, 188). In other words, it may be more useful to think of places not as an area on a map, but as constantly shifting articulations of social relations through time. Similarly, to characterize a place does not merely identify properties or qualities of it; rather the process of characterizing a place is an attempt to define it and claim a particular meaning for it. To tell a story about our food, in other words, requires us to be attentive to the places where they were grown and the social relationships that produced them.

The popularity of heirloom foods, in particular, has created new connections between people, food, and their landscapes by preserving agricultural biodiversity (Jordan 2015, 208). We find ourselves responsible for these elements as we remember them. Community gardens, urban farms, private gardens in suburban and rural areas can be ways to engage and reanimate past forms of life. More than this, for activists, seed savers, and consumers growing, buying, cooking and eating these heirloom foods can be a concrete way of preserving agricultural biodiversity. This means that the practices surrounding our shared experiences of eating can facilitate, or potentially undermine, the intentions of particular memories. Moreover, practices of commensality can fortify or undermine the resources that others need to re-experience their pasts in ways that meet their present needs and interests, including those of challenging dominant views of the past.

This opens up the possibility of telling—and listening to—a different history of the place where our food is grown. While it is important to tell stories about where our food comes from and the meanings it has for our identities, it is important to listen to the ways that our culinary heritage has also pushed people off the land and separated them from the sources of their foods. Rachel Slocum recounts such an experience when an older woman told her during an interview that she was "embarrassed for 'our culture' on how they came to be here in America" (Slocum 2007, 529). Her shame and embarrassment at the displacement and genocide of the First Nations and Native Americans helped her recognize her own accountability to others for the ways that she is (and many of us are) implicated in the ethics and politics of our own current situation. To become aware of these possibilities means that we also need to be critical of the ways that our edible memories repeat and promote certain kinds of imaginaries.

Similarly, while various sites such as farmer's markets, museum cafeterias, and festivals can be places where we follow the latest food fads, they can also be important places where memories of food are shared. As a commercial site, they can bring diverse crowds together and encourage the sampling of new foods, which may in turn be incorporated into a new tradition. In other words, in addition to the food purchased there, these markets can also bring people together in moments of contact and connection over heirloom and farm grown apples, tomatoes, and kale. Learning how to remember is a process of learning how to speak with others. As a result, farmers' markets can operate as potential sites where we can listen to a shared past that has been narrated differently or of an unfamiliar food that can become part of a familiar dish.

Community projects surrounding food can open further avenues for dialogue. The RAFT alliance in Appalachia has a mission of "[bringing] together food, farming, conservation and culinary organizations and advocates to ensure that the diverse foods and traditions unique to North America remain alive and dynamic" (Veteto 2011). This collective sees themselves as "saving the past for the future," and worry that the risk of the flint and dent varieties of corn disappearing will effectively separate them from the sources of their traditions and their identity, whether that identity is as Appalachian farmers or as indigenous Cherokee (Veteto 2011). In recognizing the diversity of traditions, food heritages, and edible memories, this group maintains that it is important to recognize the importance of white immigrants', African-Americans' and the Native Americans' different connections to the land and the ways that their relationships to the land and the food grown there are differently positioned.

Importantly, the RAFT alliance sees their work of heirloom farming and wild food harvesting in that region as an act of political and economic resistance. James Veteto, for example, suggests that cultivating heirlooms are "concrete everyday resistances, which provide counter-memories to a monoculture that [is characterized] by a society prone to negligence and forgetting" (Veteto 2011). Taking responsibility through edible memory here creates those connections between groups that can help to resist potentially oppressive practices. Through their work, they create narratives of the past through the transformation of their environment and landscape, infusing the region with accounts of how people used to live and what they used to live on. If eating food can be considered an act of resistance, heirlooms can offer one avenue for considering why people find it necessary to return to heirloom foods and why they conceive this return as an act of resistance.

More than this, the ability to share memory and listen to others' edible memories requires us to allow others' accounts of their food some influence on who we take ourselves to have been. The notion of edible memory sheds light on the ways that eating introduces and exposes one to a history that is not one's own. In fact, because the production and consumption of our food is one of the most significant ways that we remember (and forget) the past, the natural impulse to think of memory as an internal, subjective act needs to be expanded to include the social contexts in which remembering occurs. More than this, because food takes us far beyond our own locale, the practices surrounding our eating can become a site where we can learn how to thoughtfully encounter a past that is not our own, but whose meaning we nevertheless shape. Said concisely, by eating with others, and by eating others' food, we learn how to remember the past through the performances and practices of commensality.

Edible memory, I have suggested, highlights how we are responsible for the ways we shape the past. It can help us recognize how what we buy, eat, grow, and consume implicate us in a larger web of relations. This means that at least under the intentional and practiced ways of cultivating and participating in foodways, we can further understand our contributions to the meanings of the past. To move into the frame of others' pasts allows us to invest their pasts with some influence on who we take ourselves to have been. Our engagement with the values enacted through others' edible memories does not mean that we share a past, nor does it necessarily entail that we find ourselves at ease in that past. Rather, engaging with the values expressed through others' food practices can help give meaning to their past and give us an opportunity to acknowledge our own accountability to them for the ways that we are implicated in a broader, collective past.

References

Anderson, Benedict. 1991. *Imagined communities*. New York: Verso Books.

Barber, Dan. 2016. Why is this matzo different from all other matzos? *The New York Times*, April 15.

Black, Jane. 2016. The next big thing in American regional cooking: Humble appalachia. *Washington Post*. March 29.

Bourdieu, Pierre. 1989. *Outline of a theory of practice*. Trans. Richard Nice. Cambridge: Cambridge University Press.

Campbell, Sue. 2014. *Our faithfulness to the past*. Oxford: Oxford University Press.

Dimitri, Carolyn, Anne Effland, and Neilson Conklin. 2005. The 20th century transformation of U. S. agricultural farm policy. The United States Department of Agriculture. http://www.ers.usda.gov/media/259572/eib3_1_.pdf. Accessed May 25, 2016.

Guthman, Julie. 2008. Bringing good food to others: Investigating the subjects of alternative food practice. *Cultural Geographies* 15 (4): 431–447.

Jordan, Jennifer. 2015. *Edible memory; The lure of heirloom tomatoes and other forgotten foods*. Chicago: University of Chicago.

Massey, Doreen. 1995. Places and their pasts. *History Workshop Journal* 39: 182–192.

Morris, Steven. 2009. Orchards may vanish by the end of the century, conservationists warn. *The Guardian*, April 23. Accessed May 28, 2016.

Nabhan, Gary. 2010. *Forgotten fruits: Manual and manifesto*. The Renewing America's Food Tradition Alliance. http://garynabhan.com/pbf-pdf/applebklet_web-3-11.pdf. Accessed May 25, 2016.

Olick, Jeffrey, and Joyce Robbins. 1998. Social memory studies: From 'collective memory' to the sociology of mnemonic practices. *Annual Review of Sociology* 24: 105–140.

Pollan, Michael. 2008. *The omnivore's dilemma: A natural history of four meals*. New York: Penguin.

Ricoeur, Paul. 1988. *Time and narrative*. Trans. David Pellauer and Kathleen McLaughlin. Chicago: University of Chicago Press.

Severson, Kim. 2016. Museum cafeteria serves black history and a bit of comfort. *The New York Times*. November 28.

Slocum, Rachel. 2007. Whiteness, space, and alternative food practice. *Geoforum* 38: 520–533.

The Sporkful. 2016. "Other people's food" audio podcast, national public radio. New York, 21 March. Accessed March 24, 2016.

Veteto, James R. 2011. Apple-achia: The most diverse foodshed in the US, Canada, and North America. In *The place-based foods of appalachia.* eds. James R. Veteto, Gary Paul Nabhan, Regina Fitzsimmons, Kanin Routson, and DeJa Walker. The renewing America's food tradition alliance. http://garynabhan.com/pbf-pdf/AA%20APPALACHIA'S%20PLACE-BASED_FOODS.pdf. Accessed, April 18, 2016.
Zerubavel, Eviatar. 2004. *Time maps: Collective memory and the social shape of the past.* Chicago: University of Chicago Press.

Chapter 4
Building Community Capacity for Food and Agricultural Justice: Lessons from the Cuban Permaculture Movement

Justine MacKesson Williams

Abstract Researchers and activists located around the world often look to Cuba—which experienced a massive transition to small-scale, low-input agriculture in the 1990s–2000s—as a case study in sustainable agriculture adaptation. Their analyses tend to highlight external political-economic conditions, and the state policies and programs that they prompted, as the factors driving the spread of diversified, low-input farming practices. Based on 16 months of ethnographic fieldwork, this chapter argues that localized work by grassroots and non-state organizations to increase autonomy and agency has been an equally important compliment to these structural factors in the ongoing process of Cuban agricultural transformation. Specifically, it analyzes the practices and philosophies of the permaculture network of the Cuban *Fundación Antonio Núñez Jiménez de la Naturaleza y el Hombre* (FANJ) in order to provide lessons for US-based and international scholars and activists on how to build "community capacity" and "skillful disclosure" for food sovereignty and justice.

Keywords Cuba · Community capacity · Food justice · Food sovereignty · Permaculture 7500 words

4.1 Toward New Global Paradigms in Food and Agriculture

The current global food and agricultural system is heavily influenced by the visions and interests of international financial institutions, transnational corporations, and government agencies who collectively produce what scholars have called a "global food regime" (McMichael 2009; Friedmann and McMichael 1989). Although many people directly depend on agriculture for their livelihoods, and family farmers—many of them small-scale—produce 80% of the world's food supply (FAO 2015),

J.M. Williams (✉)
Department of Anthropology, University of North Carolina at Chapel Hill,
Chapel Hill, NC, USA
e-mail: JustineMWilliams@gmail.com

© Springer International Publishing AG 2017   31
I. Werkheiser and Z. Piso (eds.), *Food Justice in US and Global Contexts*,
The International Library of Environmental, Agricultural and Food Ethics 24,
DOI 10.1007/978-3-319-57174-4_4

the food regime encourages consolidation, industrialization, and integration into international markets. As a result, populations have experienced increasing economic and food access precarity. They have also increasingly mobilized to advocate for alternatives.

A movement of farmers, activists, and scholars have proposed "food sovereignty" as an alternative framework for food and agriculture systems and policies. As defined by La Via Campesina, an international alliance of peasant organizations, food sovereignty is "the right of peoples to healthy and culturally appropriate food produced through sustainable methods and their right to define their own food and agricultural systems" (La Via Campesina 2011). While there are many different visions for how food sovereignty can and should be realized, proponents generally agree that it entails prioritization of agroecology and small-scale production in order to promote thriving local economies and ecosystems.

Cuba is a fascinating context within which to consider resistance to the global food regime and the pursuit of food sovereignty because of its attention to population-wide food security (since the Revolution), and its more recent experimentations in various forms of "sustainable agriculture"[1] (since the collapse of the Soviet Union in the early 1990s). The concept of food sovereignty arrived in Cuba at a time when the island nation was undergoing a massive re-examination of its political and economic structures, including its food and agricultural system. The concept of "sovereignty," has been of acute political importance in Cuba since the Revolution, as the nation has struggled against decades of attempts by the United States to undermine the socialist leadership and structure of the country. La Via Campesina's efforts to conceptualize "food sovereignty," in the 1990s resonated with this historic Cuban struggle, as well as with Cuba's post-Soviet need to transition to more decentralized, family-scale agriculture, and to wean the system off of imported chemical and fossil fuel inputs. Thus, Cuban agricultural administrators, technicians, and small-farmers participated in early meetings of La Via Campesina and helped to bring this discourse into Cuba's own national conversations around food security and autonomy.

Small farmer organizations in North America also participated in the first meetings of La Via Campesina, and over the past several decades, many organizations in the United States and Canada promoting community food security, sustainable agriculture, and family farming have also embraced and promoted the concept of food sovereignty. In this geographic context, the movement for sovereignty has also developed alongside and in conversation with "food justice," a movement that works to "address injustices within the US food system" by confronting the structural problems—such as race, class, and gender—that limit access to food, food-producing resources (like land), and self-determination (Holt-Giménez 2015). Food justice, which emerges from the environmental

---

[1]The term "sustainable agriculture" is broad. In Cuba, various projects in environmental, social, and cultural sustainability have prioritized farming and food production philosophies ranging from organic agriculture, to agroecology, and permaculture. The general term "sustainable agriculture" is used here to encompass these connected movements.

justice movement (Holt-Giménez and Wang 2011) considers US-specific legacies of slavery and colonialism that affect the food system (Redmund 2013), but also shares many goals of the international and Cuban food sovereignty movements. This chapter is written with the intent to provide lessons to activists in North American activists working from within this context, but it should also be relevant to those working to improve food and farming systems in low to middle income countries around the world, and to those scholars reflecting specifically on the various movements and efforts underway in Cuba.

4.2   Food and Farming in Cuba: A Brief Modern History

In the 1950s, Cuban political activists and dissidents were motivated by widespread hunger, poverty, lack of land access, and inequality to mobilize and overthrow the Batista dictatorship. When revolutionaries took control of the government in 1959, they prioritized food access and distribution as a question of national concern (Jiménez Núñez 1960; Valdés 2003; Benjamin et al. 1989), and established a state procurement and distribution system to make a *canasta básica* (basic basket) of subsidized food rations available to every citizen (Wright 2009). Although the Cuban diet has not always been the most abundant or diverse, the state's commitment to evenly distributed food access has allowed it to essentially eradicate extreme hunger and malnutrition, and to rise above all other Latin American and Caribbean countries apart from Chile on the United Nation's Human Development Index (World Food Program 2015; Malik 2014).

However, this system was built around large-scale sugarcane production, which created a commodity that could be traded with the Soviet Union for imported food and other supplies. Thus, when the Soviet Union fell in the early 1990s, Cuba's entire food and economic system also collapsed. The Cuban State declared a "Special Period in the Time of Peace," and implemented austerity plans that were originally developed for times of war. During these years, Cubans reportedly reduced their daily caloric intake by 30% (Funes 2002). Many lost weight, and some became ill. The Special Period made it clear that Cuba's long-time dependence on foreign trade and industrial-scale farming had been dangerous, and it opened up new doors for sustainable agriculture advocates to push forward and develop alternative systems.

Activist growers and agronomists promoted organic agriculture and agroecology as means to boost and diversify domestic production, and the state invested in research institutions to further develop and disseminate such practices (Funes et al. 2002; Wright 2009; Benjamin and Rosset 1994). Urban community members began gardening in *tierras ociosas* (empty plots), backyards, and rooftops, and in 1994 the state started a department of urban agriculture to support the development of *organopónicos*, raised bed gardens, in every municipality across the country (Altieri et al. 1999; Premat 2009; Wright 2009). In rural areas, the *Asociación Nacional de Agricultores Pequeños* (National Association of Small Farmers,

ANAP) promoted *campesino-a-campesino* (farmer-to-farmer) knowledge exchange in order to "rescue" the traditional practices of previous generations and to develop and disseminate new, sustainable techniques.

Collectively, these efforts boosted the country's domestic food production levels and diversified many Cubans' daily diets (Altieri and Monzote 2012). Instead of losing its reputation for commitment to food security, Cuba gained new notoriety for pioneering sustainable agriculture techniques and national infrastructure for supporting them. International social movements, non-governmental organizations (NGOs), and activists began to see Cuba as evidence that ecologically and socially responsible food and agriculture projects can function to achieve not just food security, but also food sovereignty (Benjamin and Rosset 1994; Oxfam America 2001; Morgan 2006).

Many commentators have suggested that the country's success in alternative agriculture has been due to the presence of a strong and supportive state (Rosset et al. 2011; Reardon et al. 2010). Indeed, the existence of formal state institutions that support urban agriculture and agroecological research have allowed these sectors to receive financial resources and intellectual support that in other countries often goes exclusively to conventional, industrial agriculture. Additionally, the presence of mass, public organizations, such as ANAP, and local political organizations like the *Consejos Populares* (People's Councils) provide strong social bases for relatively rapid diffusion of new knowledge and implementation of local projects.

However, the Cuban state's support for sustainable, local agriculture has neither been complete nor consistent. Many individual farmers, state administrators, and government departments have remained doubtful about the possibility of supporting the national economy and food demands through low-input agriculture alone. Because the state assumes the responsibility for ensuring that sufficient quantities of food are available at the national level, rather than leaving this role to the market, the risks of failure are daunting. Since Raúl Castro took over the presidency in 2008, he has pursued the goal of food security by emphasizing efficiency and productivity over all else. Although he encourages urban and organic production in certain cases (he was the major promoter of national systems for urban agriculture even before assuming the presidency), his government is also re-investing in sugarcane and industrial production in order to increase exports (Williams 2012).

Because alternative agriculture is not yet ubiquitous in either practice or policy in Cuba, in-country proponents of low-input family agriculture and alternative environmental and livelihood philosophies have had to take on movement dynamics in order to convince both society and government functionaries that a transition away from industrial agriculture is needed in order to preserve the island's fragile ecosystem, and to encourage food sovereignty and autonomy. This has been true of the campesino-to-campesino movement, which organized small-scale, family farmers in a process of horizontal person-to-person agriculture knowledge exchange (Rosset et al. 2011), and—according to my interview participants—is also true of the permaculture movement (described further below). Thus, it is important to analyze Cuban alternative agriculture in relation to structural and policy factors, as well as through the local-level work of individuals and organizations that took advantage of new policies for de-centralized, low-input agriculture to push forward even more radical visions of ecology, society, and autonomy.

4.3 From State Policy to the Community Capacity to Disclose New Food Systems

Food sovereignty scholars and activists have generally described state policies and international economic arrangements as the primary barriers to their movement, and thus, as the potential access points for change. For instance, in pursuit of food sovereignty, La Via Campesina has rallied against trade policies of the World Trade Organization (WTO) and member organizations have voiced opposition to national policies in their home countries, such as ones ignoring the well-being of women pastoralists in Zimbabwe or allowing importation of cheap potatoes into Indonesia. However, proponents of food sovereignty are increasingly realizing the limitations of focusing efforts exclusively on policy, especially considering that states and intergovernmental organizations can co-opt or distill movement goals when food sovereignty discourse is taken out of communities and put into bureaucratic guidelines (McKay et al. 2014). It has also been noted that locally-specific historical and cultural factors—which are not easily articulated at the policy level—influence the pursuit and success of alternative food and farming models (Kerssen 2015).

  A shift to local spaces of everyday autonomy and community-based work in the pursuit of food sovereignty resonates with cultural change taking shape in 1990s Cuba. Although the state was pursuing—along with allies in Venezuela—a national vision of food sovereignty that was defined by the state's ability to provide food to all its citizens without compromising political stance and independence, members of the government were also realizing that a shift was needed to transition some responsibility for food provisioning away from centralized institutions and onto individuals, families, and local communities. In this context, food sovereignty took on a multi-scalar definition, including state sovereignty, and also the autonomous ability of people to produce their own food in ways that met their own needs and desires for agency and creative production. However, the move to decentralize food and farming necessitated less formulaic projects and policies, and more capacity in local spaces. New organizations and projects emerged in the 1990s and 2000s to fulfill this need and to help small-scale *campesinos* (peasants), *parceleros* (those with small plots for household provisioning), and even urban apartment residents engage in sustainable food production suited to their particular needs and preferences, and also to train them as promoters, facilitators, and educators in sustainable agriculture in their own regions, cities, and cooperatives.

  In the US Food Justice movement, as well, groups such as The Campaign for Food Justice Now, White Earth Recovery Project, Detroit Black Community Food

Security Network, and many others are also working to engage in education and outreach in local spaces, so that structural and policy work can articulate with on-the-ground transformation. These projects, like those in Cuba, are confronted with the question: how can movements for sustainable agriculture, food justice, and food sovereignty effectively spread their experiences, knowledge, and skills to larger numbers of people?

Spinosa et al. (1999) notion of disclosure provides a useful framework for understanding why localized coordination among individuals is a necessary accompaniment to political and structural transformation. They argue that history is changed, and that "new worlds" of action and meaning are created, when people come together in "disclosive spaces." In such spaces, people notice disharmonies within society and coordinate actions that will lead toward new policies, programs, or cultural formations. Spinosa et al. stress that people cannot make a difference just by thinking together; they must be motivated to take actions that will bring new "worlds" into material existence. They call the process of doing so "skillful disclosure." Following their analysis, it would not be sufficient, for instance, for a group of policymakers to simply imagine a world of sustainable, productive farms, and produce legal changes that could allow it to occur. People must also come together in spaces where they can collectively reproduce this vision and work toward bringing it into existence. Disclosure is necessary for food sovereignty and food justice movements, therefore, in order to point out elements of the food and agricultural system that remain inconsistent with these goals, and to conceptualize a pathway forward.

But what enables such spaces to exist and encourages people to come together in them? One way of considering whether the cultural or social atmosphere for disclosure exists is through the concept of "community capacity." As defined by the Aspen Institute, community capacity is the "combined influence of a community's commitment, resources and skills that can be deployed to build on community strengths and address community problems and opportunities" (Rural Economic Policy Program n.d.). The concept emerged in the 1990s as an alternative to conventional developmental paradigms. Rather than providing aid or forcing structural adjustment on non-western nations and communities, the idea was to support the development of communities who were able to identify their own problems and work proactively to address them.

Like any term, community capacity can be re-interpreted and co-opted based on the intentions of those who deploy it. It can be used, for instance, by international agencies to re-brand conventional development programs that attempt to cultivate individuals who will seek out and act according to western paradigms of progress and development. In these cases, communities are considered to have "capacity" only when acting toward the same goals that hegemonic institutions view as beneficial for development. However, the term can also be used as a framework for identifying grassroots participation and self-determination. For the purposes of this chapter, "community capacity" is used to describe the ability of localized groups of people to work toward new realities, take advantage of existing beneficial state structures, and advocate for political change based on the existence of strong social connectivity, local spaces for coordinating actions, and infrastructure for supporting them. Rather than simply echoing a linguistic trend of development agencies, the term also resonates with the preferred vocabulary of Cuban sustainable agriculture promoters, who use the term *capacitación* (capacitation) to refer to people's ability to successful carry out and promote alternative agriculture and commonly emphasize the need to develop *capacidades* (capacities) for sustainable agriculture. As one leader of the campesino-to-campesino movement for agroecology described during an anonymous interview regarding spreading agroecology in order to transform the country's agriculture: "Above all else, the challenge was to identify capacities, and to develop them."

4.4 Researching Alternative Agriculture in Central Cuba

Although alternative, sustainable agriculture remains confined to the minority of Cuban agricultural land, the movement to expand it continues. The simultaneous patchiness and endurance of these projects raises the questions: what allows the movement to continue reaching new people and building commitment? And on the other hand, what is constraining it from gaining ground? Is this simply a question of a contradictory policy framework, or something more?

Between 2011 and 2015, I conducted 16 months of ethnographic research with urban growers, rural farmers, and agricultural administrators and technicians in Cuba, with a focus on a central Cuban province that has been a center for several alternative agriculture projects and movements.[2] The resulting analysis—based on 50 in-depth semi-structured interviews as well as extensive participant observation and informal interviews with at least 100 additional participants—found the largest challenge in the contemporary alternative agriculture movement of the province to be activating and supporting sufficient promoters and supporters of alternative agriculture to continually share information, guidance, and knowledge with more people. In other words, although there were many "pockets" of capacity for alternative agriculture promotion and instances of "disclosure," they were not yet available to all farmers or would-be food producers. Additionally, the research identified the permaculture network of the *Fundación Antonio Núñez Jiménez de la Naturaleza y el Hombre* (Antonio Núñez Jiménez Foundation for Nature and Man, FANJ) as the most active organization (during the fieldwork period) working to overcome the challenge, provide community for alternative agriculture knowledge and skills development, and generate larger networks of committed alternative

---

[2]To protect the privacy of the people who participated in this research I have chosen not to reveal the name of the specific province or city where this fieldwork was conducted.

agriculturists. The following section situates the relevance of their work, describes the philosophies and practices of permaculture, and considers how others may learn from these experiences in the pursuit of food sovereignty and justice in Cuba, North America, and beyond.

4.5   Becoming Sustainable Agriculturists

Both within and outside of Cuba, it is commonly reported that Cuba began to farm sustainably not because the government or people chose to, but because dwindling access to international imports left them with no other choice. Some researchers have speculated that farmers might, then, return to using chemical inputs if given the opportunity (Nelson et al. 2009). However, the participants of this study (who associated with alternative agriculture organizations in Cuba) voiced strong ethical commitments to alternative forms of agriculture, and indicated that they would not return to chemical usage knowing what they do now. Many of these growers began farming because of economic pressures. Some began to work on urban farms after 1993, when the state mandated that *organopónicos* (raised bed farms) be implemented in all cities, and required most of their operators to farm without chemicals in order to avoid public health consequences of spraying chemicals in cities. Others came from *campesino* backgrounds, but had moved on to other professions, and returned to family farms when the Special Period weakened the economy and access to food. However, when asked how they became "agroecologists" or "permaculturists" (the two identities embraced by sustainable agriculturists in this research site), none cited these events; these moments may have prompted changes in practice, but they did not convert them to the alternative ways of thinking they now identify with.

Rather, the majority of informants identified specific individuals, and the groups they were connected to, as bringing them into a new world and way of thinking. For instance, many identify Esteban[3] a local agricultural engineer, former of employee of Urban Agriculture, and founder of a local organic agriculture group for talking to them about alternative farming philosophies and inviting them to participate in his group. Others recall a particular neighbor, friend, or colleague who was already involved in alternative agriculture and introduced them to it, or a chance encounter at which they happened to meet someone and strike up a conversation.

Research participants who came to identify as sustainable agriculture practitioners in the 1990s thru early 2000s mentioned a variety of groups and individuals as important to their early education and *capacitación* (capacitation) in sustainable agriculture. However, the majority of post-2005 entrants described someone

---

[3]Following ethnographic conventions and the terms of informed consent that were described to participants prior to interviews, all names here are pseudonyms, in order to protect the anonymity of informants.

connected to the *Fundación Antonio Núñez Jiménez de la Naturaleza y el Hombre* (Antonio Núñez Jiménez Foundation for Nature and Man, FANJ) permaculture network as key to their personal transformation. Almost all participants indicated in interviews that FANJ, a Cuban non-governmental organization, is now the most important organization providing them support and continuing to promote sustainable agriculture.

For instance, Alfredo, who farms coffee, vegetables, fruits, and some grains on his family farm explains that he began to learn about agroecology when the local office of the small farmers' association (ANAP) organized a meet-up group for farmers interested in agroecology. However, he later connected with FANJ and learned about permaculture, and he says that now, "FANJ has the most protagonism in sustainable agriculture." He believes this is true because of their work in outreach, but also because of the particular philosophy he believes permaculture offers:

I see in permaculture more opportunities than in agroecology. Agroecology gives you the opportunity to conserve soil, you see? It teaches you to conserve soil and seeds following the campesino tradition. But permaculture takes you above and beyond. It includes the conservation of soils, but also allows you to better *aprovechar* [take advantage of/work with] nature and live with it… It's a life philosophy.

For Alfredo and many others, national policy changes prompted their entrance into low-input farming, and national, centralized, state-affiliated associations helped them to begin learning about alternative forms of production, but FANJ, which offers a broader philosophical discourse, and works intensively in local spaces to support community, disclosive conversation, and new philosophical visions, offers greater transformative potential.

Growers working in urban contexts and also practicing permaculture also pointed out differences between the state Urban Agriculture Department and FANJ, both of which they are members of. "It's a completely different thing," José Luís, who works for the Urban Agriculture Department, but also promotes permaculture described. He and numerous others remarked during interviews that, "Urban Agriculture is too linear," referring both to the departmental guidelines that require farms to follow the same layout of long, straight rows, and their rigidity in enforcing these guidelines. "It's very interesting," José Luís said with a twinkle in his eye, "That all of the Triple Crown *organopónicos* [in this city] are run by permaculturists." The Triple Crown is the highest rating that the state can award to productive, diversified urban farms. That all urban farms in this municipality with this ranking were run by permaculturists indicates the importance of belonging to both a supportive state infrastructure (such as is provided by the department of urban agriculture) and another group that can provide the capacity to thrive within it. To understand how FANJ operates, the next section turns to an examination of their work and philosophy.

4.6   FANJ and the Adaption of Permaculture

In 1993, a solidarity brigade travelled to Havana from Australia to promote a practice they called permaculture. State agencies were not convinced to adapt the philosophy, but a year later, Cuban naturalist, geographer, and self-proclaimed nature lover Antonio Núñez Jiménez took advantage of a new Cuban law to found his own NGO (FANJ), and his staff took up permaculture as part of their organizational mission and philosophy. They defined *permacultura criolla* (creole or Cuban permaculture) as "a system of principles and methods useful for designing human settlements" (Cruz et al. 2006), and began to offer workshops and courses in and around Havana to already-practicing urban farmers, as well as to individuals interested in self-provisioning with home gardens. In 2003, FANJ hired a promoter to work from a central Cuban province and the organization began promoting permaculture to a similar cohort of individuals there, plus rural agriculturists seeking to change their farming styles.

Permaculture was originally developed in the 1970s by Australians Bill Mollison and David Holgrem. They described it as a creative design process that, following a set of social and ethical principles, guides practitioners to imitate natural patterns in the construction of sustainable habitations (Mollison 1991). The philosophy and the practices associated with it became popularized amongst niche communities in Australia, North America, Europe, and the United Kingdom. From there, it travelled to the Global South, including locations throughout Asia and Latin America.

Although permaculture has emerged in a unique context in Cuba, it is worth noting that in the English-speaking world, permaculture is commonly associated with a "hippie" lifestyle. Even as sustainable agriculture, agroecology, and organic agriculture enter into mainstream US discourse, permaculture remains at the fringes, and is often characterized as unscientific (Ferguson and Lovell 2013). Moreover, the North American permaculture movement has increasingly come under fire in the food justice community for its lack of diversity and cultural representation. Reflecting on her experience as a Black woman in permaculture, one food justice advocate says she has wondered, "How is it that these permaculture principles speak to me, yet the movement does not"? (Baxter 2015).

In Cuba, a more diverse body of people practice permaculture, and Cuban permaculturists who have interacted with their North American counterparts note a distinct difference in the geographic manifestations of the practice. They point out that although some US permaculturists practice in isolation and spend significant money on permaculture equipment and materials, that they use it to enact low-resource farming or gardening from a community- and socially-informed basis. They liken it to the Cuban propensity to *inventar,* or creatively devise solutions to problems using available resources and skills (Gold 2015; Del Real and Pertierra 2008).

4.7 Permaculture Principles and Practices

As do organizations in other countries, FANJ teaches permaculture as a practice that revolves around eleven ethical principles (Cruz et al. 2006):

1. An element (e.g. the herb oregano) supports various functions (e.g. providing a harvestable cooking ingredient and warding off pests);
2. A function (e.g. bringing water to the garden) is supported by various elements (e.g. rainfall collected in a tank, plus a local stream);
3. You should look for opportunities rather than see problems;
4. Cooperate instead of compete;
5. Use energy efficiently;
6. Close cycles in order to minimize waste;
7. Obtain benefits;
8. Use natural resources without depleting them;
9. Use border areas and value the marginal;
10. Guarantee diversity; and
11. Respect natural cycles and succession.

These principles are presented at Cuban permaculture design courses, *interiorizado* (interiorized) during group activities, and verbally cited when permaculturists come together to plan or discuss the construction or progress of a new permaculture farm/garden. They provide rules of thumb, ethical orientation, and the base of a shared discourse that permaculturists use to communicate with each other and others about their farming/gardening practices. Esteban, a leading permaculture promoter, described the philosophy and principles as such:

> Permaculture is a methodology to understand and to work... or you could call it a way to manage content. You come to better manage the content, whatever kind of content it is. Say it's money. Depending on the form in which you manage this money, you will end up with more money, or you will lose it all. So when you apply this methodology, or the principles – which aren't exactly the same in all of the world – you are going to end up being more functional, or more efficient from the energetic point of view, the economic point of view, and the spatial point of view. When you see that permaculture is a new form to analyze what you are doing, and you analyze everything that you do, [permaculture] becomes a science. It's a powerful science. If boys and girls were taught to think and reflect, and they didn't have to memorize what someone else thought, the world would advance more quickly.

To act on the principles and philosophy of permaculture, practitioners draw on farming and gardening techniques that are also used in agroecology, organic agriculture, and the traditions of small-scale farming communities. They include composting, intercopping, the integration of animal and plant-production, biological pest management, and the use of vertical space (such as trellising) to intensify production in small spaces. As they determine what to grow and how to carryout these techniques across available space, permaculturists are taught to identify what they call "elements in the system" (including anything ranging from wind, the sun, or a stream, to neighbor's chickens, or car exhaust from a nearby highway), consider how they will affect the farm, and then organize their plant and animal production into various "zones." Using this set of techniques, principles, and practices, permaculturists report that they are able to better *aprovechar el espacio* (take advantage of space) and diversify their production in ways that can satisfy both their family's self-provisioning and the market.

4.8   Creating Community Through Permaculture

Though permaculturists often come to FANJ with the intention to learn practices that will benefit their individual farms or households, they encounter a social world that revolves around strong social bonds within the network, improved familial solidarity, and a sense of interconnection with the broader community. For many practitioners, this becomes very appealing, and ultimately sparks their commitment to the network. Moreover, they consider it to be a key element of FANJ's success in encouraging people to adapt sustainable agriculture techniques and supporting flourishing farms. As Xiomara, an urban farmer and permaculturist, says: "We feel as if we were a family. The work that we do as promoters of permaculture in [this province] is known all around the world precisely because of this."

When a person joins FANJ as a permaculturist, they are assigned to a local neighborhood or town group. Each group has a volunteer leader that is responsible for maintaining lines of communication, passing on news about upcoming events of interest, organizing local meetings, and bringing together the group to support someone when they need help on their farm or at their house. As, Caridad explains of her local permaculture group:

> We'll say, *caballeros* [gentlemen/women], the last Saturday of the month we'll go to Carmina's system. If there is anything she needs help with, we'll help her. If not we'll sit around and *compartir* [share] with each other. We'll share the information the foundation has sent us… if there is going to be an advanced course or a youth course for example. If there is going to be an exhibition or an event. If we are going to have visitors. We get together and talk about *nuestras cosas* [our things] and laugh a bit [translated from Spanish by the author].

Like Xiomara, Caridad says that her local group is like a family to her. For some permaculturists, their relationships grow to extend beyond organized meetings to frequent informal get-togethers. Gustavo says of a friend and fellow permaculturist "We are like brothers. If a week passes and I don't see Julio I'll say 'Hey! Julio hasn't been by [the farm]!'"

The feeling of close connection between permaculture practitioners not only facilitates friendly visits and opportunities for permaculture discourse to be repeated and reinforced, but also fosters a sense of mutual responsibility. It underscores the commitment permaculturists have to share time, materials, and other resources with each other. As another permaculturist describes: "We always support each other, help each other, share with each other. When someone needs something, someone else helps. We visit each other's *patios* (back yard gardens)."

In addition to emphasizing relationships between permaculturists, the permaculture philosophy also encourages an examination of all relationships affecting a person, and therefore their permaculture "system" (food production zone and home). Gender equity, for example, has been identified in national permaculture workshops as a countrywide problem and barrier to permaculturists' well being. Thus, meetings organized by FANJ often include discussion of gender relations and reflections on the causes of gender-based violence, or, more mundanely, expectations of labor-division with the household. For example, Esteban confesses that before he affiliated with FANJ and became a permaculturist, he would wait at home for his wife to prepare dinner, even if she had to work later than he did. Now, he says he will always start cooking if he is the first to arrive home, and explains that he understands that he is not doing her a "favor," but is contributing his fair share to "the well-being of all of the household."

Cuban Permaculturists see the family as the central "zone" of their system, but they are also concerned with how their system interacts with wider systems. Thus, FANJ teaches that each permaculture system—or farm—is also a "subsystem" of the larger community. To put this idea into practice, some permaculturists form alliances within their neighbors. Gladys and her co-workers, for instance, who run a municipal urban farm, regularly offer classes to students in neighboring primary schools, high schools, and the local university. They also run a composting program, which sends student volunteers to neighbors' homes to collect kitchen scraps, which ultimately become the rich organic matter that produces their abundant flower and vegetable crops. Participants are encouraged to come to the farm to access low cost vegetables and herbs.

4.9 Capacity, Disclosure, and Food Sovereignty in Permaculture

Gladys's composting program is indicative of the Cuban permaculture movement's conception of both food sovereignty and capacity to disclose it. FANJ describes both national and individual food sovereignty as goals that must be achieved through "local participation." In this formulation, food sovereignty is seen as achievable when individuals are able to actively work to resolve local food provisioning, agricultural, and ecological problems. The availability of organic matter —topsoil—is a problem for many would-be growers. The permaculture network was not the first or only to identify and consider it, but the local network they had enabled a disclosive space for collaborative problem discussion and resolution that propelled Gladys toward a solution. Moreover, as she shares this experience with other members of the network, some have moved on to start similar compost collection practices for their own gardens. In each instance, they are affecting their capacity to change the material reality of their farm or garden ecosystem.

The phrase "*no crear dependencia*" (don't create dependency), a slogan of the permaculture movement, is also seen in efforts like the composting program. Permaculturists stress the importance of working in cooperation with each other to find autonomy from external agencies or entities that might derail or slow their work. This is also seen in FANJ's seed saving project; in order to ensure a constant supply of diversified seeds and protect permaculture systems from the instability of state seed markets and their connection to international economies, permaculturists grow, harvest, and save their own seeds, and get together formally twice a year, and informally as needed to share and swap them.

Additionally, permaculture's emphasis on the interconnectivity of all "elements" in a permaculture system encourages permaculturists to be observant of and to discuss problems they witness in their surrounding ecosystem and culture. For instance, at a national gathering, permaculturists living in a coastal neighborhood of the central province complained that their local political leaders were doing nothing to protect pollution and litter from entering local mangrove swamps. In moments like these, the permaculture movement reveals its active critique of norms and institutions that permaculturists do not believe to be sufficient. Such critique is revealed, as well, in the movement's previously mentioned differences with the Urban Agriculture Department. However, rather than resigning themselves to passive complaint, permaculture activists in Cuba encourage direct engagement. Following the lamentation of these coastal dwellers, a woman from a coastal area in another part of the country spoke out that her permaculture group had started a committee to patrol against pollution and educate the local community on how to avoid contributing to it. The conversation facilitator jumped into suggest that the original speakers must do the same in their community—and "take responsibility" for the problem, themselves. In these projects of norm shifting, the permaculture movement prioritizes the "demonstration" of alternatives, rather than engagement in conflictive debate. The same tactic is followed in disagreements with Urban Agriculture Department officials, who permaculturists are working to slowly per-suade to change guidelines through test plots where they can show the productivity and success of new growing methods.

The sense of agency and individual/community-sovereignty that actions like these provide to permaculturists is particularly meaningful in contemporary Cuban society, where intense bureaucratization has caused many people to feel immobi-lized in their everyday lives (Gold 2015). Héctor, who has a very small, suburban farm describes this feeling of frustration in an encounter he had with ANAP. He wanted to cut down one of the many trees he had planted on his own land to build a new wooden goat-pen, and was required to secure permission and equipment through ANAP. Héctor says he passed through a convoluted and protracted paperwork process before finally being granted permission many months later. "An association should represent you and help you to do things, not block you!" he insisted. While dealings with large-scale institutions in this period of limited resources and staffing can cause Cuban growers to feel stymied, they describe that the experiences they have in permaculture as helping them to feel "motivated," "hopeful," and "satisfied;" feelings which support their ongoing pursuit of social and ecological transformation and disclosure.

Disclosure also requires that groups have the capacity to grow and expand outward, so that change is spread beyond fringe communities. To do this, FANJ and its permaculture network employ a variety of institutional and informal practices to recruit new members. When a potential permaculturist is identified, the permaculture community begins by working to *sensibilizar* (sensitize them to) them to permacultural thinking through informal social events and site (farm/garden/home) visits. These events are characterized by casual conversation and ice-breaking activities, which allow people to get to know each other. FANJ, through its permaculture promoters, will then invite new participants to a cost-free, intensive course, which formally constitutes the participant as a trained "permaculturist," and entitles them to attend all future FANJ events, including seed swaps, work days, presentations, and occasional advanced courses. When no formal courses have been planned, permaculturists work, instead, to informally train interested individuals through self-organized, local workshops and exchanges.

In both informal and formal spaces of education, the Cuban permaculture movement uses a pedagogical philosophy that draws on values of horizontal knowledge exchange and learning *en la practica* (in practice) as promoted by the international campesino-to-campesino movement and Paolo Freire's philosophies of popular education (Freire 2000). Based on this, they suggest that the best way to learn is through direct conversation, horizontal knowledge-transfer, and hands-on learning. For instance, Héctor once organized a workshop on soil management for members of his church congregation. During the event, he led participants around his small farm, showed them his worm compost, methods of crop rotation, and other farm features. Then, he took everyone to a clearing where he had gathered fallen leaves, various weeds and plant cuttings, food scraps from his household, and waste from a composting toilet, and he proceeded to lead them in building a compost pile. "I know about the science of soil," he told me afterwards, reflecting on what he had and had not shared with participants during the workshop, "but I didn't think that was the most important. I wanted them to observe and have the practice."

Through such exchanges, the permaculture network has slowly spread a new form of ecological and social consciousness for food production across the island. Beginning with only six practitioners in the early 2000s, FANJ reported in 2015 that 123 people from the central province have now passed permaculture courses and 67 additional people have joined the network through informal outreach and workshops. Similar numbers have been documented in other parts of the country. While these numbers may be small in comparison to the wider population of farmers and would-be gardeners who could join alternative agriculture movements in Cuba, they are significant for a small NGO.

Although FANJ and the permaculturists focus on localized manifestations of food sovereignty, they also conceptualize these in relation to national goals. Members of the permaculture movement have calculated that their small production systems are sufficiently productive that if multiplied across available Cuban land, they could support national self-sufficiency in food (in calories at least, if not in all commonly consumed products) (Casimiro-González 2007).

4.10   Lessons for Food Sovereignty and Justice Movements

Cuban food and agricultural workers, farmers, and activists have faced the challenge of extreme material scarcity, limited inputs, transitioning state institutions, and heavily depleted soil in their efforts to achieve ecological sustainability and food sovereignty. The creation of policies, infrastructure, and institutions supporting low-input, decentralized forms of "sustainable agriculture," opened the doors to new projects and experimentations. However, to take advantage of them, and to begin the still-ongoing work of "capacitating" individuals and creating cultural change in society at large, Cuba has needed additional organizations able to work in local contexts to create discussion and make room for agency, therefore building the capacity to implement and disclose new practices. In other parts of the world, additional and/or locally specific iterations of these challenges are experienced. In the United States, for instance, we also face soil depletion, but are met with excessive access to inputs, and the additional challenge of historical structures of racism, classicism, and sexism that challenge equitable access to the resources needed produce food, and create food sovereignty. Compared to Cuba, US groups have very limited government programs to which they can turn to for support in accessing land or beginning to produce healthier and more diverse foods. However, food justice activists and scholars can draw important lessons from Cuban permaculturists who have also had to supplement and go beyond state structures in order to seek transformation in their food and farming systems.

This chapter has examined Cuba's permaculture movement—through an ethnographic analysis based in a central Cuban province—not in an attempt to argue that permaculture is inherently more transformative or conducive to food sovereignty and justice than "agroecology," "organic agriculture," or any other methodology of sustainable agriculture, but to consider why this particular iteration of it has been so influential in affecting the research participants, and causing them to adopt long-term commitments to sustainability and food system change. It has argued that permaculture principles create a shared form of discourse that draw Cuban permaculturists together as a "family," creating a community in which people continue to discuss problems, build local, participatory capacity for resolving them, and begin to disclose new visions for food autonomy and both local and national sovereignty.

The network has enabled physical spaces and opportunities for practitioners to come together regularly to reflect on their shared philosophy, and to plan future activities. Through a strong social support structure, it keeps people connected to these spaces on an ongoing basis. It is characterized by a shared discourse that ties permaculturists together as a distinctive group who share a collective desire to continue promoting the discourse to new people. The network and the geographic communities to which it expands demonstrate an interest in and ability to take on new local environmental and food-related projects.

Food Justice proponents in North America may be facing a distinct set of problems from those encountered and prioritized by Cuban permaculturists, but by drawing on FANJ's example, and taking seriously the need to compliment efforts for structural change with localized efforts to create social networks and shared values, they may also be able to create communities with capacity to discuss problems, disclose them, and generate locally-appropriate, autonomous actions for creating new food and farming systems.

Acknowledgements The author wishes to thank the National Science Foundation, the UNC Graduate School, and the UNC Institute for the Study of the America's for their generous support of various phases of this research and writing. She is also grateful to the Fundación Antonio Núñez Jiménez de la Naturaleza y el Hombre for providing visa sponsorship, support, and mentorship during her fieldwork.

References

Altieri, Miguel A., Nelso Companioni, Kristina Cañizares, Catherine Murphy, Peter Rosset, Martin Bourque, and Clara I. Nicholls. 1999. The greening of the 'barrios': Urban agriculture for food security in Cuba. *Agriculture and Human Values* 16 (2): 131–140. doi:10.1023/A: 1007545304561.

Altieri, Miguel A., and Fernando Funes Monzote. 2012. The paradox of Cuban agriculture. *Monthly Review* 63(8). http://monthlyreview.org/2012/01/01/the-paradox-of-cuban-agriculture/.

Baxter, Kirtrina. 2015. A black woman practicing permaculture. *Permaculture design: Regenerating life together.*

Benjamin, Medea, Joseph Collins, and Michael Scott. 1989. *No free lunch: Food and revolution in Cuba today.* San Francisco: The Institute for Food and Development Policy.

Benjamin, Medea, and Peter Rosset. 1994. *The greening of the revolution: Cuba's experiment with organic agriculture.* New York: Ocean Press.

Casimiro-González, José Antonio. 2007. *Pensando Con La Familia En La Finca Agroecológica.* Habana: Cubasolar.

Cruz, Maria Caridad, Roberto Sánchez Medina, and Carmen Cabrera. 2006. *Permacultura criolla.* La Habana, Cuba: FANJ.

Del Real, Patricio, and Anna Cristina Pertierra. 2008. Inventar: Recent struggles and inventions in housing in two Cuban cities. *Buildings & Landscapes: Journal of the Vernacular Architecture Forum* 15: 78–92.

FAO. 2015. *The state of food and agriculture, social protection and agriculture: Breaking the cycle of rural poverty.* Rome: Food and Agriculture Organization of the United Nations.

Ferguson, Rafter Sass, and Sarah Taylor Lovell. 2013. Permaculture for agroecology: Design, movement, practice, and worldview. A review. *Agronomy for Sustainable Development* 34 (2): 251–274. doi:10.1007/s13593-013-0181-6.

Freire, Paulo. 2000. *Pedagogy of the Oppressed, 30th Anniversary Edition.* 30th Anniversary edition. New York: Bloomsbury Academic.

Friedmann, Harriet, and Philip McMichael. 1989. Agriculture and the state system: The rise and decline of national agricultures, 1870 to the present. *Sociologia Ruralis* 29 (2): 93–117.

Funes, Fernando. 2002. The organic farming movement in Cuba. In *Sustainable agriculture and resistance: Transforming food production in Cuba*, 1–26. Oakland, Calif.: Food First Books.

Funes, Fernando, et al. (eds.). 2002. *Sustainable agriculture and resistance: Transforming local food production in Cuba*. Oakland: Food First Books.

Gold, Marina. 2015. *People and state in socialist Cuba: ideas and practices of revolution*. New York: Palgrave Macmillan. http://www.palgrave.com/us/book/9781137539816.

Holt-Giménez, Eric. 2015. Food justice and food sovereignty in the USA. *Nyeleni Newsletter*, September.

Holt-Giménez, Eric, and Yi Wang. 2011. Reform or transformation? The pivotal role of food justice in the U.S. food movement. *Race/Ethnicity: Multidisciplinary Global Contexts* 5 (1): 83–102.

Kerssen, Tanya M. 2015. Food sovereignty and the Quinoa boom: Challenges to sustainable re-peasantisation in the southern Altiplano of Bolivia. *Third World Quarterly* 36 (3): 489–507. doi:10.1080/01436597.2015.1002992.

La Via Campesina. 2011. Organisation: The international peasant's voice. February 9. http://viacampesina.org/en/index.php/organisation-mainmenu-44.

Malik, Khalid. 2014. *2014 human development report*. New York: The United Nations Development Programme. http://www.undp.org/content/undp/en/home/librarypage/hdr/2014-human-development-report.html.

McKay, Ben, Ryan Nehring, and Marygold Walsh-Dilley. 2014. The 'state' of food sovereignty in Latin America: Political projects and alternative pathways in Venezuela, Ecuador and Bolivia. *The Journal of Peasant Studies* 41 (6): 1175–1200. doi:10.1080/03066150.2014.964217.

McMichael, Philip. 2009. A food regime genealogy. *The Journal of Peasant Studies* 36 (1): 139–169. doi:10.1080/03066150902820354.

Mollison, Bill. 1991. *Introduction to permaculture*. Tasmania, Australia: Tagari.

Morgan, Faith. 2006. *The power of community: How Cuba survived peak oil*. The Community Solution.

Nelson, Erin, Steffanie Scott, Judie Cukier, and Ángel Leyva Galán. 2009. Institutionalizing agroecology: Successes and challenges in Cuba. *Agriculture and Human Values* 26 (3): 233–243. doi:http://dx.doi.org.libproxy.lib.unc.edu/10.1007/s10460-008-9156-7.

Núñez Jiménez, Antonio. 1960. *Reforma Agraria En La Revolución Cubana*. Habana: Departamento de Relaciones Públicos.

Oxfam America. 2001. *Cuba: Going against the green*. Boston and Washington, D.C.

Premat, Adriana. 2009. State power, private plots and the greening of Havana's urban agriculture movement. *City and Society* 21 (1): 28–57.

Redmund, LaDonna. 2013. *Food+justice=democracy*. Manhattan: Tedx.

Rosset, Peter, Braulio Machín Sosa, Adilén María Roque Jaime, Ávila Lozano, and Dana Rocío. 2011. The Campesino-to-Campesino agroecology movement of ANAP in Cuba: Social process methodology in the construction of sustainable peasant agriculture and food sovereignty. *The Journal of Peasant Studies* 38 (1): 161–191.

Rural Economic Policy Program. n.d. *Measuring community capacity building: A workbook-in-progress for rural communities*. The Aspen Institute. https://www.aspeninstitute.org/sites/default/files/content/docs/csg/Measuring_Community_Capactiy_Building.pdf.

Reardon, Simón, Juan Alberto, and Reinaldo Alemán Pérez. 2010. Agroecology and the development of indicators of food sovereignty in Cuban food systems. *Journal of Sustainable Agriculture* 34 (8): 907–922.

Spinosa, Charles, Fernando Flores, and Hubert L. Dreyfus. 1999. *Disclosing new worlds: Entrepreneurship, democratic action, and the cultivation of solidarity*. Reprint edition. The MIT Press.

Valdés, Orlando. 2003. *Historia de La Reforma Agraria En Cuba*. Habana: Editorial de Ciencias Sociales.

Williams, Justine M. 2012. *Agroecology versus productivism: Competing discourses on the future of Cuban agriculture*. MA, Chapel Hill: Anthropology Department at the University of North Carolina.

World Food Program. 2015. WFP Cuba brief. www.wfp.org/countries/cuba.

Wright, Julia. 2009. *Sustainable agriculture and food security in an era of oil scarcity : Lessons from Cuba*. London ; Sterling, VA: Earthscan.

Chapter 5
It's Not Just About Us: Food as a Mechanism for Environmental and Social Justice in Mato Grosso, Brazil

Marisela A. Chávez

Abstract The Landless Workers Movement (MST) is one of the most important social movements in the world for the implementation of agrarian land reform. Their fight for access to land has been based on the premise that land should serve a "social function." Since its birth in the 1980s, the MST has settled more than one million people in Brazil on approximately 35 million acres of land (an area about the size of Paraguay). Many of the settlements across the country have demonstrated a commitment to move beyond social justice by combining environmental justice into their discourses and activities, and pinning their struggle with the fight for food sovereignty. This ethnographic research explored the different ways that environmental discourses and activities are being incorporated into the movement by describing the experience of the 12 de Outubro settlement in the state of Mato Grosso. Interviews with members of 12 de Outubro reveal that by implementing alternative agricultural methods like agroecology and agroforestry, they believe they are able to restore and protect the land that they acquire, while working towards food sovereignty. Secondly, they hope to demonstrate that their struggle for access to land is not just for individual benefit, but rather, that by growing healthy food sustainably and by developing a cooperative that benefits the entire community, land is fulfilling the social function that it should. Finally, they believe that their activities have connected them to the larger urban community through the establishment of CANTASOL, a solidarity commercialization system, extending awareness about food, the environment, and social justice into the urban sphere.

Keywords Food sovereignty · Social movements · Environmental justice

M.A. Chávez (✉)
University of Kansas, Lawrence, KS, USA
e-mail: chavmarisela@gmail.com

© Springer International Publishing AG 2017
I. Werkheiser and Z. Piso (eds.), *Food Justice in US and Global Contexts*,
The International Library of Environmental, Agricultural and Food Ethics 24,
DOI 10.1007/978-3-319-57174-4_5

5.1 Introduction

*By politicizing market culture, and its material consequences, these struggles
reformulate the meaning and content of social change. Those deemed casualties of
progress become agents or vehicles of critique of the normalizing claims of
development. Their critique is not so much in development's terms (success or
failure), but in terms that are infused with the particular values and meanings
through which they engage in struggle for rights, access, and representation*
McMichael 4.

In Brazil, the landless poor have been questioning the amount of land belonging
to very few in an ever more resounding way, placing themselves at the center of the
land debate during the past century. Since the early 1980s when the Landless
Workers Movement (MST) became formalized, at least one million people have
been resettled in about 35 million acres of land (an area about the size of Paraguay).
This social movement has spread to every state, and though they share the same
ideological platform, their commitment to the implementation of the agrarian
reform and reaching social justice has manifested itself in different ways. The types
of activities and development that has taken place at settlements has varied greatly.
However, there are indications that that is changing. At the national level, the MST
has adopted a discourse that is supportive of sustainable agriculture and committed
to food sovereignty, it has even established a couple of national schools where
members can be trained and obtain degrees in agronomy and civic organization.
Examples of MST settlements that have adopted this model can be seen all over the
country. But perhaps a particularly striking example comes from the state of Mato
Grosso,[1] the "Midwest of Brazil," one of the most dynamic agricultural frontiers of
the Amazon.

Mato Grosso is the only state in Brazil that holds within its borders three distinct
biomes—Cerrado (savanna), Pantanal (wetlands), and Amazon rainforest.
Intensive, mechanized agriculture and cattle ranching have been major drivers of
deforestation in this region, threatening not only these delicate ecosystems, but also
the livelihoods of the people that inhabit them. During the period of colonization of
the Amazon region, the military dictatorship encouraged the settlement of Mato
Grosso and its neighboring states through a series of public and private settlement
projects. These projects incentivized the migration of large numbers of people.[2]
Within a span of 50 years, the landscape of the state has been transformed dramatically due mainly to the expansion of intensive agriculture[3] and cattle

---

[1]Total area of 903,378.292 km^2.

[2]Between 1960 and 1970 the estimated population was about 117,608 people, but by 1980 that
number had jumped to 433,545 (IBGE, *Censos demográficos de 1980, 1970, 1960*). According to
the 2010 census, the total population in Mato Grosso was 3,035,122 people (IBGE).

[3]Currently, agriculture makes up about 40% of the state's GDP (IBGE). Mato Grosso has steadily
increased its production of soybeans over the past seven years and it currently leads as the state

ranching.[4] In the last couple of decades, the huge investments for agricultural production (particularly soy). By 2013, the total land area utilized to cultivate soy was 7.9 million hectares, or about 51.8% of the total land area (IBGE).

Because of the profitability of big agriculture and some mining projects, land is highly contested in Mato Grosso and the practice of *grilagem*[5] has been widespread. So much so, that one study demonstrated that at one point, the land titles exceeded the total state land area (Wolford and Wright 20). The amount of land owned or operated by the big agriculture companies (most of them multinational) has also created a lot of tension. It has brought with it cultural shifts, and a growing disparity of what and how people think about food production in the state. In some circles, the "agroboys"[6] exemplify progress, while in others, they are resented because they represent land inequality issues.

It is in this context of tensions over land and big investments in intensified agriculture that one MST settlement, 12 de Outubro,[7] has decided to challenge the dominant agricultural models surrounding it in defense of the agrarian reform and the environment. The level of organization and range of activities taking place at 12 de Outubro, coupled with the intentionality in which its programs are connected— from farming to student engagement—made it an exciting research location. I approached the research expecting to gain some more understanding of the manifestations of food sovereignty in the community, and walked away with examples of how people negotiate with these ideas and implement them at different scales. The experience of 12 de Outubro demonstrates the ways in which MST members talk about and live out their vision for agrarian reform. By engaging in sustainable agriculture and building strong connections to the closest urban center, they are using food as a vehicle through which they can be heard and understood. Most importantly, it enables them exercise their allegiance with the food sovereignty movement and experiment with different agricultural models to minimize

---

(Footnote 3 continued)

with the highest total production. In 2013, the state was producing 23,417,000 metric tons of soybeans.

[4]The total area used for cattle ranching was 24 million hectares, with a total of about 29 million heads of cattle.

[5]Illegal possession of land and illegal land titling. The word itself refers to drawing up a title and placing it in a box with crickets so that it can appear aged to help individuals legitimatize their illegal possession of land.

[6]"Agroboys" is a local term used for the culture of large landowners that have adopted U.S. ranchers fashion and practices. Most of this has been influenced by Texan cowboys and can be traced back to the close connections between Texan ranchers and the expansion of intensified farming in Mato Grosso.

[7]The 12 de Outubro settlement was chosen as a study site for this project after I conducted a survey of all of the existing MST settlements and camps in Mato Grosso with help from the Pastoral Land Commission in Brazil. Subsequent interactions with MST members and local NGOs at regional meetings provided some insight on the activities that were taking place at 12 de Outubro which led me to visit and then return for a longer stay to conduct the field work.

their impact on the environment. It's about food, creating awareness, and protecting the environment.

5.2 Bringing Food to the City

Every Thursday evening, several families come together to sell their goods in the city. They have spent the last couple of days collecting and organizing the produce and other goods from the settlement. It is a coordinated effort, a concept they are familiar with and practice widely in everything they do in the settlement. There is excitement in the air, lately they have made more and more friends in the city. They have regular customers, people stop by and ask them questions. People recognize their banner and the red flag waving by the table. There is a lot of activity—people coming and going, picking up the produce they have ordered or stopping by to see if there is anything they might be able to pick up. The eggs and chicken go first, "we have a lot of demand for that!" Rosa says proudly. "It's because it tastes different, we let the chickens roam around, they are healthier than I am!" she laughs. They are proud of what they have been able to accomplish. In only a year, they have gone from just growing food for themselves to forming a budding community supported agriculture program. New faces stop by.

Is this grown close by?

"At 12 de Outubro," Rosa replies, with a huge smile on her face.

Rosa represents just one of the 150 families that occupied part of a large estate some 50 km outside of Sinop, Mato Grosso in 2007. They arrived armed with some basic essentials and the hope to have piece of land they could call their own someday. It took two years of living in make-shift camps before INCRA[8] redirected them a neighboring fazenda and granted them the land. The 4000 ha were divided up evenly amongst the families, granting them each a 20 ha plot, and leaving the remaining 1000 ha for a community operated agroforestry extractive reserve. The settlement is one of the symbols of victory for the MST in the state of Mato Grosso. It lies in the heart of the state's most concentrated soybean production area, which makes their push for alternative forms of agriculture to sustain their communities that much more interesting (and quite often, contentious).

It is not easy to get there, the only access is through a dirt road, marked only with a small sign on the side of the highway. The settlement inherited the buildings that had belonged to the estate—barns, a couple of small houses, a mill, and some other

---

[8]The *Instituto Nacional de Colonização e Reforma Agrária* (INCRA, National Institute of Colonization and Agrarian Reform) is a federal entity has the largest land jurisdiction in the region, it owns land is responsible for land titling. It was formed in 1970 to resettle people at 100,000 families in undesignated or illegally occupied lands, it was also assigned the responsibility of overseeing the colonization of the Amazon during the 1970s.

basic infrastructure. The school and the old barn adjacent to it, are the central hub of the community. This is where most of the meetings take place, where people congregate. There are only a few businesses, a small store where basic provisions can be purchased (like sugar, milk, hygienic products, dry goods, and household cleaning items), and next to it a small bar that is owned by the same family. Its an important informal meeting space where people exchange ideas, discuss issues, and share news, especially after larger settlement meetings that take place across the street in the old barn. A web of dirt roads connect the plots and the homes, as well as small lake and the forested communal area. It is a quiet community.

5.3 Building the Cooperative

On the first day that I arrived to the settlement, I was invited to attend a COOPERVIA[9] and CANTASOL[10] meeting. CANTASOL is a project that emerged in the settlement in 2013 in response to a proposal by two UNEMAT[11] students who were representing the university's Canteiros[12] initiative. The students, Paulo and Daniele, and their supervising faculty member proposed a project[13] that would benefit both the *assentados* and the university students, who were needing to complete service hours for their degree programs. It was simple, why didn't they partner together to start a community supported agriculture program? They had learned that the settlement was already producing a variety of organic produce and goods sustainably, why not expand it to generate some supplemental income for their families? They could design a website[14] to market and ensure the sale of their products online, which would increase their customer base, and give them access to customers in the city. The profits would be a benefit for the settlement, and the surrounding area would benefit by having access to more sources of organic produce. After some discussion and voting with the rest of the community, they agreed and CANTASOL was formalized.

---

[9]*Cooperativa de Produtores Agropecuários da Região Norte de Mato Grosso* or Northern Region Cooperative of Agricultural Producers.

[10]*Sistema Canteiros de Comercialização Solidária* (CANTASOL) or Canteiros Solidarity Trade System project.

[11]*Universidade Estadual do Mato Grosso* or Mato Grosso State University.

[12]Canteiros is an initiative developed by students and professors at UNEMAT who seek to break past what they considered to be the exclusion and elitism of academia by encouraging the exchange of ideas with the community and engaging in service-learning projects. The objective is to find ways to strengthen the working class, not to maintain inequalities.

[13]The idea was modeled after another organization, Instituto Ouro Verde, who had already established a successful community supported agriculture program like this in Alta Floresta, Mato Grosso.

[14]The CANTASOL website can be found at: http://cantasol.org.br/portal/.

The website went at the beginning of 2013, and the first sales took place just a week later. The students arranged to have a space at the entrance of the university in Sinop, where the orders could be picked up by consumers. Sales average between R \$500 and R\$800 per week. The supplemental income is welcomed by all of those involved, though the community engagement in a solidarity system that they can all benefit from is celebrated the most (field notes). For many of the consumers, the online CANTASOL portal is their first point of interaction with the settlement, and sometimes, with MST members in general. The program is accomplishing several things, it is guaranteeing direct sale of goods between the producers and the consumers, it is promoting agroecological production, it is building a solidary economy, and it is creating awareness both about food and the people who produce it. In this case, it is building a bridge between marginalized peoples and other stakeholders in the region.

There are two teams that work together to keep the operation running smoothly, the 'city team' and the 'settlement team'. The city team is composed of Daniele and Paulo, as well as the professor that supervises their involvement in the project. Their primary responsibilities are managing the website, delivering the products at the UNEMAT campus in Sinop, building and managing relationships with consumers, marketing, and accounting. The settlement team is composed of the cooperative that sell their goods through the program, and the youth that take part in the on-site management of the program.

Every Saturday a general meeting is held with everyone involved in the program. There, the cooperative producers turn in a list of items that they have available for sale and the inventory is updated online. Then, consumers have from Sunday to Tuesday to place their orders online. At the end of the day on Tuesdays, the city team gives the settlement team a report on the orders that were placed through the website. Over the next two days, the settlement team is responsible for reaching out to the producers, and collecting and organizing the orders. On Thursdays, the orders are delivered to the city team at the university where the consumers pick them up and confirm payment. Finally, at the Saturday meetings the producers receive final payment for the goods that they sold.

The whole process is built on the values of the MST—land, solidarity, education, and cooperation. From the point of the view of the settlement, it is an opportunity for them to extend their net to make organic produce available to more people at fair prices. Not all MST members have rural roots, some joined the movement from large urban areas. They know all too well the challenges for the urban poor to have access to healthy produce. This topic came up often at the general meetings when they discussed adjustments in the prices of goods. They strongly voiced that prices need to be accessible because the urban poor have just as much right to access organic food as the city's elite (field notes).

5.4 Producing the Goods

Currently, there are about 30 families that benefit directly from the CANTASOL project, which means that they are making weekly sales consistently (field notes). Their products include produce (fruits and vegetables), eggs, chickens, pork, jams, candies, organic cleaning products, and bread. Though the products that are sold are not all organic, a majority of them are.[15] On the website, the products that are organic are labeled as such, and at the time of sale they are differentiated by a different colored label and price (field notes). Though the numbers fluctuate, they have about 300 consumers that regularly purchase their products and there is optimism that that number will keep growing. They have also organized a couple of farmer's markets[16] in Sinop to provide the consumers with the opportunity to interact with the producers and market their products.

The producers in the settlement practice *agricultura familiar* or family agriculture. Each family has a plot where they are able to plant and cultivate for their own consumption or to sell (as is the case for those involved in the cooperative or that sell independently). Some people have rented out their plots for cattle grazing, while others utilize them for their own livestock. Utilizing the agroecological system is highly encouraged, and the settlement counts with a few individuals who have been trained and certified in agroecology by the national movement's agronomy school.[17] There are also workshops and classes that occur frequently supported by their partner organizations (like CPT, UNEMAT, and Instituto Ouro Verde) that share best practices, like how to make liquid biofertilizers, compost, and design agroforestry and agroecological management systems, among other things (field notes).

In addition to their plots, most of the families have vegetable and herb gardens around their homes. They cultivate manioc, lettuce, squash, beans, chicken, eggs, and fruits. I visited the gardens and plots of the each of my informants, but relied on their description of the type of system that they used to classify them. I wanted to know how they *articulated* their practices, rather than imposing my own terminology. In these interviews, people talked about how they produced and why in different ways. However, there was a common root—awareness of the negative impacts of agrochemicals both for the environment, and for their own health and the desire to learn more about nature's own models. In different places in the settlement (especially in the communal buildings) there were images which established a clear connection between the use of agritoxins and health. Signs like: "Agritoxins Kill. Agrarian Reform Creates Employment and Healthy Food, Agribusiness = Agritoxins." Beyond just establishing the connection with health, it also pushes forward the

---

[15]About 25 of the 30 families that participate produce their goods organically (field notes).

[16]*Feiras da agricultura familiar e reforma agraria* or family agriculture and agrarian reform fairs.

[17]The MST national movement has a school called Florestan Fernandes in São Paulo, and a research center in Rio Grande do Sul where a few of the individuals from the settlement have spent some time to learning about the different aspects of these alternative agricultural models.

discussion about the types of practices engaged in by agribusiness to produce their goods. Furthermore, it tries to create an awareness that differentiates the practices supported by the movement at the local, state, and national levels.

The settlement is surrounded by soybean plantations, which can potentially have some negative repercussions on the livelihoods of the *assentados*, especially because they do not subscribe to the monoculture system in which those plantations are set up (both in their use of agrochemical and GMO seeds). According to one account, some producers involved in the CANTASOL project have been negatively impacted by the agritoxins that are sprayed on the some of the soybean plantations nearby.

The *assentados* work very hard to differentiate themselves from the urban community. Especially when it comes to their interactions with the environment and the kinds of inputs that they utilize in their farms. Those who utilize a small amount of inputs or that utilize environmentally friendly ones, are proud to talk about it and give demonstrations on how they work. For some, this is their first foray into agriculture as adults, coming back to the *campo* after having lived and worked in the city as laborers, mechanics, etc. for most of their lives.

5.5   Engaging Youth

The CANTASOL program was developed with the idea that eventually, the settlement would operate it independently. Because of that, it is integrated into the curriculum of many of the workshops that the settlement's youth participate in conjunction with the *Novos Talentos*[18] (New Talents) classes at the settlement school.[19] The school at the settlement was built by MST members, with the collaboration of other settlements and some funding from the state. In addition to the normal pre-K to eighth curriculum that is imparted there every day, a series of other evening and weekend classes take are held to increase participation and promote activism[20] within the settlement. The classes are team taught by professors from UNEMAT, teachers, and volunteers from the settlement. Anyone in the settlement

---

[18]*Novos Talentos* is another program that was established in partnership with the UNEMAT Canteiros initiative.

[19]Even though it is a state sponsored school, the community plays a very active role in carefully choosing who is qualified to teach there. The leadership at the settlement takes many steps to ensure that the education imparted there aligns to the movement's own ideals. They are focused on nurturing critical citizens, people who can think and make decisions collectively.

[20]The pedagogic approach at the assentamento school is reminiscent of the Freiren model that was implemented in the CEBs. The school's regular k-8th grade curriculum and the Novos Talentos classes follow the same bottom-up leadership development in which solidarity and cooperation are favored over individual or elitist relationships. Because of the early connections between the Brazilian liberation theologians of the Catholic Church, the CPT, and the landless, it is not surprising to see this as the model being implemented especially because it promotes liberation and collective thinking.

can participate, it is not just limited to school-aged children. The three main classes that are linked to the CANTASOL program are audio-visual technologies,[21] agronomy,[22] and civic engagement.[23] They build on the skills learned in each class to inform the participation of the food production process, and participants are given the training to assume leadership and administrative positions within the project itself.

A good example of building of encouraging collaboration amongst different groups of students is through mandala designs. First, the students in the agronomy class learn about the different kinds of fruits and vegetables that can be grown in the region. Then, in art class they learn about mandala designs and do activities to experiment with that. Back in their agronomy class, they are introduced to the concept of mandala agriculture and asked to adapt their mandalas designs to create a mandala agriculture design. They are asked to think about the types of fruits, vegetables, and legumes as well as animals that can be raised and cultivated within this type of system. Finally, at the civic engagement class, several designs are selected and merged to put into practice. The students are asked to choose the elements of each design that they like the best in order to come up with a collective design (field notes). In each step of the process, the students are encouraged to think creatively and collectively. But beyond that, they are introduced to concepts and given the opportunity to materialize them, putting into practice what they have learned. They deliberately enable students to look at agriculture from a different vantage point, one shaped by the principals of agroecology.

For the university students involved, the involvement of the settlement's youth has been one of the most positive outcomes of the program because of it's a powerful educational tool.[24] Most importantly, it helps the youth to prepare for the future, learning how to manage and improve the institutions of the community (the cooperatives, the associations, the school, etc.). When they participate in these

---

[21]In the audio-visual technologies classes the students learn how to take photos, conduct interviews, film, and utilize computer software to produce short documentaries that explore a variety of themes. The most important aspect of this is that the topics they explore are all related to their own realities, they interview their neighbors and document the issues that they face.

[22]The agronomy class is designed to teach the *assentados* about the importance of agriculture, expose them to different philosophical lines in agriculture, and provide them with tools and training to farm sustainably. They learn how to grow and cultivate crops by experimenting with some of the different systems that they are learning about.

[23]The civic engagement class is designed to teach team-building, communication, critical thinking, and community collaboration skills.

[24]As part of one her classes, one teacher took the students to the university to show them how the delivery/pick-up of the products worked. One of the students did not understand why some products were sold in a bundle and others by the kilo. So, the teacher grabbed one of the bundles of green onions and counted them out one by one. She explained how light they were and that if sold by the kilo, the producer would not be able to profit very much. Then, she demonstrated by placing the onions on the scale and asking the students to do the math to determine what the profit would be utilizing one approach versus the other (field notes).

spaces, they gain important experiences of the potential of transforming conditions of life, and the type of work that these communities engage in.

5.6 Food Sovereignty in Practice

"For us, an agrarian reform does not have any meaning if it doesn't have a notion of environmental conservation, of taking care of the air that we breath, the water, restoring the land and farming it sustainably" (field notes). The experience of 12 de Outubro can teach us a lot about the social, economic, and environmental importance of engaging in sustainable alternative forms of agriculture. As a whole, the MST movement has been developing pedagogy based on ideas of social justice and food sovereignty, and disseminating it throughout the various organizational levels. The notion of food sovereignty within the movement varies by location, but the campaigns developed at the national level have been aimed at motivating its members to engage in sustainable agricultural production practices like agroecology. They have been training its members in the foundations of agroecology, even offering certification and degree programs. For the *assentados* involved in the program at 12 de Outubro, food sovereignty is about living with dignity, feeding themselves in a way that helps them take better care of the land that they worked so hard to obtain. They inherited an estate that was very degraded but they have noticed changes taking place in the landscape since they started implementing agroecology and agroforestry methods. They have recovered natural springs, and have even observed some bird and mammal species that have begun to reappear in larger numbers, especially bird species (field notes). They have also recognized the productivity of alternative agricultural models vs monoculture. They like the quality of the products that are derived from a more diversified system. But it is a matter of economics too. They understand that producing a single type of crop in high concentration can lead to susceptibility of pests and diseases in crops, especially if you don't have the additional expensive inputs needed to avoid that (fertilizers, pesticides, etc.). It is just not cost-efficient, they say, for family farmers.

The major differences were in the *ways* that people articulated their subscription to food sovereignty. Some framed it as more environmentally friendly practices to food production, others linked it more to the integrity of people and the environment, others to having more autonomy. For them, it's about justice—for people and the environment. They believe that looking at agriculture through a different lens helps connects people to the understanding that it's not just about people respecting the environment, but also about people having a different relationship with each other.

5.7   Linking Rural and Urban Spaces

Perhaps one of the most interesting aspects of this settlement, is its objective to establish and maintain a strong connection to the urban community in Sinop. The partnerships that they have made with UNEMAT, have opened up clear and purposeful paths to connecting the *assentados* to the inhabitants of Sinop. This opens up the possibility to exchange ideas and practices, and to share frameworks with an extended community. Those involved with the CANTASOL project observe that it has helped to connect the *assentados* to each other (by purchasing from each other the goods that they produce), and it has also contributed to establishing a connection with people who live in the city of Sinop. For example, at least twice per week, the people who purchase the products in the city have some type of interaction with the producers from the settlement. In those exchanges, there is opportunity for dialogue about what is going on in the settlement and the challenges that they face. These opportunities can help change the preconceptions that people have of members of the MST.[25] In the media, they are often portrayed as criminals and vagabonds.

For the settlement, food has given them the ability to demonstrate the positive things that they are accomplishing, to get the chance to show the reality of the settlement. Linking the urban and rural spaces enables them to introduce more people to the concept of food sovereignty and communicate the importance of making people aware to the fact that access to healthy food is not just an issue for marginalized communities, but a challenge that everyone faces. They say, "it's not just about us," meaning that concern about what we eat and how it's produced is a cause that everyone should rally around, not just marginalized communities.

The Novo Talentos project is also an important link because of the exchange of students and teachers between the settlement and UNEMAT, and the audiovisual materials that are produced and shared amongst the constituents of both groups provide good platforms for creating awareness and fostering dialogue. These partnerships are foundational in the transformation of preconceptions that the inhabitants of Sinop have in regards to the MST. Furthermore, they are important ways for members of the MST to continue establishing open dialogue and collaboration.

---

[25]Daniele pointed out that before she got involved with the Cantasol cooperative project, she had never been exposed to the MST very much other than the comments she heard from her uncles, "I didn't really understand what the movement was, or what kind of work they do because I didn't have very much proximity to them because my uncles were always very prejudiced against them. After I met them and started working my admiration became very big. I managed to change the mind of many of them. Here in Brazil that organization suffers a lot from prejudice. And after, when I began to wear the MST shirt, I felt in my own skin what they go through on a daily basis— at the university, on the street. That day at the fair, people stopped to buy things, but when they saw the flag of the movement they did not like that very much" (field notes).

5.8 What's Next? Conclusions and Challenges Ahead

My observations and interviews at 12 de Outubro some nuances about food production practices and the settlement's regional engagement. It also shed some light about food sovereignty in practice, and connections between agrarian citizenship and alternative agricultural models.

In the end, the experience of 12 de Outubro reveals that it's about individual choices—the choices that people are able to make about what they plant and how they make their livelihood. It highlights the things that individuals are *able* to negotiate with and the support systems that are needed to provide the right vehicles for that. It is about the importance of food as a medium to address social and environmental injustice, both in the access of food (creating opportunities to have access to organic food) and in the cultivation of it (through alternative agricultural that aligns more closely to nature's own processes).

More broadly, I observed that by adopting a food sovereignty stance, the MST has begun a process of transforming itself as not must merely a socio-political movement, but an environmental-socio-political force in Brazil. There were several indicators of this, but nothing conclusive. Among them, the literature produced by the movement, especially materials that I found in their website and in their different social media sites. But perhaps the most important, and that which I had the most contact with, were the dialogues from the interviews in which individuals often pointed to the importance of combining the environmental, social, and political. Especially when individuals communicated that their fight for land and agrarian reform was inseparable to environmental causes because they depend on the environment for survival. Of special consideration were also the dialogues of individuals indicating that because of this understanding, they have become more aware of environmental issues. Though not often communicated clearly, there were some indications of an awareness that at the very least, the MST was not a movement that solely focused on agrarian reform, but rather, that it was including discussions of food and the environment as centerpieces to their objectives.

They are building connections through food. The CANTASOL project has increased their interactions between the people who live in Sinop and the *assentados*, by providing urban consumers access to locally-grown foods, these *assentados* are not just linking their food systems, but they are also extending conversations about food sovereignty into the urban sphere. In the process, they are hoping to change the perceptions that people have of them in order to facilitate continued dialogue and collaboration.

MST members at 12 de Outubro are directly and indirectly pushing forth a discussion about food as mechanism to address social and environmental justice in the region. They are accomplishing this through their day-to-day-activities, especially in their promotion of alternative agriculture as means to protect an environment that has been exploited before. Most importantly, they are accomplishing this by connecting injustices to people (lack of services, exploitation, land to grow food) to the environment. Both of these ideas culminate in the CANTASOL program.

Despite their success, the movement still faces serious challenges. When I visited the stand at the university recently, I ran into a young girl that I recognized. I had taken a photograph of her at one of the farmer's markets that the settlement had organized in Sinop. She told me her story. She is 17 years old and decided to leave the settlement because she didn't see a future for herself there. She wants to live in a city, to have access to college and a steady income, she doesn't want her future to depend on farming. Yet, she is struggling to find work and is facing the possibility of having to return. Cases like these are a reminder that ideals of the MST movement will be continuously tested. More than 30 years ago, thousands of people came together to fight for land, placing their hope in creating something that they could leave behind for the next generations. They won and since then, they have been counting victories across the country and working hard to ensure the future of their families. But does that mean that the next generation will want the same things? The movement has invested a lot of time and money into the creation of schools and training programs for its youth, to instill an understanding of the struggle for land, the importance of agrarian reform and social justice. But is it equipped to face the pressures of a rapidly changing society? Will its youth continue to embrace its foundations or will they seek opportunities outside of what the movement can offer them?

These are the challenges that the MST faces, but places like 12 de Outubro offer the possibilities of what can be accomplished when strong partnerships are forged within and outside of the movement. The settlement has embraced food as a vehicle for social and environmental justice by connecting education with activism and egalitarian development.

References

Altieri, M.A. 1987. *Agroecology: The scientific basis of alternative agriculture.* Boulder: Westview Press. Print.

Altieri, Miguel A. 2009. Agroecology, small farms, and food sovereignty. *Monthly Review* 61 (3): 102–113. Print.

Altieri, Miguel A., and Victor Manuel Toledo. 2011. The agroecological revolution in Latin America: Rescuing nature, ensuring food sovereignty and empowering peasants. *The Journal of Peasant Studies* 38 (3): 587–612. Print.

Altieri, Miguel A., Fernando R. Funes-Monzote, and Paulo Petersen. 2012. Agroecologically efficient agricultural systems for smallholder farmers: Contributions to food sovereignty. *Agronomy for Sustainable Development* 32: 1–13. Print.

Amaggi. Web. http://amaggi.com.br/?lang=en. Accessed March 30, 2015.

Anderson, Molly D. 2013. Beyond food security to realizing food rights in the US. *Journal of Rural Studies* 29: 113–122. Print.

Anderson, Molly D. 2008. Rights-based food systems and the goals of the food systems reform. *Agriculture and Human Values* 25: 593–608. Print.

Branford, Sue, and Jan Rocha. 2002. *Cutting the wire: The story of the landless movement in Brazil*. London: Latin American Bureau. Print.

CANTASOL. Sistema de Canteiros de Comercialização Solidária. Web. http://cantasol.org.br/portal/. Accessed July 2014.

Carrol, C.R., J.H. Vandermeer, and P.M. Rosset. 1990. *Agroecology*. New York: McGraw Hill Publishing Company. Print.

Chappell, Michael, and Liliana A. LaValle. 2011. Food security and biodiversity: Can we have both? *Agriculture and Human Values* 28: 3–26. Print.

Comissão Pastoral da Terra (CPT). Canuto, Antônio, Cássia Regina da Silva Luz, and Edmundo Rodrigues Costa eds. 2014. *Conflitos no Campo-Brasil 2014*. Goiâna: CPT. Web. http://www.cptnacional.org.br/index.php/component/jdownloads/finish/43-conflitos-no-campo-brasil-publicacao/2392-conflitos-no-campo-brasil-2014?Itemid=23. Accessed March 20, 2015.

Corrêa Filho, Virgílio. 1969. *História de Mato Grosso*. Rio de Janeiro: Instituto Nacional do Livro. Print.

Desmarais, Annette A. 2007. *La Vía Campesina: Globalization and the power of peasants*. Ontario: Fernwood. Print.

Desmarais, Annette A. 2002. The Vía Campesina: Consolidating and international peasant and farm movements. *The Journal of Peasant Studies* 29 (2): 91–124. Print.

de Souza, Eduardo Ferreira. 2004. *Do Silencio á Satanização: O Discurso de Veja e o MST*. São Paulo: Annablume. Print.

Edelmen, Marc, Tony Weis, Amita Baviskar, Saturnino M. Borras Jr, Eric Holt-Giménez, Deniz Kandiyoti and Wendy Wolford. 2014. Introduction: Critical perspectives on food sovereignty. *The Journal of Peasant Studies 41*(6): 911–931. Print.

Ehrlich, P.R., A.H. Ehrlich, and G.C. Daily. 1993. Food security, population, and environment. *Population and Development Review* 19 (1): 1–32. Print.

Fairbairn, Madeleine. 2010. Framing resistance: International food regimes and the roots of food sovereignty. In *Food sovereignty: Reconnecting food, nature and community*. Oakland: Food First Books. Print.

Forman, Shepard. 1971. Disunity and discontent: A study of peasant political movements in Brazil. *Journal of Latin American Studies 3*: 14–16. Web.

Friedmann, Harriet. 1987. International regimes of food and agriculture since 1870. In *Peasants and peasant societies*, ed. T. Shanin. Oxford: Basil Backwell. Print.

Friedmann, Harriet. 1989. Agriculture and the state system: The rise and fall of national agriculture, 1870 to the present. *Sociologia Ruralis* 29 (2): 93–117. Print.

Gardner, Morgan. 2005. *Linking activism: Ecology, social justice, and education for social change*. New York: Routledge. Print.

Holt-Giménez, Eric. 2009. From food crisis to food sovereignty: The challenge of social movements. *Monthly Review* 61 (3): 142–156. Print.

Holt-Giménez, Eric, and Annie Shattuck. 2011. Food Crises, food regimes and food movements: Rumblings of reform or tides of transformation? *The Journal of Peasant Studies* 38 (1): 109–144. Print.

Instituto Brasileiro de Geografia e Estadísticas. Web. http://www.ibge.gov.br/english/. Accessed December 15, 2014.

James Jr., H.S. 2003. On finding solutions to ethical problems in agriculture. *Journal of Agriculture and Environmental Ethics* 16: 439–457. Web.

Jansen, Kees. 2014. The debate on food sovereignty theory: Agrarian capitalism, dispossession and agroecology. *The Journal of Peasant Studies* 1–20. Print.

La Vía Campesina. 1996. The right to produce and access to land. Position of La Vía Campesina on Food Sovereignty presented at the World Food Summit, November 13–17, Rome.

La Via Campesina International Peasant's Movement. What is La Via Campesina? Web. http://viacampesina.org.

MacLachlan, Colin M. 2003. *A History of modern Brazil: The past against the future*. Wilmington: Scholarly Resources Inc. Print.

McMichael, Philip (ed.). 2010. *Contesting development: Critical struggles for social change*. New York: Routledge. Print.

McMichael, Philip. 2009. A food regime geneology. *The Journal of Peasant Studies* 36 (1): 139–169. Print.

Movimento dos Trabalhadores Rurais Sem Terra. http://www.mst.org.br/.

Neto, Elias Nasrala, Francisco Antonio de Castro Lacaz, and Wanderlei Antonio Pignati. 2014. Health surveillance and agribusiness: The impact of pesticides on health and the environment. Danger ahead! *Ciencia & Saude Coletiva 19*(12): 4709–4718. Print.

Peet, Richard, and Elaine Hartwick. 2009. *Theories of development: Contentions, arguments, alternatives*. New York: The Guilford Press. Print.

Perfecto, Ivette, John Vandermeer, and Angus Wright. 2009. *Nature's matrix: Linking agriculture, conservation and food sovereignty*. London: Earthscan. Print.

Pestana Barros, Carlos, Ari Francisco de Araujo Jr., and João Ricardo Faria. 2013. Brazilian land tenure and conflicts: The landless peasants movement. *Cato Journal 33*(1): 47–75. Print.

Pimentel, D., and M. Pimentel. 1996. *Food, energy, and society*. Niwot: University Press of Colorado. Print.

Pretty, J.N., J.I.L. Morison, and R.E. Hine. 2003. Reducing food poverty by increasing agricultural sustainability in developing countries. *Agriculture, Ecosystems & Environment* 95 (1): 217. Print.

Ricketts, T.H. 2001. The matrix matters: Effective isolation in fragmented landscapes. *American Naturalist* 158: 87–99. Print.

Rosset, Peter. 2011. Food sovereignty and alternative paradigms to confront land grabbing and the food and climate crises. *Development* 54 (1): 21–30. Print.

Rosset, Peter, Raj Patel, and Michael Courville (eds.). 2006. *Promised land: Competing visions of agrarian reform*. Oakland: First Food Books. Print.

Saure, Sérgio and Sergio Pereira Leite. 2011. Agrarian structure, foreign land ownership, and land value in Brazil. Global Land Grabbing Conference. Web. http://www.iss.nl/fileadmin/ASSETS/iss/Documents/Conference_papers/LDPI/14_Sergio_Sauer_and_Sergio_Pereira_Leite_-_ENGLISH.pdf. Accessed January 25, 2015.

Schanbacher, William D. 2010. *The politics of food: the global conflict between food security and food sovereignty*. Santa Barbara: Praeger Security International. Print.

de Schutter, Olivier. 2011. Agroecology and the right to food. Rep. no. A/HRC/16/49. Geneva: United Nations. Report Presented at the 16th Session of the United Nations Human Rights Council. Web.

Seufert, V., N. Ramankutty, and J.A. Foley. 2012. Comparing the yields of organic and conventional agriculture. *Nature* 485: 229–232. Print.

Smith, T.Lynn. 1964. Land reform in Brazil. *Luso-Brazilian Review* 1 (2): 3–20. Web.

Thompson, P.B., R.J. Matthews, and E.O. van Ravenswaay. 1994. *Ethics, public policy, and agriculture*. New York: Macmillan Publishing Co. Print.

Vandermeer, John. 1995. The ecological basis of alternative agriculture. *Annual Review of Ecology and Systematics* 26: 201–224. Print.

Vandermeer, John, and T. Dietsch. 2003. The fateful dialectic: Agriculture and conservation. *Endangered Species Update* 20 (4): 199–207. Print.

Vandermeer, John, and Ivette Perfecto. 1997. The agroecosystem: A need for the conservation biologist's lens. *Conservation Biology* 11 (3): 591–592. Print.

Wittman, Hannah. 2010. Mobilizing agrarian citizenship: A new rural paradigm for Brazil. In *Contesting development: Critical struggles for social change*, ed. Philip McMichael. New York: Routledge. Print.

Wittman, Hannah, Annette Aurélie Desmarais, and Nettie Wiebe (eds.). 2010. *Food sovereignty: Reconnecting food, nature and community*. Oakland: Food First Books. Print.

Wolford, Wendy. 2010. *This land is ours now: Social mobilization and the meanings of land in Brazil*. Durham: Duke University Press. Print.

Wright, Angus, and Wendy Wolford. 2003. *To inherit the earth: The landless movement and the struggle for a new Brazil*. Oakland: Food First Books. Print.

Chapter 6
Save the Whale? Ecological Memory and the Human-Whale Bond in Japan's Small Coastal Villages

Seven Mattes

Abstract Whales are a common property and a potential natural resource for the taking. Of the many resources they can provide humans, their flesh, or "whale meat," has become controversial in the past three decades. The controversy lies largely with the whale as a prominent charismatic mega fauna. Whales have become a symbol and source of environmental activism, floating in the middle of a highly contested political and ideological struggle. Japan stands at the center of the international whaling dispute, refusing to accept the global anti-whaling norm. Since the 1982 moratorium, Japan has put in an annual request to the IWC each year to create an exception of the moratorium for a number of small coastal whaling villages to carry out traditional practices. They are continually denied, despite the centrality of the whale in these cultures. As conservation efforts, dams, and other modern alterations relocate humans from their traditional lands or prevent humans from interacting with key species, we are increasingly discovering that affected communities lose more than access to natural resources, but that key parts of the culture itself is forced to be left behind, if not forgotten entirely over time. This chapter explores what is lost when whaling is removed from small coastal whaling villages of Japan—addressing how global anti-whaling discourse may save whales, but harm human-whale relations in Japan.

Keywords Japan · Whaling · Ecological memory

"A whale on the beach is wealth for seven villages," says an old Japanese proverb. Throughout the world whales are considered common property—invoking the need for regulation on an international scale to avoid the problems inherent in the commons (Hardin 1968). Thriving within and around human-derived national boundaries, yet existing as a living natural resource, whales have become a symbol and source of environmental activism, finding themselves in the middle of a highly contested political and ideological struggle. Japan has stood at the center of the

S. Mattes (✉)
Department of Anthropology, Michigan State University, East Lansing, MI, USA
e-mail: mattes.seven@gmail.com

© Springer International Publishing AG 2017   67
I. Werkheiser and Z. Piso (eds.), *Food Justice in US and Global Contexts*,
The International Library of Environmental, Agricultural and Food Ethics 24,
DOI 10.1007/978-3-319-57174-4_6

whaling debate for decades. Even with the plethora of information and misinformation within the volatile discourse surrounding Japanese whaling, there are victims of the struggle who are obscured under the dust cloud of international debate—the small coastal whaling villages of Japan and the human-whale relationships that exist therein. Whaling regulations' focused only on the global level disregards the traditional centrality of the whale in these community's distinct cultures.

Problematizing prevalent perceptions of Japan as a homogenous, modern industrial nation I ask why these villages are excluded from the same indigenous rights to hunt whale meat as other "non-Western" and "traditional" communities. It is important to note that this chapter is not intended as an argument in support of whaling, but as a space to sort through the misconceptions the international whaling discourse has propagated while highlighting the significance and value of the human-animal relationships involved in regards to sustainable subsistence practices. Taking into consideration the complexities of hunter-prey relationships, particularly those with an extensive socio-historical foundation, and pulling from Shelter's (2007) concept of ecological memory, I ask what is lost when dismantling the relationships these communities have held with whales over the course of centuries? As we increasingly live in a world in which humans are separated from nature, sometimes in the name of "preservation" or "conservation", we forget the important of interaction and "being with" non-human others (King 2010).

This chapter begins with a brief introduction to the socio-historical background of whaling in Japan. Next, it introduces the entrance of the whale as a mascot of the environmental movement and corresponding regulations of various whale species. Introducing and applying the framework of ecological memory to the traditional whaling communities of coastal Japan, I suggest that understanding the significance of human-animal relationships in sustainable subsistence practices may shed light on alternative options that preserve a way of life that culturally values a threatened species.

6.1 Misconceptions of Japanese Whaling

The international whaling controversy is ripe with misconceptions. These misconceptions regarding whales, traditional whaling practices, and the diversity of Japanese culture are highly influential in the decision to not grant aboriginal subsistence whaling rights to the traditional whaling villages of Japan. The most prominent misconceptions are within the generalization of "Japanese whaling," often discussed as though it is a singular practice. To the contrary, "whaling" encompasses a variety of hunting strategies involving a wide range of whales.

There are Baird's beaked whales, minke whales, bottlenose dolphins, and pilot whales living near the coast of Japan during the hunting season for practitioners of small-type coastal whaling. Of the listed species, all are too small to be regulated by the International Whaling Commission (IWC), the organization who dictates global policies regarding whale hunting, except for the minke whales. While environ-mentalists commonly decorate signs and t-shirts with, "Save the Whale," the failure to specify exactly which whale they are referring to leads to what Kalland and Moeran (1992) calls the belief in the "super whale," or

> A whale which is at the same time the largest animals on earth (true of the blue whale), has the largest brain (the sperm whale), has a large brain-to-body-weight ratio (the bottlenose dolphin), sings nicely (the humpback whale), is friendly (the gray whale), but endangered (the bowhead and the blue whales), and so on (8).

Each whale species has distinct patterns, sizes, predation styles, and statuses on the endangered species list. In other words, there is no one whale to save—in Japan least of all.

Accompanying the various species of whales in Japan's hunting purview are three types of whaling: small-type coastal whaling (STCW), large-type coastal whaling (LTCW), and pelagic whaling. When environmentalists protest Japanese whaling, they are often referring to LTCW and pelagic whaling, rather than STCW. These three types differ in years practiced, strategy utilized, and the location of the hunt. LTCW hunts larger whales, specifically sperm and baleen whales. They use larger boats, around 700 tons, whereas STCW boats average 70 tons. The LTCW hunts occur not only close to shore, as with STCW, but can take place throughout the ocean, capable of staying out at sea for a month at a time. The smaller STCW boats return to shore every night and never venture further than 150 miles from shore. However, both STCW and LTCW depend on coastal processing (Kalland and Moeran 1992, 95). Pelagic whaling differs from both the above types in that the entire operation—hunting and processing—takes place in large ocean vessels, allowing the whaling to occur far from a Japanese base over an even longer period of time.

Similarly, Japanese whaling is not generalizable due to its long and diverse history, with the direct hunting of whales occurring as early as the 16th century (Kalland and Moeran 1992). Whaling as a livelihood tended to remain centered on select coastal villages up until the war period, which heavily influenced the whale consumption. During World War II, pelagic and LTCW whaling was halted due to the destruction of the easily targeted whaling vessels being brought down by enemies in open waters. During this time, STCW increased due to the decrease in the other whaling types. Post-war large-scale starvation led to an increased need for a source of protein. Whale meat constituted as much as 47% of protein intake during the post-war period in Japan, a large percent that remained high even years after the war, constituting 23% of protein intake in 1964 (Akimichi 1988). When Japanese adults recall the eating of whale meat, it is often of this wartime nour-ishment (Arch 2016). However, the negative effects of this increase in whaling became noticed internationally. "By 1960, Japan had surpassed Norway as the leading whaling nation, but it was obvious to all that resources were being seriously depleted and that regulations were necessary" (Kalland and Moeran 1992, 90).

"Japanese whaling" is often discussed in general terms, particularly by those in opposition to the practice. However, the variation among whales, whaling practices, and whalers is vast. This chapter is focused on STCW and the whalers and whale species within the traditional whaling villages, whose stories and interactions are marginalized in the generalized terminology.

6.2 Regulation and the "Save the Whale" Campaign

The generalized associations of the whale, and of whaling practices—highly divided from the reality of whaling in traditional settings—served as a platform on which regulation was encouraged. International whaling regulation began with the creation of the International Whaling Commission (IWC) in 1946, for the purpose of providing "...for the proper conservation of whale stocks and thus make possible the orderly development of the whaling industry" (IWC 2016). Japan became a member in 1951. However, the 1960s and 1970s began a period of stricter and stricter whaling regulations imposed by the IWC, imposing limits and moratoriums on specific species of whales based on their level of depletion. While the IWC's original conception was for the promotion of sustainable whaling, the organization's orientation shifted into one fighting for whales' rights and a global "save the whales" campaign became the new agenda (Hirata 2008). Furthermore, this new agenda focused on the whale's welfare as the first and foremost priority, rather than the whaling industry—including the people who relied upon the animals as a livelihood. The whale became the poster child not just for the IWC, but also for the increasingly popular environmentalist movement, including prominent NGOs such as Greenpeace. Freeman (1990) explains that the whale was chosen as an environmental symbol because of their worldwide recognition combined with widespread misinformation about their biology, "The more mysterious or mythic the animal can be made to appear, the more interest its name or appearance generates. Information that embellishes the mythic properties of the creature will include various anthropocentric characterisations thereby increasing the public's concern for, and identification with it" (Freeman 1990).

Kalland and Moeran (1992) suggests that it was the public protests that came as a result of exploitation of whales in the Antarctic that led to their widespread attention by environmentalists. He states that this exploitation, "...has led to a public outcry, headed by environmentalists, that all whales are on the verge of extinction". This myth of the vulnerable concept of the whale insisted that they must be protected. Furthermore, due to their unique characteristics of being both mammal and marine life, Kalland and Moeran proposes that they are, "...'betwixt and between' and hence singled out for special attention, allowing them in some ways to take on the characteristics of a 'totem' for many environmentalists" (Kalland and Moeran 1992, 6). Advocates for the whale created the aforementioned "super whale", encompassing all positives traits to a generalized symbolic whale to serve as the mascot for their cause. The reality of whales, whaling and the people involved in the industry—such as the Japanese whalers in small coastal whaling villages—were largely unspoken of in the path to save this newly fetishized creature of the sea.

6.2.1 The Moratorium

This symbolic position, distant from many of the populations (and the diverse ideologies regarding whales therein) it affected, served as one of the factors that led to worldwide regulations on whaling throughout the 1970s and 80s.[1] Regardless of these regulations, which were to ensure sustainable whaling practices, the IWC implemented a 1982 decision to place a moratorium on all commercial whaling. "The moratorium was highly controversial as it was never endorsed by the IWC's Scientific Committee, which had not identified any need for a blanket moratorium on all whale species" (Hirata 2008, 185). This moratorium was based largely on the Western ideological beliefs that were driving the IWC's decision-making—beliefs that conflicted with those of whaling cultures that were affected by these regulations. Thus, there was opposition to this decision, particularly by Japan and Norway. The moratorium halted whaling for the 1984-6 seasons regardless of objections by these prominent whaling nations. This ban was meant to be a temporary five-year moratorium, followed by an assessment to determine the future of commercial whaling, "but it has become a de facto permanent ban on commercial whaling, even though in 1992 the IWC Scientific Committee reported that many whaling stocks had recovered..." (Hirata 2008, 186). Currently, approximately 2000 whales a year are hunted between Japan, Norway, Iceland, and Aboriginal Subsistence whaling. This is the end of nearly a century decline from pre-moratorium numbers, in which over 30,000 whales were hunted on a year basis (prior to the 1960s). In 2014, Japan reportedly hunted 196 whales, and an average of 480 per year in the four years prior (WDC 2016).

Twenty years later and the battle over the moratorium continues. Anti-whaling sentiment has become the norm internationally. Japan is, therefore, a renegade for opposing this international anti-whaling convention and continuing whaling for research/scientific purposes (Morikawa 2009; Hirata 2008). The reaction to Japan's insistence on scientific whaling, and hope for an end to the moratorium, is vocally negative. Japan's Institute of Cetacean Research (ICR) states,

---

[1]The Marine Mammal Protection Act of 1972, halting the taking of marine mammals in American and international waters and the importation of these animals into the United States (Hirata 2008) was followed by the successful and highly acclaimed Endangered Species Act of 1973 incorporated whales, as did the UN Convention on International Trade in Endangered Species of Wild Fauna and Flora (CITES), which restricted any international trade regarding species who are potentially harmed by trade. Regulations directly specifying whales were simultaneously being implemented. As early as 1972, the UN Conference on the Human Environment in Stockholm proposed a ten-year whaling moratorium on all commercial whaling. This proposal influenced the IWC to further restrict whaling. First, they established the New Management Procedures, which separated whale stocks into three categories. This categorization provided the basis for the IWC to create distinct quotas for each type of whale, rather than implementing a total moratorium. Further specifications in 1976 separated whales by the species to determine distinct quotas, in regards to their endangerment. Finally, in 1979 all pelagic whaling was banned by the IWC and the Indian Ocean was established as a cetacean sanctuary (Hirata 2008).

Emotionally based anti-whaling positions calling for an end to all commercial and research whaling irrespective of the abundance of whale stocks are unhelpful in resolving difficult international negotiations and have led to the current dysfunctional nature of the IWC characterized by its institutionalized and polarized rhetoric and confrontational conduct of its meetings (ICR 2016).

A IWC meeting in 2007, "Conference for the Normalization of the International Whaling Commission" reiterated these concerns, citing issues such as, "Disregard for the principle of science-based policy and rule-making; Disrespect of cultural diversity related to food and the ethics; Increasing emotionalism concerning whales; Institutionalized combative/confrontational discourse that discourages cooperation" (ICR 2007). The IWC's basis for the continuance of the whaling moratorium, the conference's notes portray, was not based on science, but polarizing ideals. These polarizing ideals directly affected traditional Japanese whaling villages.

6.2.2 Indigenous Whaling and Cultural Diversity

Despite these strong regulations, the IWC permits Arctic communities to whale based on their historical whaling traditions. Regardless of regional parts of Japan's historical whaling practices, they are denied a similar status. At the annual IWC meetings, Japan continues to argue for the importance of granting a number of small coastal STCW communities the right to whale due to the centrality of the whale in their cultural tradition (Hirata 2008). The rights of these few small coastal villages to whale, while important on their own, have become highly political. It has ceased to be simply a scientific issue, but one of cultural values and national pride for Japan (Morikawa 2009). These traditional whaling villages have thus become a pawn in the middle of a global debate to end the moratorium on commercial whaling.

The romanticism associated with indigenous groups, combined with ignorance of Japan's diversity and whaling history, are perhaps key reasons for the discrepancy in regulations. Kalland and Moeran (1992) comments on why Eskimos can hunt one of the world's most endangered whales while Japan is not allowed to hunt minke whales, who can easily survive sustainable hunting, "'It's alright for them,' goes the underlying, but unstated, line of reasoning. 'They live in the wild and are close to nature'" (Kalland and Moeran 1992, 6). There is a perception of Japan as a homogenous, urbanized, modern society. They stand as a first world, wealthy country having no need for a seemingly archaic past subsistence practice. To the contrary, Japan is a highly diverse nation, with communities ranging from the hectic salary-men of Tokyo to rural farmers in the mountains. These generalizations of Japan as a homogeneous nation serve to hide the reality of the financial and cultural losses that have been sustained following the whaling moratorium in those villages that once specialized in this subsistence practice.

Whether the scientific studies conducted by Japan are legit and necessary is not an issue that will be discussed here (see Morikawa 2009; Watanabe 2009).

However, manufacturing a discrepancy between Japanese traditional whaling vil-lages and the Arctic communities in terms of a cultural-historical tradition of whaling is arguably a misconception by the IWC. The messy discourse on the Japanese whaling debate often disregards the significance of the STCW villages, scoffing at the notion of giving these villages aboriginal whaling rights—viewing them as a mere ploy to further an end to the moratorium. This may be the case, yet the whaling culture embodied in these villages holds significance and the impact of the removal of the whale from their lives is apparent.[2]

This chapter takes the whaling culture and human-whale relations within these regions seriously, specifically for the purpose of addressing significant cultural loss in the midst of global conservation efforts. The preservation of these communities, both economically and culturally, is not regarded as an important endeavor by the IWC in light of the larger, global anti-whaling ideals and concerns (see Abel 2005; Hirata 2008). From the perspective Japan places in the media, these whaling communities struggle as globally focused ideological imperialists impose alterna-tive perceptions of the whale, nature, and whale-human relations, on their culture (Hirata 2008). It is the culturally diverse views of acceptable whale-human, and nature-culture relationships, which stand at the center of the ongoing ideological debates occurring in the IWC and in public discourse surrounding whaling. As one Japanese women, Sachiko S. Zushi, comments on the whaling ban, "'I feel the Japanese view of nature and the Western view of nature are fundamentally different. We Japanese consider humans as part of nature, but the Western people do not seem to think that way. I wonder where the human's place is in the Western view of the world'" (Zushi 1988, 116).

Zushi may be correct in that alternative views of the nature/culture relationships, and human/animal relationships, is a key factor in the whaling controversy. There is a strong dominant belief in the so-called "Western" world that the physical envi-ronment defined by geography textbooks is merely a stage on which our human cultural actions take place (Ingold 2000, 51–2). This is part of a nature-culture dualism that permeates not only interactions with nature, but non-human animals as well. Humans stand firm on the cultural end of the binary, whereas non-human entities such as plants, dirt, and cows, are part of nature—absent of those special features that supposedly elevate humanity above the "beasts". Non-human animals were viewed as "beast machines" for centuries, creatures not anymore capable of feeling pain than a standard clock. It has only been in the last two centuries that this mindset has changed. Animal welfare has slowly become a topic of discussion and

---

[2]It is important to note that the line between socio-political storytelling, and general manipulation of the facts, and the reality of documented and published whaling culture reported here continues to be questionable (e.g., Watanabe 2009; Morikawa 2009). The Japanese government has overemphasized the centrality of the whale on a national scale, for instance, whereas whaling culture (food, spiritual significance, etc.) is more limited to traditional whaling regions. Published works that rose out of the center of this debate should be called into question given the political culture surrounding their publications. For the purposes of this discussion, these published works and accounts will be addressed as legitimate.

non-human animals are increasingly recognized as having the capability to think and, as a main driver of welfare initiatives, suffer (Kalof 2007, 138–9). However, these same philosophies are not echoed in Japan's history, which has a tradition of viewing some animals in a similar light to that of human beings, allowing them to be deeply integrated into their social world.

While the West is familiar only with their own history of animal perceptions and human-animal relationships, decision-makers perhaps know little about Japan's. As Japan is a first world nation, as many countries in the West, it may be assumed that they share similar beliefs regarding nature-culture relations—especially when compared to that of aboriginal communities who have been granted the right to whale by the IWC. This has the potential to lead to misunderstandings concerning Japan's reasoning for remaining pro-whaling. The Western argument that whales should be specially protected, as they are rational and intelligent creatures, falls on deaf ears to a nation that does not necessarily judge an animal's worth by their distance from the West's "beast machine". Whereas the West bases their perceptions of the whale, and what is right and wrong therein, on their own socio-historical background, Japan has its own equally valid and strong beliefs regarding the natural world and the beings therein (see Imanishi 2013).

6.3 Ecological Memory and the Nature-Culture Divide

Japanese pro-whalers argue that the absence of the whale in traditional whaling villages results in the loss of far more than a long-established local occupation—but also historically and spiritually-rooted human-whale relationships. As conservation efforts, dams, and other modern alterations relocate humans from their traditional home or prevent humans from interacting with culturally-relevant species for tourism or conservation purposes, we are increasingly discovering that those societies lose more than access to natural resources. In these circumstances, key parts of the culture itself is forced to be left behind, if not forgotten entirely over time. Significantly, this loss can be linked the lack of acknowledgement of alternative perceptions of the nature-culture divide or hunter-prey relationships.

In *Imagining Serengeti: A History of Landscape Memory in Tanzania from Earliest Times to the Present* (2007), Shelter discusses the cultural tragedy that can ensue when removing a society from their traditional landscape in the name of environmentalism. National parks are common sights of environmental justice conflict as indigenous people (associated with "culture" by decision-makers) have been forced out to conserve "nature". As Figueroa and Waitt (2010) explains, "… the indigenous populations related to National Parks have typically been forced out or displaced, sometimes ignored as nomadic, often too easily perceived as absent from the area because of colonial-induced blind-spots created by ideological principles of nature-human separation in the nationalistic breaking of frontiers" (136). These cultural conflicts result in the loss of traditional lands in the name of ecological conservation, which is often separated from concerns of indigenous justice (Gottlieb 1994).

This occurred with the Massai, hunter-gatherers' who have lived partially within what is now the Serengeti National Park for two thousand years. Defined as a "wilderness" area and applauded as a great conservation effort on a global scale, the Massai were removed from the area that became Serengeti National Park. Shelter explains that the Massai "...view of the landscape has not been a part of the global conversations of other people who care about the Serengeti". Despite the area existing as part of their traditional homeland, "Western Serengeti peoples have been dismissed as recently arrived poachers within a landscape envisioned as empty of people. Yet, for as long as we have memory, the western Serengeti has been a profoundly humanized landscape with the stories, hopes, and challenges of its people deeply embedded in its rocks and hills, polls and streams, vistas and valleys" (Shelter 2007, 1).

Shelter uses the term "core spatial images" throughout this work to represent the reoccurring images within the landscape found in oral narratives of the people who resided in what has become this famous park. These images are associated with the history, cosmology, and social relationships within a culture in a particular environmental context. The cultural memory is implanted on the landscape, existing as part of the culture itself.

While the Massai were removed from their landscape, their cultural history remained within the boundaries of the now human-less park. When Shelter arrived on site and elicited their memories of the land, they jumped at the opportunity to become re-acquainted with their past. "Elders wanted to take me out to see the places themselves and to walk over the landscape as they told the stories, providing more information about people, events, and ideas associated with the places" (Shelter 2007, 19). As they walked, particular areas served as a mnemonic device for remembering their oral history. Shelter discovered that the landscape itself is saturated with meaning, "...oral traditions encode social relationships and identities by employing a spatial imagery that includes landscapes and topography" (20). A person's social identity alters how they view spatial areas in their surroundings and this identity is then reflected within oral narratives. Shelter states, "How people order their memories within a particular spatial construct or landscape, then, depends on their own social identity—their socially shared, situational definition of self in relation to others—that both shapes and is shaped by the landscape" (21). The Massai not only lost their land, but a key part of their cultural and social identity that was embedded in the landscape itself when they were removed from their traditional homeland. Their story reveals how closely their lives were, and remain, intertwined with the "nature" that was taken from them.

This same cultural intertwinement is experienced within alternative perceptions of hunter-prey relationships. The Cree, native hunters of northeastern Canada, interpret the stopping and staring of a caribou as an offering up of themselves to the hunter as a result of the hunter offering respect and even superiority to the animal (Ingold 2000, 13, 48). "They say the animal offers itself up, quite intentionally and in a spirit of good-will or even love towards the hunter. The bodily substance of the caribou is not taken, it is *received*" (Ingold 2000, 13). The Cree interpretation is a result of their understanding of non-human animals, in which, "…they are partners with humans in an encompassing 'cosmic economy of sharing'" (Ingold 2000, 48). Personhood, for the Cree, is simply an inherent part of being alive. This personhood of non-humans includes the personality and experience of living that humans themselves possess, including their own way of experiencing and understanding the world. The hunters' interactions with the caribou are, thus, not as humans versus nonhumans as with many Western hunters, but as a social interaction full of meaning on both sides of the weapon.[3]

Similarly, Northern indigenous communities share rich cultural relationships with the whales they hunt. The Inuit closely link their traditional beliefs and practices to whales, who are "common themes in the songs, legends, toponymy, art, dance, and the thoughts of Inuit everywhere…" (Freeman et al. 1998, 38). The Inuit cite whale hunting and the many rituals surrounding the practice as a link to their ancestors and an honoring of their mutually respectful relationship with whales and other prey they hunt. "Long ago, Inuit came to understand that animals provide people with the necessities of life in this world, and people, by observing the appropriate rituals and etiquette when animals are killed and consumed, are able to ensure that animal populations will remain healthy and ever-present" (40). The Inuit-whale relationship is one of deep respect—with rituals and practices reflecting this reverence. "The whale is more than food to us. It is the center of our life and culture. The taking and sharing of the whale is our Eucharist and Passover. The whaling festival is our Easter and Christmas, the arctic celebration of the mysteries of life" (55). As the caribou and the Cree, the Inuit state that their prey have no objection to their death at an Inuit's hand, and may even volunteer to be killed by a particular person. Due to this close and long-standing relationship with the whale, Inuit and other Alaskan indigenous communities are currently permitted to hunt a specified quota of traditional whale species under the title of Aboriginal Subsistence Whaling. The presence of the whale and the ability to carry on the ancestral traditions and the "core spatial images" within are recognized as central to preserving their culture.

6.4 Whaling Culture of Japan

This disheartening study of the Serengeti National Park, the beliefs of the Cree, and the human-whale relationship of the Inuit finds similarities in the small coastal whaling villages of Japan. The hunting of the various whale species (referred to as "whaling" from here on) in Japanese waters has existed in Japan dating back as far as 9000 years. Jōmon-aged pottery has been found with vertebral whale imprints,

---

[3]Where the caribou stands on this cycle of exchange is presently unknown, but worthy of future multispecies inquiry.

alongside a large quantity of cetacean remains at early settlements. These archae-ological findings serve as evidence revealing not only the consumption of whale meat, but also the inclusion of whales in Jōmon culture (ICR 2004). Nearly thirty years have passed since most forms of commercial whaling has existed as a viable and legal livelihood, resulting in the end of not just a way of life, but of cultural traditions built on the tradition of whaling. Due to the longstanding practice of whaling in these areas, a distinctive whaling culture had developed.

Akimichi (1988) defines whaling culture as, "...the shared knowledge of whaling transmitted across generations. This shared knowledge consists of a number of different socio-cultural inputs: a common heritage and worldview, an understanding of ecological (including spiritual) and technological relations between human beings and whales, special distribution practices, and a food cul-ture" (76). This common heritage is accentuated with myths, folk tales, and other narrative events that communicate historical events and ways of life for whaling villages throughout history, continuously reiterating the current generations' iden-tity as a part of this longstanding community. Knowledge of whales, their biology, their migration habitats, as well as hunting strategies are similarly passed on gen-eration to generation. These narratives and the continual whaling culture as a whole serve to confirm the essential dependence on the species they hunt to survive, as "...those who participate in a whaling culture are very aware that their entire social fabric depends on the continued existence of the whale and of whaling. This consciousness gives rise to the concept of an ecosystemic relationship between the two" (Akimichi 1988, 76).

This distinct whaling culture exists in those areas in which STCW whaling became the prevailing livelihood. The moratorium did not only remove the eco-nomic wellbeing of these communities, but disconnected them from the heart of their traditional culture. These areas correspond with the villages that Japan hopes to gain aboriginal rights to whale. The full integration of whales into the Japanese social world in these villages remains illustrated in the form of distinctive food culture, religious beliefs and ceremonies, livelihood, community, identity, and the mourning of the potential and total loss of all.

Japanese cuisine is an integral part of regional and national identity. While Japan is a relatively small nation, the mountainous geography has resulted in a large number of distinct regions, represented by dialectical, religious, and cultural vari-ants. At present, no cultural aspect is more touted in tourism advertisements than regional cuisine. "Regional variations in whale dishes among the STCW commu-nities, and other regions of Japan are pronounced, and serve to strengthen local identity" (Akimichi 1988, 4). This cuisine utilizes all parts of the whale just as it had it the past, shipping the parts unused to those regions that do include them in their regional specialties. Mutsuko O. of Osaka, a contributor to the 1988 compi-lation *Women's Tales of Whaling: Life Stories of 11 Japanese Women who Live with Whaling*, speaks passionately about her popular Osaka restaurant, which succeeded once she introduced traditional whale cuisine that hearkens back to Osaka's past. Mutsuko (1988) comments on the ethnocentrism of another country dictating what the Japanese should eat, stating, "'This is like telling your next door neighbor not to eat his dinner just because you don't like his food. That is awfully rude, isn't it?'" (Mutsuko 1988, 96). The loss of the food culture, particularly in whaling villages that relied upon fresh whale for tourism, is far from only a loss capital. Mutsuko ends, "America might be powerful enough to force Japan to give up whaling, but they cannot take our food culture away" (Mutsuko 1988, 98).

The regional whaling cuisine is not only a source of community pride and identity, but it has become incorporated into various celebratory functions, such as religious festivals. Shinto and Buddhism are the two main religions of Japan. While both retain their own individuality, the rituals, festivals, and general beliefs pertaining to these religions are often intertwined in everyday Japanese life. Whaling is no exception. In these traditional whaling villages, both religions have shrines, deities, rituals, festivals, and beliefs associated with whales and whaling practices. While fresh whale is valued and preferred among most villages, as early as 1988 frozen whale meat was being stock piled to preserve for use in the traditional celebrations due to the moratorium (Akimichi 1988, 4).

Distinctive whaling rituals have developed in these small coastal villages that represent both the interconnectivity of whales to their local culture and the perception of the human in the natural world. The Japanese understanding of the soul differs from that of the Judeo-Christian conception, for example. Whereas in Judeo-Chrsitian tradition, souls arguably belong to humans alone, the Japanese give the potential for souls to both animals and objects alike. Indeed, whales are not seen to have only souls, but are often believed to be gifts from the deities, or *kami*, themselves. "The Japanese perceive an interdependent world between animals and humans and reciprocal relations between these realms. The taking of a life is at the same time a giving of a life and the depth of this feeling is apparent in the energy and time expended in atonement and gratitude for it" (Akimichi 1988, 53). Kalland and Moeran adds, "To repay the whales for sacrificing their lives, whalers have furthermore to take care of their souls, or else these whale souls can turn into 'hungry ghosts' which might cause illness, accidents, or other misfortune" (Kalland and Moeran 1992, 152). To ensure a safe journey and good catch, whalers and their families visit whale shrines, which exist in prevalence in whaling villages, some dating as far back as the 17th century, and carry out religious rituals (Akimichi 1988, 53).

One of these rituals is a Buddhist memorial service for those whales caught in the hunting season. A Buddhist sutra is read for both the whale's soul and for those that passed at sea. While people interpret these services differently, "The temple priest in Arikawa, for example, performs the memorial service in the belief that the whale will be released from rebirth in this world and enter Paradise as a 'Buddha,' though he supposed that some people might believe that the whale will be reborn as a human" (Kalland and Moeran 1992, 154). In select villages, it is the practice to treat whale souls in the same manner as human souls, complete with giving the whales posthumous names and including them in Buddhist death records. Further, to some whalers, the rituals are to express gratitude for their sacrifices as opposed matters pertaining to an afterlife.

Women carried out the religious rituals "with great fervor," according to Kalland and Moeran, when their husbands were out on a hunt (1992, 152). He expresses these religious rituals in terms of their loss,

> With the moratorium in force, there is no longer any need to go to temples in order to pray for the souls of killed whales or to shrines to give thanks for good catches and for whalers' protection at sea. Whalers' wives, too, do not need to go on pilgrimages or pray for good catches and safe voyages in front of their house altars, and the morning rituals are considerably shorter (Kalland and Moeran 1992, 188).

On a community level, these rituals serve to "...give the local residents both the feeling of a common heritage and meaning to their lives" (Kalland and Moeran 1992, 152). Festivals are held yearly to further reaffirm their identity as a whaling community and acknowledge their dependence on the whale. The rituals serve as another form of reciprocity to their success as a village.

In these whaling villages, the whale stands at the center of religious beliefs and rituals, bringing together members of the village for rituals that benefit the entire community. This naturally reaffirms the village residents as an interlinked community, connecting them through their unique traditional whaling culture. With the implementation of the moratorium, the community not only loses the glue that holds it together, but the complex relationships between humans and whales disintegrate.

6.5   Conclusion

As with the Massai and the Serengeti, the Cree and the caribou, and the Inuit and the whale, the whale body and traditional practices carried out surrounding him or her is ripe with "core spatial images" that connect the current generation to the past— calling on religious, technological, and mythical traditions that have been carried out for centuries. The relationships between these communities and the whales they hunt are part of the social fabric that holds these communities together. The whaling moratorium has turned the whale-human relationships in these communities into increasingly dated memories—resulting in the potential loss of significant cultural traditions that valued the species of whales with whom they once interacted.

The human-animal relationship has a role in sustainable practices. Western notions of a separation of nature and culture all too often remove the human from the "nature" being saved. This dualistic view is not only harmful to societies with alternative relationships with nature, as we saw with the Massai, but potentially harmful to the natural entities marked for saving. These traditions valued and respected the various species of whales in their region, their continued well-being as sustainable resources, and the relationships held with her kind, honoring her through ritual, festivals, and celebrated cuisine. Rules and ritual surrounding the hunter-prey relationship are often based on the maintenance of sustaining the prey on whom they rely and based on generations of local knowledge regarding the species at hand. Without the reliance on the whale as prey, what will become of the human-whale relationships in these villages?

Regulations have an important place in both international and domestic whaling practices. However, in continuing a blanket moratorium on whaling based on Western ideological beliefs and perceptions (as opposed to scientific findings), the IWC ignores the human, and significant human-animal entanglements, involved in the controversy. Acknowledging the cultural significance of ecological memory, the diversity of human-nature interactions, and the respect inherent in many traditional subsistence practices can potentially aid, rather than harm, conservation practices.

References

Abel, Jessamyn R. 2005. The Ambivalence of Whaling: Conflicting Cultures in Identity Formation. In *Japanimals: History and Culture in Japan's Animal Life*, ed. Gregory. M. Pflugfelder, and Brett L. Walker, 314–341. Michigan: University of Michigan, Center for Japanese Studies.

Akimichi, Tomoya. 1988. *Small-Type Coastal Whaling in Japan: Report of an International Workshop.* Japan Social Sciences Association of Canada Fund to Promote International Education Exchange and Boreal Institute for Northern Studies.

Arch, Jakobina. 2016. Whale Meat in Early Postwar Japan: Natural Resources and Food Culture. *Environmental History* 21: 467–487.

Figueroa, Robert M., and Gordan Waitt. 2010. Climb: Restorative Justice, Environmental Heritage, and the Moral Terrains of Uluru- Kata Tjuta National Park. *Environmental Philosophy* 7 (2): 135–163.

Freeman, M.M.R. 1990. *Political Issues with Regard to Contemporary Whaling.* Tokyo, Japan: Institute of Cetacean Research.

Freeman, M.M.R, L. Bogoslovskaya, R.A. Caulfield, I. Egede, Igor I. Krupnik, and M.G. Stevenson. 1998. *Inuit, Whaling, and Sustainability.* Altimira Press.

Gottlieb, Robert. 1994. *Forcing the Spring: The Transformation of the American Environmental Movement.* Washington DC: Island Press.

Hardin, Garret. 1968. The Tragedy of the Commons. *Science* 162 (3859): 1243–1248.

Hirata, Keiko. 2008. Japan's Whaling Politics. In *Norms, Interests, and Power in Japanese Foreign Policy*, eds. Yoichiro Sato, Keiko Hirata, 175–209. Palgrave Macmillan.

Imanishi, Kinji. 2013. *A Japanese View of Nature: The World of Living Things.* Routledge Curzon.

Ingold, Tim. 2000. *The Perception of the Environment: Essays in Livelihood, Dwelling and Skill.* New York, NY: Routledge.

Institute of Cetacean Research (ICR). 2016. http://www.icrwhale.org/. Accessed May 2016.

Institute for Cetacean Research (ICR). 2007. Chair's Summary. *Conference for the Normalization of the International Whaling Commission.* http://www.jfa.maff.go.jp/j/whale/w_document/pdf/iwc_houkoku.pdf. Accessed May 2016.

Institute for Cetacean Research (ICR). 2004. *The 2nd Summit of Japanese Traditional Whaling Communities: Ikitsuki, Nagasaki, Reports and Proceedings.*

International Whaling Commission (IWC). 2016. *History and Purpose.* https://iwc.int/. Accessed March 2016.

Kalland, Arne, and Moeran, Brian. 1992. *Japanese Whaling: End of an Era?* Curzon Press Ltd.

Kalof, Linda. 2007. *Looking at Animals in Human History.* London: Reaktion Books.

King, Barbara J. 2010. *Being With Animals: Why We Are Obsessed with the Furry, Scaly.* Harmony: Feathered Creatures Who Populate Our World.

Morikawa, Jun. 2009. Whaling in Japan. Oxford University Press.

Mutsuko, O. 1988. Nobody Can Take our Food Culture Away. In *Women's Tales of Whaling: Life Stories of 11 Japanese Women who Live with Whaling*, ed. Junichi Takahashi. Tokyo, Japan: Japan Whaling Association.

Shelter, Jan Bender. 2007. *Imagining Serengeti: A History of Landscape Memory in Tanzania from Earliest Times to the Present*. Athens Ohio: Ohio University Press.

Stevens, Fisher. (Producer), and Psihoyos, Louie. (Director). 2009. *The Cove [Motion Picture]*. United States: Participant Media.

Takahashi, Junichi. 1988. *Women's Tales of Whaling: Life Stories of 11 Japanese Women who Live with Whaling*. Tokyo, Japan: Japan Whaling Association.

Watanabe, Hiroyuki. 2009. *Japan's Whaling: The Politics of Culture in Historical Perspective*. Trans Pacific Press.

WDC. 2016. Whale and Dolphin Conservation. http://us.whales.org/.

Zushi, S.S. 1988. A Housewife Questions the Whaling Ban. In *Women's Tales of Whaling: Life Stories of 11 Japanese Women who Live with Whaling*, ed. Junichi Takahashi. Tokyo, Japan: Japan Whaling Association.

Part II
Food Justice and the Built Environment

Chapter 7
A Vignette from Our Kitchen Table

Lisa Oliver King

Abstract In this vignette, Lisa Oliver King discusses how one activist group includes emancipatory, theoretical education and understanding to build capacity for groups to tackle more concrete projects.

Keywords Food Justice · Urban Food Systems · Community Capacity · Praxis

Our Kitchen Table (OKT) addresses the health issues resulting from food insecurity and environmental health issues, which inequitably impact the African American and income-challenged residents of Southeast Grand Rapids, Michigan. As a group of women living in or having ties to these neighborhoods, members of OKT created the Food Diversity Project (FDP), a cooperative food-growing model in which members educate participants and address the root causes of these problems. Currently funded by the W.K. Kellogg Foundation, it is a community-led advocacy project that involves like-minded partners, utilizes popular education activities, analyzes public policy, and mobilizes citizens to network as backyard and community gardeners. OKT's activities are designed for implementing long-term solutions to food insecurity and environmental toxins in that they deal with the systemic causes of poverty. The FDP is designed to institute food justice and serve as an alternative to many initiatives that provide Band-Aid solutions to poverty related issues.

OKT typically consists of ten core members, mainly women, who are concerned with food sovereignty, reducing health disparities, community activism, and politics. Drawing upon their members' expertise, OKT organizes in response to community residents interested in building toxic-free neighborhoods and a self-sustaining food system. To achieve this objective, in 2008 OKT began

L.O. King (✉)
Our Kitchen Table, Grand Rapids, MI, USA
e-mail: admin@oktjustice.org

© Springer International Publishing AG 2017   85
I. Werkheiser and Z. Piso (eds.), *Food Justice in US and Global Contexts*,
The International Library of Environmental, Agricultural and Food Ethics 24,
DOI 10.1007/978-3-319-57174-4_7

facilitating a social network of household and community gardeners who share their excess harvest and resources and teach others how to grow food. They refer to this capacity-building model as the Food Diversity Project (FDP). One of its major goals is to reduce the high rates of "food insecurity"—defined as little to no access to fresh, healthy, culturally appropriate, or affordable food—as experienced by residents living in the Southeast. The Southeast disproportionly consists of resource depleted and economically distressed neighborhoods in Grand Rapids.

The FDP takes an interactive popular education approach for explaining the relationship between economic and health disparities with social injustice. The goal of popular education, a model developed by Paulo Freire, is to develop peoples' capacity for social change through a collective problem-solving approach and critical analysis of social problems. OKT provides support for food growers while teaching them how to be agents in shaping social change. Participants collectively build their own capacity through resident-led activities in which they learn how growing their own food challenges the inequality impacting their livelihoods, supplements income, and has health benefits.

The Food Diversity Project's application of the popular education method involves several capacity-building activities related to promoting food security, avoiding exposure to lead in the soil, and education on sustainable subsistence activities. Interactive workshops provide education on canning and seed saving, composting, starting plants from seeds, organic growing methods, growing winter crops, and selecting seeds for plant diversity. Walking and bicycle tour workshops are organized around both household gardens and naturally growing food in the neighborhoods. During these tours, examples of topics include lead in the soil; herbal gardening for health and culinary purposes; and urban foraging for edible fruit, nuts, and weeds that grow in public spaces within the city. Cooking demonstrations are another type of workshop, in which peer educators prepare food with both familiar and unfamiliar types of locally grown produce. At times, renowned food activists facilitate these workshops. Such guests have included Ladonna Redmond and Bryant Terry.

OKT also works closely with individual families and hosts the Southeast Area Farmers' Market (SEAFM). Over the 2015 growing season, the FDP provided support for 24 families. Participants completed an educational series, a biochemist tested their soil for lead, and a gardening coach assisted in maintaining and providing advice on their gardens. Together, the families produced 3,125 lb of produce, herbs, and edible flowers. The SEAFM is located in income-challenged neighborhoods that lack full servicing grocery stores and farmers markets. Typically, 90% of the vendors are women of color who grow food out of their household gardens, which are located in these neighborhoods. The FDP consists of numerous other undertakings that include supporting community gardens, partnering with schools for gardening with youth, and collaborating with healthcare organizations and other partners for various initiatives.

Popular education involves raising critical consciousness. The FDP achieves this objective through media and discussions as to how public policy can cause lasting change. To educate the public, OKT uses media for sharing participants' stories and upcoming events with local newspapers and online publications. For the same purposes, OKT produces zines as well as curricula for schools and health organizations. Further, OKT maintains a Facebook page and website (https://oktjustice. org/). Members have also appeared on radio shows as guests and have produced public service announcements. In regards to policy, the FDP includes an educational series, and the capacity building activities integrate discussion about food related policy at the city, state, federal, and international level. Participants are notified of marches or rallies organized to change policy.

Overall, conversations center on critical topics as they relate to social inequality and food justice. The FDP has a deeply ingrained food justice perspective for challenging the systemic inequality causing health disparities in the area. OKT achieves this objective through encouraging people to grow their own food in a sovereign fashion while educating community residents on how to prevent high exposure to environmental toxins. The FDP began as a series of discussions around the kitchen table and has become the core initiative of OKT, an organization nationally recognized for the FDP. The FDP is an exemplary model in which other places nationwide could replicate and borrow its major tenets for achieving food justice.

Chapter 8
Introduction to Food Justice and the Built Environment

Shane Epting

Abstract The need for the comingling of theory and practice in food justice is particularly striking in the context of the built environment, where reactive policies without a theoretical framework and theories which do not take into account the surprises found in application have both failed many times in the past. Shane Epting calls for transdisciplinary collaboration to find paths toward food justice.

Keywords Food Justice · Transdisciplinarity · Urban Food Systems · Community Capacity · Praxis · Philosophy of the City

8.1 Food Justice and the Built Environment

In 1681, Sor Juana Inés de la Cruz held that people can learn philosophy in the kitchen, quipping that Aristotle would have had better insights if he observed how food changes while cooking (de la Cruz et al. 2009). Despite her facetious tone, she was arguing that people could understand metaphysical principles from observing the everyday physical world. In addition to this lesson, examining humankind's relationships with food can teach us about urban justice, showing how it intersects with life in the city.

To achieve this end, learning how food gets from the industrial farm to the grocery store means understanding a confluence of policy, technology, farming, cooking, and marketing. Consider that investigating what goes into a typical loaf of store-bought bread reveals many food-justice issues. Although most ingredients are safe, high fructose corn syrup (HFCS) is common. It is associated with obesity and

S. Epting (✉)
University of Nevada, Las Vegas, USA
e-mail: shane.epting@gmail.com

© Springer International Publishing AG 2017   89
I. Werkheiser and Z. Piso (eds.), *Food Justice in US and Global Contexts*,
The International Library of Environmental, Agricultural and Food Ethics 24,
DOI 10.1007/978-3-319-57174-4_8

diabetes. While too much HFCS can affect any person, recent research suggests that its consumption can disproportionately harm African-American populations (Saab et al. 2015). Due to the US Government's corn subsidy policies, it serves as a ubiquitous, inexpensive sweetener, though such practices have received several criticisms (Fields 2004; Finkelstein and Zuckerman 2010; Franck 2013). Azodicarbonamide, a synthesized chemical used in plastics, also makes the list, showing up in breads and beer (Dennis et al. 1997).

While the above account is not exhaustive, it suggests that seemingly innocuous food matters surround us, blending into the daily humdrum. Scholars and activists show how such matters are morally problematic in myriad ways, often overlapping with other social concerns. For example, Guthman (2011), in *Weighing In: Obesity, Food Justice, and the Limits of Capitalism*, illustrates how issues such as obesity and food justice also include considerations for urban planning. In built environments, thinking about food systems means bringing several socio-politico dimensions into view, basic requirements for gaining a comprehensive understanding of such affairs.

When examining these concerns in today's cities, we learn about the obstacles that hinder residents' control over their food security. Lisa King Oliver, for instance, provides an account of Our Kitchen Table (OKT), a nonprofit organization that serves southeast Grand Rapids, Michigan. This group addresses several dimensions of food injustice, from policy issues to the public health concerns mentioned above, along with organizing community gardeners and educational services. These efforts make up their "Food Diversity Project," an undertaking that embodies the ideals present in the food-justice literature. Through demonstrating how activists can embed such theories into their practices, OKT exhibits practical ways to transform some elements of their local food system.

As a small group of women, OKT educates through hands-on experience, taking residents on neighborhood tours to teach them about food-related toxic dangers and urban horticulture, along with cooking and canning presentations. As part of a networked community, OKT partners with local families to host the Southeast Area Farmers' Market, producing 3125 lb of fruits and vegetables, along with herbs and edible flowers. Their efforts illustrate that resident groups can exercise autonomy, contributing to food-system reform. Although such actions have limits, they count as steps toward achieving food justice in the built environment, despite their humble nature.

To be fair, groups such as OKT are dealing with an unjust food system that became part of the urban landscape over a lengthy period, and any positive impacts should be celebrated as progress. For instance, Samantha Noll's chapter, "Food Sovereignty in the City: Challenging Historical Barriers to Food Justice," shows that several elements and policies during the nineteenth century created the sharp division between urban and rural settings. In turn, for many contemporary urban centers in the US, food production and city life remain separate.

For example, as numerous urban centers in the US were developing, people linked animals with disease and filth, providing reasons to remove them from the city; they were also associated with lower-class status (Gamber 2005). With automobiles replacing horses as transportation, the manure required for farming was not readily available. This condition, coupled with the rising cost of real estate, helped push agriculture into rural settings. Such events encouraged the separation of urban and rural living. Achieving food sovereignty, Noll argues, means that this divide, along with the mindset that accompanies it, must go. Groups such as OKT demonstrate how to work toward this goal. They represent a unique approach for dealing with a multifaceted food-access issue in Grand Rapids. Due to distinctive circumstances of food-justice issues in other cities, population centers must develop specific solutions that can alleviate harm. One could argue that there is a pressing need to provide immediate succor in dire instances. This notion suggests that people could benefit from short-term solutions, even if such measures leave unjust systems in place. While this attitude could garner criticism, it is gaining momentum.

Considering the problematic nature of food justice in built environments, one could make a case that we should closely examine solutions to ensure that they do not smuggle in additional concerns. Emily Holmes and Christopher Peterson's chapter, "Race, Religion, and Justice: From Privilege to Solidarity in the Mid-South Food Movement," for instance, show how the structure of food-justice movements can raise moral concerns, despite the best intentions of the people involved. They spotlight a hidden aspect that pervades several social justice concerns, such as white people determining and securing the conditions for justice on the behalf of minorities. Best intentions aside, such issues call into question the ethical dimensions of social movements, showing that reform efforts must go beyond customary models to determine inclusive measures.

In turn, approaching food sovereignty in built environments requires a multi-dimensional approach that combines theoretical insights and workable solutions. Discovering this combination is an onerous task, but trans-disciplinary projects that involve academics and professionals are successfully emerging.[1] Consider, for example, that public schools have become associated with food justice, and partnerships between scholars and teachers exhibit beneficial ways to mitigate harm.

To gain insights into how to achieve such feats, Sarah Riggs Stapleton reveals that working with educators can uncover outlier food issues that standard policies cannot address. In "Views from the Classroom: Teachers on Food in a Low-Income Urban School District," Stapleton examines how a group of teachers deal with food issues on their campuses. Collectively weighed against state or national policies, these narratives exhibit that food-justice concerns in urban schools require intimate knowledge of students' lives, perspectives that perhaps only teachers can provide.

Stapleton illustrates how participatory-action research exposes students' problems that impede education. For instance, through researching with a teacher, Person Cole, she uncovers how providing healthy lunches during schools is essential for bolstering classroom instruction, but the presence of vending machines

---

[1]"Trans-disciplinary" in this context suggests that the involved efforts go beyond the academy to include the private sector, but it could also include government agencies.

on campus undermines such efforts. Through demonstrating that students will succumb to the temptation of processed foods, this case provides a glimpse into the kinds of concerns that public-private partnerships raise. Cole's study shows that food-justice issues in urban schools require inclusive, participatory efforts.

Although the example above addresses a "normal" food problem in public schools, other instances such as Melissa Washburn's backpack feeding program show that education-food issues extend beyond the classroom. The problem is that students do not simply lack food for only the school day, but they lack nutritional foods in general. Through securing a grant to provide needy high school kids with an additional snack, she was able to alleviate some of the harm that poverty causes. One can argue that Washburn was going beyond her duty, holding that her job begins and ends with instruction-related activities. This is fair criticism, but one could counter this claim through arguing that good teaching involves caring about all of the factors that support or hurt education. In turn, Washburn could justify her approach.

Elisabeth Mari's chapter, "Healthy Food on Wheels: An Exploration of Mobile Produce Markets through a Food Justice Lens," illustrates that mobile produce markets can provide temporary relief for people who lack access to fresh fruits and vegetables. Yet, residents lacking access to produce is only one of many food-justice challenges that cities face. Through examining enterprises in eight cities, she identifies how every urban setting has distinct issues, meaning that there is not a one-size-fits-all solution to food security. The fact that these issues exist suggests that government leaders should work toward long-term relief. The complex character of such issues means that academic and grassroots partnerships could benefit how we understand the necessary conditions that are required to mitigate harm.

In the cases explored above, there are two common themes. The first is that they concern events in the 'real-world,' occurring at present. These instances should hold prominence because they can discover or improve solutions, as the authors examined above make evident. Emphasizing this point does not mean that we should completely ignore systemic issues, only to highlight a specific condition that is relevant to a nascent area of applied work.

Second, such instances are solution-focused. Seen in concert, the remarkable feature of these examples is that they can serve as guides to progress for cities facing similar problems. While concerns will vary across cities, examining successes provides urban planners, policymakers, and resident groups with tested ways to approach food issues in built environments. For instance, including participatory elements into school lunch policy as noted above provides a starting point for conversation. Although such measures are not "silver-bullet" solutions to urban food problems, they do have a record of success that could benefit similar cases.

The notion of trans-disciplinary research present in cases such as Oliver's OKT and Stapleton's backpack program suggest that combining viewpoints from academia and grassroots movements has a two-fold advantage for addressing urban food justice. One benefit rests on the inherent goal-focused nature of such writings, promoting urban food justice through its documentation. That is to say, through reporting how community groups and informal leaders such as teachers can discover solutions to complex food issues, they illustrate a way forward.

Such measures remind us of the importance behind justice scholars who research specific affairs such as food policy and practice. They can exhibit how theoretical-yet-practical steps can alleviate harm that socio-politico conditions cause. This point leads us to the second benefit, an advantage for the academy. Consider, for instance, that cases such as OKT can give researchers a means to gauge the effectiveness of the theories that guide policy decisions. In the example that Stapleton provides, overlooked details such as vending machines can sabotage administrator's efforts to prime students for learning with nutritious foods. This point indicates that real-world examples can show researchers how well theoretical approaches work or fail in practice for food justice cases in the built environment. In turn, the scholarly work of researchers and the lived-world experience of activists can merge, illustrating paths to food-justice progress.

References

de la Cruz, Sister, Juana Inés, Electa Arenal, and Amanda Powell. 2009. *The Answer: Including Sor Filotea's Letter and New Selected Poems*. New York City: Feminist Press at CUNY.

Dennis, M.J., R.C. Massey, R. Ginn, P. Willetts, C. Crews, and I. Parker. 1997. The Contribution of Azodicarbonamide to Ethyl Carbamate Formation in Bread and Beer. *Food Additives & Contaminants* 14 (1): 101–108.

Finkelstein, E.A., and L. Zuckerman. 2010. *The fattening of America: How the economy makes us fat, if it matters, and what to do about it*. Hoboken, NJ: Wiley.

Fields, S. 2004. The Fat of the Land: Do Agricultural Subsides Foster Poor Health?. *Environmental Health Perspectives*, 112 (14): A820.

Franck, C., S. M. Grandi and M. J.Eisenberg. 2013. Agricultural subsidies and the American obesity epidemic. *American journal of preventive medicine*, 45 (3): 327–333.

Gamber, W. 2005. Away from Home: Middle-Class Boarders in the Nineteenth-Century City. *Journal of Urban History* 31: 289–305.

Guthman, J. 2011. *Weighing In: Obesity, Food Justice, and the Limits of Capitalism*. Oakland, CA: University of California Press.

Saab, K., J. Kendrick, J. Yracheta, M. Lanaspa, M. Pollard, and R. Johnson. 2015. New insights on the risk for cardiovascular disease in African Americans: the role of added sugars. *Journal of the American Society of Nephrology* 26: 247–257.

Chapter 9
Food Sovereignty in the City: Challenging Historical Barriers to Food Justice

Samantha E. Noll

Abstract Local food initiatives are steadily becoming a part of contemporary cities around the world and can take on many forms. While some of these initiatives are concerned with providing consumers with farm-fresh produce, a growing portion are concerned with increasing the food sovereignty of marginalized urban communities. This chapter provides an analysis of urban contexts with the aim of identifying conceptual barriers that may act as roadblocks to achieving food sovereignty in cities. Specifically, this paper argues that presupposed commitments created during the birth of the modern city could act as conceptual barriers for the implementation of food sovereignty programs and that urban food activists and programs that challenge these barriers are helping to achieve the goal of restoring food sovereignty to local communities, no matter their reasons for doing so. At the very least, understanding the complexities of these barriers and how they operate helps to strengthen ties between urban food projects, provides these initiatives with ways to undermine common arguments used to support restrictive ordinances and policies, and illustrates the transformative potential of food sovereignty movements.

Keywords Urban agriculture · Food sovereignty · Food justice · Food policy · Environmental ethics · Food ethics

Local food initiatives are steadily becoming a part of contemporary cities around the world and can take on many forms (Holt-Gimenez 2011; DeLind 2011; Martinez et al. 2010).[1] While some of these initiatives are concerned with providing consumers with farm-fresh produce or improving urban population's access to foodstuffs in urban areas, a growing portion of movements are concerned with

---

[1]Note: this chapter draws on and expands some of the ideas in my article "History lessons: What urban environmental ethics can learn from nineteenth century cities." *Journal of Agricultural and Environmental Ethics* Volume 28(1): 143–159.

S.E. Noll (✉)
The School of Politics, Philosophy, and Public Affairs,
Washington State University, Pullman, WA, USA
e-mail: snoll@haverford.edu

© Springer International Publishing AG 2017
I. Werkheiser and Z. Piso (eds.), *Food Justice in US and Global Contexts*,
The International Library of Environmental, Agricultural and Food Ethics 24,
DOI 10.1007/978-3-319-57174-4_9

increasing the food sovereignty of marginalized urban communities. Here food sovereignty should be understood as a broader understanding of food, where food is bound up with human rights concerns or justice issues.[2] In contrast to programs that make use of industrialized food systems, food sovereignty movements work hard to increase local community control of the production, processing, and distribution of food, as this is seen as a necessary condition for liberating communities from oppression (Werkheiser and Noll 2014; Schanbacher 2010). With this in mind, urban food initiatives often make use of various strategies, from direct community action, to lobbying for local policy and ordnance changes, to achieve their various short term and long term goals.

The purpose of this chapter is to provide a theoretical analysis of urban contexts with the aim of identifying conceptual barriers that may act as roadblocks to achieving food sovereignty in cities. Specifically, in this paper, I argue that (1) presupposed commitments created during the birth of the modern city could act as conceptual barriers for the implementation of food sovereignty programs and (2) that urban food activists and programs that challenge these barriers are helping to achieve the goal of restoring food sovereignty to local communities, no matter their reasons for doing so. While increasing food sovereignty in urban areas includes policy, direct action, and community activism, purposefully challenging historical conceptual barriers may help to remove roadblocks to food sovereignty.

In addition, the theoretical analysis presented in this chapter could help urban food practitioners the following practical ways: (1) It could help cultivate cooperative relationships between a multiplicity of urban food projects guided by disparate goals (2) help undermine seemingly reasonable arguments (such as nuance and public health claims) commonly used to justify strict ordinances and policies concerning food production in the city, by exposing their biased histories, and (3) provide alternative visions of urban spaces as centers of agricultural production that could serve as examples and/or models for future development. At the very least, understanding the complexities of barriers to urban agriculture and how they operate helps to illustrate the transformative potential of food sovereignty movements.

9.1 Urban Agriculture

In the current literature, urban agriculture is generally characterized as part of the local food movement that focuses on food production in urban areas or cities. Here, "urban areas" or "cities" should be broadly defined to include built areas often held

---

[2]It should be noted here that the terms "food sovereignty" and "food justice" are often used interchangeably, as both signify food related movements that accept a wide range of justice concerns, including but not limited to increasing community control of food systems. For this reason, both terms will be used interchangeably in this chapter.

in opposition to the country, natural areas, or the wilderness (Light 2003).[3] In this context, urban agricultural initiatives are often understood as projects that achieve particular ends, such as a novel way to provide populations living in city centers with a fresh supply of foodstuffs (Angotti 2015), a method to support sustainability efforts by limiting our ecological footprint (Huang and Drescher 2015), a way to connect urbanites with the natural world (Light 2003), and an activity to promote community connections (Delind 2011). Thus, while common urban farming endeavors (such as farmers' markets, community gardens, etc.) can all be placed under the umbrella of urban agriculture, they are guided by disparate goals and motivations (Delind 2011; Werkheiser and Noll 2014). For example, international organizations, such as the FAO, conceptualizes urban food initiatives as a way of increasing food and nutrition security around the world, as urban populations are increasingly facing global food price inflation (Food and Nutrition Security). In local contexts, community supported agriculture projects, such as the Kimberton CSA in Pennsylvania, are frequently guided by the goals of providing local populations with biodynamic, organic, and/or sustainably produced fruits, vegetables, and animal products, supporting the local economy, and training future farmers (Kimberton 2017). Community gardens, such as the Hunter Park Garden House Project in Michigan and The School Garden Project of Lane County in Oregon, are guided by the goals of providing food education, strengthening community ties, and bettering the health of local populations (Hunter Park Garden House 2017, The School Garden Project 2016).

9.2 Food Sovereignty in Cities

Within this wider urban food landscape, food justice or sovereignty movements, such as the Detroit Food Justice Taskforce, are committed to the specific goal of increasing local community control of food systems (Detroit Food Justice Task Force). In fact, key justice commitments and an expanded conception of "food" play important roles guiding food sovereignty movements in city contexts

---

[3]This paper, specifically, utilizes Andrew Light's (2003) broad definition of "cities" found in his seminal essay on the urban blind-spot in environmental philosophy. As such, the paper roughly defines "urban areas" or "cities" (the terms will be used interchangeably) as built spaces connected to both the past and the future, as they are continually being developed in such a way as to act as physical manifestations of community priorities and values (Light 2003). In addition, it is important to note that urban areas are often opposed, or form dualisms, with other areas. For example, the city is opposed to the country and nature is often opposed to culture. This definition is deliberately vague, as its aim is to capture the essence of built areas, while still respecting the unique cultural and historical manifestations found in individual contexts. Due to this vagueness, it should be noted that the insights identified in this paper may not apply to all cities, but still of may be of use at a future date. As the historical section draws heavily on work in North America, this essay may be of particular use to practitioners working in similar contexts.

(Werkheiser and Noll 2014). The following definition of food sovereignty in Declaration of Nyéléni illustrates this point perfectly:

> Food sovereignty is the right of peoples to healthy and culturally appropriate food produced through ecologically sound and sustainable methods, and their right to define their own food and agriculture systems. It puts the aspirations and needs of those who produce, distribute and consume food at the heart of food systems and policies rather than the demands of markets and corporations… It ensures that the rights to use and manage our lands, territories, waters, seeds, livestock and biodiversity are in the hands of those of us who produce food. Food sovereignty implies new social relations free of oppression and inequality between men and women, peoples, racial groups, social classes and generations.

In the above passage, food sovereignty's definitions of "food" moves beyond the limited conception of "food" as a commodity or food-stuff to include a wide range of social justice concerns, such as matters of food autonomy, ecological preservation, food system and policy issues, future generations, and gender and racial equality. Here the conception of food that guides food sovereignty movements is bound up with personal, spiritual, and social identity (Werkheiser and Noll 2014; Pimbert 2008; Whittman et al. 2011). As such, it is difficult if not impossible to separate food issues from wider social issues. In addition, food is seen as a way of addressing a plethora of human rights or justice concerns beyond increasing access and thus employs an expanded conception of justice (Flora 2010). In urban contexts, food sovereignty initiatives address a wide range of justice issues often faced by marginalized communities in urban contexts, such as the realities of living in a food desert (Schafft et al. 2009), biases in food systems and food policy (Taylor and Ard 2015), and the lack of urban spaces for food production (Angotti 2015). As such, while food sovereignty based programs frequently work towards many of the goals guiding other urban agricultural programs, a key impetus for food justice is addressing human rights and justice concerns in urban contexts by increasing community control of food systems.

9.3 The Context of the City: Challenges Faced by Urban Agriculture Projects

While "urban food" captures a wide array of food related projects guided by various values, these movements often face similar challenges unique to the urban context, such as soil contamination, the high price of land, and regulatory hurdles, when attempting to start farming enterprises in cities. In fact, one of the greatest barriers faced by urban agricultural programs are policy and zoning regulations (Huang and Drescher 2015; De Zeeuw et al. 2000; LeJava and Goonan 2012). Depending on the specific context, current zoning ordinances can make it difficult, if not impossible, to acquire sites for community gardens and other production activities, thus hindering the implementation of agricultural programs (Castillo et al. 2013; Roehr and Kunigk 2009; Voigt 2011). Indeed, factors such as whether or not your operation is community based or commercially focused can greatly impact whether or not you can successfully integrate your enterprise into the cityscape. This is particularly the case concerning food animals, as municipalities often prohibit raising livestock (such as chickens, goats, and pigs) in urban areas due to public health concerns, such as that they could spread diseases, become a nuisance, or attract pests (Huang and Drescher 2015; Pollock et al. 2012).

The above list of concerns is not meant to be exhaustive but rather to curiously illustrate how the context of the city creates unique challenges for urban food projects. Land-use policy, zoning regulations, food system choices, and most importantly for this chapter, food initiatives do not happen in a vacuum, but are grounded in specific environmental, historical, and socio-economic contexts (Mcgirr and Batterbury 2016; Mitchell 2003). These contexts are not value free but are molded by the values of previous generations (Lyson 2004; Thompson 2015) and interwoven with basic commitments and assumptions that often help to guide decision making, such as particular conceptions of disease (Atkins 2012b; Barnes 1995; Howell 2012), what a city is (Noll 2015), and the perceived "right" relationship is between humans and animals (Brantz 2011; Clutton-Brock 2011; Kalof 2007).[4] The next section of this chapter takes up what Rose (2003) calls "the project of delimiting and determining the governing features of everyday social existence" in the context of the city (p. 462). Specifically, it explores how (a) the context of the city and (b) actors and stakeholders in this context include or accept key assumptions and commitments that play a role in shaping the current challenges faced by urban agriculture initiatives today.

For instance, as argued below, the seemingly reasonable justifications for prohibiting livestock urban areas just listed, such as public health and nuisance concerns, where historically guided by sustained efforts to limit the food sovereignty of working class communities and guided by a specific vision of how cities should be structured. Today we often assume that cities, in general, are densely developed population centers largely divorced from the natural world, in general, and food production, in particular (Light 2001; Thompson 1994). However, this was not always the case. Prior to the nineteenth century, urban areas were often sites of intense agricultural production, with food production, processing, and distribution occurring in the urban neighborhood. The next section teases out metaphysical commitments created during the birth of the modern city that are now embedded in the cityscape and illustrates how they (1) help to inform the policies and regulations above and (2) act as conceptual barriers for the implementation of food sovereignty

---

[4]For the purposes of this chapter, a "commitment" should be understood as an underlying feature of social existence (Rose 2003) or a basic concept that a person holds concerning what something "is" (Inwagen 2013; Noll 2015). Such commitments are influential in all areas of life, including the development of personal identity (Ricoeur and Blamey 1995), how we treat various groups, such as other those of other cultures, animals, and the environment, and in scientific inquiry (Haraway 1989; Harding 1993). There is a plethora of work in humanities and social sciences devoted to teasing out the social, symbolic, and cultural aspects of food (Mcgirr and Batterbury 2016). This chapter builds on this work, by identifying historical commitments that may act as conceptual barriers to food sovereignty.

programs. It will end by arguing that urban food initiatives that challenge these barriers are helping to achieve the goal of restoring food sovereignty, no matter their specific goals.

9.4 An Analysis of Nineteenth Century Cities[5]

During the nineteenth century, urban areas around the world experienced widespread changes that culminated in the development of what today is recognized as the basic structure of the modern city (Atkins 2012a). It should be noted here that the literature on the urban landscape is vast and covers a wide range of topics (Steinhoff 2011). For the purposes of this chapter, the following analysis primarily focuses on teasing out conceptual and structural changes concerning (1) the relationships between humans and domesticated animals in urban areas and (2) urban agriculture practices. It pays attention to how the shifting contextual landscape culminated in drastic changes to urban agriculture and animal husbandry practices and to the power dynamics helping to prompt these changes. As will be discussed, cities during this time saw drastic changes to urban food production, processing, and distribution. Specifically, shifting understandings of key concepts combined to cleanse the city of livestock and the resulting loss of cheap fertilizer (in the form of manure) forced urban crop production to move out of the city.

9.5 Animals in the City: Shifting Human & Domesticated Animal Relationships

The roles of domesticated animals in urban landscapes greatly shifted during the nineteenth century, so much so that, by the dawn of the twentieth century, cityscapes were effectively cleansed of all but pet animals and those predominantly living outside human control (Atkins 2012a; Brantz 2011; Noll 2015). In contrast, inhabitants in pre-modern urban areas included various types of agricultural and work animals—animals who often spent their lives and eventually died in urban environments. Specifically, the shifting understandings of (1) civilization and domestication and (2) disease and filth combined with class based discrimination to culminate in the removal of all but a handful of non-human animals from the city streets. Concerning the first point, key factors impacting human-animal

---

[5]The analysis of Nineteenth Century Cities presented in this chapter is, by necessity, short, as the work needs to fit within the limited scope of an essay. However, for a more detailed account, readers may find McNeur's (2014) book *Taming Manhattan: Environmental battles in the antebellum city* or Atkin's book (2012a) *Introduction. In Animal cities: Beastly urban histories* helpful.

relationships during the nineteenth century were (a) an increasing emphasis and reliance on science and rationality, coupled with a growing faith in progress, and (b) the emergence of new concepts of "civilization" and "domestication" built upon the key dualisms of wild/tame and primitive/cultivated (Brantz 2011; Noll 2015). These labels signify different relationships between humans and non-human animals (Palmer 2011), as wildness came to emphasize "the absence of a relation and a disposition that is markedly not 'tame,' while a domesticated animal is one that is both controlled by humans and has been made dependent upon humans in various degrees" (Noll 2015). Due to these conceptual developments, "civilized" spaces, such as the home and the city, that were traditionally viewed as a place where various animals (such as agricultural and work animals) were welcomed (Edwards 2011; Pascua 2011), were increasingly conceptualized as a place where only the most "tame" or controlled animals (mostly companion animals) could safely enter (Clutton-Brock 2011). In the context of the city, this translated into the upper class ideal of the home as place where an increasingly private and animal-free space was now understood as a marker of status (Gamber 2005). As will be discussed further, most urban dwellers during the nineteenth century lived in boarding houses or with their food animals. For these reasons, they never reached this ideal and were negatively judged by the upper classes.

A consequence of this shift was an increase in claims that agricultural and other working animals were "nuisances" and, therefore, should be further removed from city common areas, such as streets and parks. In fact, these charges played an increasingly prominent role in de-animalizing the city. Like the home, urban areas were re-conceptualized as places of civilization and thus no longer acceptable habitats for many non-human animals classified as "wild," or at least not "tame" enough for civilized spaces (see Michelfelder 2003; McNeur 2011; Mizelle 2011). In this context, "nuisance" claims should be understood as a way of reinforcing boundaries, as these arguments ultimately rested upon the charge that non-humans have transgressed human-animal boundaries in some way. Indeed, as intimated above, such claims fueled by the new concepts of "civilization" and "domestication" exacerbated class-strife helped to ignite the lengthy campaign to remove pigs from New York City at the end of the nineteenth century. The next section of this chapter uses this case study to illustrate how the above changes combined with class strife to de-animalize the city and thus reduce the food sovereignty of immigrant populations in urban contexts.

9.6 Food Justice and a History of Class Strife

In pre-nineteenth century cities, it was common for people of all classes to own food and working animals, such as pigs (Noll 2015). However, due to various cultural changes of the time, such as the new ideal of the home and rising land prices, middle-class and wealthy New Yorkers increasingly opted to purchase food at markets instead of growing gardens and raising livestock (McNeur 2011). While the upper classes distanced themselves from food production, working class communities, such as those of African, Irish, and German American communities, relied on raising pigs and other food animals as an integral part of securing enough food for their families. These city neighborhoods were areas where humans, pigs, and various domesticated animals shared both city common space, such as streets, and living space. However, "the cultural ideal of the home cleansed of working animals distanced the upper class from most animals (all but pets and horses), connected raising livestock to the lower class, and essentially shifted this class' perception of pigs from that of a useful animal to nuances or ones that transgress human-animal boundaries" (Noll 2015, p. 148). This shift, in combination with a sharp increase in population during the time period and the resulting push to "gentrify" working class neighborhoods ignited a major legal battle over the proper place of pigs in the city commons and the urban home.

The result was a violent clash between wealthy New York citizens, who argued that city pigs were a nuisance, impeded progress, violated the delicate sensibilities of cultured ladies, and a health hazard, and working class citizens, who argued that food animals were useful for cleaning city streets, as pigs eat garbage accumulated in urban areas, and necessary for providing income and food to their families (McNeur 2011, 2014; Noll 2015). Indeed, tradesmen at "piggerys" often collected food waste during this time and boiled it down to both sell to various manufactures and to fatten up pigs that were then sold to local butchers (McNeur 2014). After lengthy legal battles, city officials and police officers clashed with working class communities (the producers of pigs processed at "piggerys") over the forceful removal of pigs from the neighborhoods around Central Park in what is now called New York City's "Piggery" War of 1859 (McNeur 2014). This war ended in the removal of pigs from urban common areas and city homes and new land use laws strictly limiting animal agricultural production in the city proper.

Importantly for this chapter, these changes also resulted in the marked reduction of food sovereignty in city neighborhoods, as working class communities were now increasingly reliant on markets to purchase food-stuffs. In essence, this argument was not about pigs but a deeper conflict concerning divergent visions of the proper use of urban public space (McNeur 2011), food production in the city, and human-animal relationships. The emerging concepts of civilization and domestication guided by the key dualisms of wild/tame and primitive/cultivated combined with class prejudices to cleanse urban environments of all animals considered nuisances by the wealthy. A New York Times writer captures the interplay of these concepts with class discrimination perfectly, when he describes a working-class neighborhood as an assemblage of "shanties in which the pigs and the Patricks lie down together while little ones of Celtic and swinish origin lie miscellaneously, with billy-goats here and there interspersed" (McNeur 2015, p. 1). Beyond the specific context of New York City, such arguments were used to remove a wide array of working animals from the city sphere, including, interestingly, stray dogs and dogs used in dogcarts, which were the primary means of transportation for those unable to afford a horse (Howell 2012; Kalof 2007). Thus "nuisance" claims are not value neutral but come out of often class-charged conflicts concerning divergent visions of how urban areas should be structured and used.

9.7 Disease and the Cleansing of Human Spaces

In addition to nuisance claims, basic assumptions concerning disease, how diseases are transmitted, and dirt or "filth" also played a part in the de-animalization of urban areas (Noll 2015). Atkins (2012b) argues that major changes of widely held conceptions of disease transmission helped to accelerate the process of de-animalizing urban areas. During the nineteenth century, populations commonly believed that foul odors could transmit diseases and that the increasingly chaotic and polluted city environments could cause illness (Atkins 2012b; Barnes 1995; Coleman 1982; McNeur 2011). In fact, "several significant elements of the pre-germ theory etiology of tuberculosis survived intact through the late nineteenth century… Among these elements are filth, stench, and overcrowding, all symptomatic of the underlying pathology of the city" (Barnes 1995, p. 25). In the early part of the nineteenth century, household livestock were a common site in the city streets, slaughter houses were located in urban neighborhoods, and manure from working animals, such as horses, filled the gutters (Atkins 2012b, p. 85). However, the above assumptions associated with disease, dirt, and disease transmission manifested into the increased sanitary policing of urban spaces that culminated in the use of new legislation to address issues of smell coming from horse manure, drains, slaughter houses, and trash (Atkins 2012b; Stallybrass and White 1986).

Indeed, as we just explored above, claims that food animals were a health hazard helped bolster arguments for removing working animals from city environments (McNeur 2011). Such claims, coupled with rising real estate costs, were also used to push slaughter houses and piggerys from city contexts and (DeMello 2011) increasing fears of rabies were used as justifications to remove stray dogs from city-centers (Howell 2012) and to make the use of dogcarts cost prohibitive for poor and working class families (Kalof 2007). Commonly held beliefs concerning the transmission of disease, and the roles that dirt and manure play in this transmission, again manifested along class-lines, with increasing legislation intended to police industries and behavior predominately undertaken in urban working class neighborhoods. In fact, while this analysis largely focuses on relationships between animals and humans in urban areas, McNeur (2011) argues that conflicts, such as the "Piggery" War of 1859 were as much about censoring and changing the behavior and habits of the city's working class and reducing their control over food production, as they were about human-animal relationships. Shifting conceptions of disease helped to bolster claims that the animals of the working class were "nuisances" and thus should be removed from city contexts. This, in turn, impacted the ability of various communities that were dependent on these animals to produce food and other goods and thus impacted their food sovereignty.

9.8 Vegetable Production in the City

The de-animalization of city landscapes had serious impacts in other areas of urban agriculture practices, especially in the area of vegetable production (Noll 2015). In fact, Atkins (2012c) argues that above changes, coupled with automotive technology replacing the horse, helped to undermine food production and the sustainable relationship that made this food production possible between the city-center and the area surrounding the city (between 10 and 50 miles, depending on the context) known as the "charmed circle" (p. 53). In particular, vegetable production in the city greatly benefited from the manure being produced by the plethora of food and work animals housed there, as this abundant resource was continually being used as a cheap source of fertilizer. The fertilizer was used in the peri-urban areas surrounding the city to grow crops that fetched a low price and that could easily be transported, such as beans, potatoes, and grain, and in horticultural operations in the city center to grow delicate and high value crops that were in high demand, such as broccoli, celery, and asparagus (Atkins 2012c, p. 54).

This network between the city proper and peri-urban areas was commonly known as the "manured region," or the radius of agricultural prosperity being sustained by the city's steady supply of the cheap fertilizer. As the amount of manure being produced in city centers decreased due to de-animalization, the prices paid for the product increased. These economic factors combined with rising land costs and the conceptual shifts above to effectively remove most vegetable production from the city sphere. The result of this exodus of crop production from the city landscape was the removal of food production activities to areas far outside cities, the increasing dependence of cities on global food systems, an "extinction of experience" of interactions between humans and the natural world and the disappearance of local knowledge related to agriculture from urban landscapes (Barthel et al. 2010, 2015; Pyle 1978; Miller 2005).

Thus, the current urban landscape bears little resemblance to pre-nineteenth century cities. Historically, cities were sites of intense agricultural production, with food production, processing, and distribution occurring in the larger urban landscape and within specific neighborhoods. As argued above shifting understandings of (1) civilization and domestication and (2) disease and filth combined with class based discrimination to culminate in the removal of all but a handful of non-human animals from the city streets. This, in turn, impacted the ability of various communities to produce food, undermined the "charmed circle," resulted in the "extinction of experience" and an increasing reliance of global food systems. While the conception of the city as an area largely divorced from food animals and crop production is largely assumed today, the analysis of the birth of the modern city outlined above illustrates how the current general structure of the modern city is the result of sustained efforts to limit the food sovereignty of working class communities and guided by a specific vision of how cities should be structured.

9.9   The Context of the City: Challenges Revisited

When viewed from this position, current urban agricultural initiatives are not new and novel approaches to growing food or increasing food security in cities but a challenge to modern conceptions of civilization (and the dualisms in this concept) and re-instantiation of an older vision of cities, where individual communities again have increasing control of the production, processing, and distribution of food. Many of the social barriers faced by urban food initiatives, such as policy, zoning regulations, and municipal ordinances, in so far as they limit the population's ability to grow food, can be understood as vestiges of the de-animalization efforts of the nineteenth century. This is especially the case concerning ordinances that ban food animals from urban areas, as municipalities often prohibit raising livestock (such as chickens, goats, and pigs) due to the historically-charged concerns that they could spread diseases, become a nuisance, or attract pests (Huang and Drescher 2015; Pollock et al. 2012). Rather than being value neutral, such claims are made in a context where these justifications were directed at the poor and used to limit the food sovereignty of working class communities, as a whole, and working class producers, in particular. For this reason, I argue that all urban food initiatives, no matter the values guiding them, that help to amend these ordinances at least indirectly contribute to food sovereignty initiatives, in that they are helping to change the social structure of the city. Let us turn to a survey of municipal laws concerning back yard chickens as an example of my argument.

9.10   Backyard Chickens: A Case Study

Raising backyard chickens has become increasingly popular in urban areas for a multiplicity of reasons, such as the desire to increase personal control over food, a reaction to industrial farming practices, and a way to address a broken food system (Bouvier 2012; Wooten and Ackerman 2011). In response to the increased desire of those living in cities to raise chickens, local governments across the United States are in the process of amending or considering amending their ordinances to allow chickens in urban areas (Bouvier 2012). However, even in the face of this wide-spread push to allow chickens, city leaders commonly raise a similar list of concerns, such as that chickens may be a nuisance, could decrease property values, emit greenhouse emissions, cause odors, pass along diseases, and that "chickens do not belong in cities" (Bouvier 2012, p. 9). The above concerns coupled with the wide-spread desire for backyard chickens has resulted in a highly contentious controversy over regulation. While each of the above concerns do not stand up to scrutiny and are being adequately addressed by backyard chicken enthusiasts, what is important for this paper is that the list contains references to the commitments that contributed to the historical de-animalization of modern cityscapes, such as nuisance claims, issues concerning disease, filth, and odors, and conceptions of the city as a place of civilization and thus no place for a chicken. Just as in the historical case-study concerning pigs in New York, the same concerns are currently being used as justifications for keeping city landscapes free of food animals. Thus, one could argue that commitments created during the birth the modern city may still be acting as conceptual barriers to the implementation of urban food projects.

However, in contrast to historical clashes between classes on the issue of urban animal husbandry, today advocates for backyard chickens and animals are making substantial progress changing ordinances and laws to allow for the reintroduction of a wide range of livestock and food animals in urban areas, such as bees, goats, and chickens (Wooten and Ackerman 2011). Indeed, we're even seeing elected officials in the historically charged context of New York City coming out in support of legalizing pig pet ownership. For example, State Senator Tony Avella is quoted as saying that "people die on (illegal) construction sites ... and yet our threatening enforcement is against the person who has a small pig. There is a total inconsistency in the way the city deals with certain activities" (Fenton 2013). When viewed from an ahistorical position, there may seem to be an inconsistency in the laws, as they've largely succeed in removing animals from working-class neighborhoods. Today advocates for urban livestock come from wide range of socio-economic classes and pursue the reintroduction of food animals for a variety of reasons, such as to increase food security, personal health or environmental reasons, or to increase local control of food systems (Bouvier 2012). This diverse lobbying base is increasingly successful in undermining the above challenges and bringing about change at the legislative level. Thus, no matter their goals, this case study illustrates how urban food advocates and programs are challenging barriers that once contributed to the reduction of food sovereignty in urban neighborhoods and thus is helping to make food sovereignty possible.

This does not necessarily mean, however, that changing ordinances or laws to allow for a greater variety of food animals and agricultural production projects in the city will automatically translate into increased food sovereignty for the population, as a whole, or for traditionally marginalized communities, in particular. There are several other factors that are necessary to help increase the food sovereignty of populations, such as strong community networks, community members that are committed to the goal of increasing food sovereignty, access to affordable land, the availability of soil tests and farming expertise, the potential for acquiring grants, etc. In fact, the availability of social capital plays a major role in determining whether or not specific communities have the resources necessary to pursue food sovereignty and traditionally marginalized communities often historically suffer from a lack of these resources. However, even after acknowledging the important role that historical and contextual factors play, I argue that removing legal barriers that limit local food production is a necessary first step in working towards food sovereignty, but is by no means sufficient for reaching this goal.

In this vein, changing local ordinances and laws will not automatically "achieve" food sovereignty in urban areas. Food sovereignty includes a wide range of social justice concerns, such as matters of food autonomy, ecological preservation, determining what duties we owe to future generations, etc. (Pimbert 2008;

Whittman et al. 2011). Changes in ordinances cannot address all of these wide-ranging concerns, as they largely lie outside of the legal scope. Nor will all of these concerns be of equal importance to all communities working towards the goal of increasing food sovereignty. This chapter is making the more modest claim that removing legal barriers to urban food production will help to open up the space necessary for local communities to play an important role in determining their specific food landscape. This does not mean that there will be no conflicts or that agricultural production should be prioritized in all situations. As with conflicts of interest associated with greenspace initiatives (Walker 2012), such changes serve the purpose of helping to open up a space for communities to be at the table when important food system and land use decisions are made.

It should also be noted here that, as a counter-argument, one could argue that urban projects do not contribute to food sovereignty, as the above changes in agricultural production are due to (1) economic developments, rather than social and to (2) improving knowledge in the realm of public health. Concerning the first point, one could argue that Atkin's (2012c) historical analysis of the undermine of the charmed circle and thus of vegetable production, makes use of Von Thunen's predictive model[6] describing how peri-urban areas organize food production, which looks at economic and not social factors. Regardless of conceptual shifts, the technological developments that displaced the horse combined with rising land values to impact the market prices of manure costs. This, in turn, made it economically unfeasible to grow crops in city areas, thus pushing agricultural production to cheaper land outside of the city or, in the language of Von Thunen's model, a ring with lowering production intensity. Thus, at least in the case of vegetable production, social factors did not play an important role in undermining urban food systems.

In reply, I argue that this counter argument is only applicable to the vegetable production section and not the larger argument that takes shifting human-animal relationships into account, as the larger analysis does not use von Thunen's model. Even when accepting this limited scope, however, the critique does not hold, as, even accepting the facile and dubious separation of the economic and social spheres, spatial organization is predominantly the result of historical social relationships, as much as it is influenced by economic factors (Barthel et al. 2015). In fact, von Thunen's model itself has been critiqued, with scholars arguing that Thunen's prescribed boundaries between different types of production (in this case milk production) "are as much the product of economic margins as they are the product of political outcomes" (Barthel et al. 2015, p. 81). In the complicated contexts of the city, a multiplicity of factors influence both the structure of urban

---

[6]Von Thunen's influential thesis was that when there is a lone market located in an urban center, "crops with high transportation costs and intensive uses of land would be produced near the market than would other types of crops... Distance determined land value and transportation costs and therefore the margin of profit from a particular enterprise needed to be sufficient to pay these costs" (Barthel et al. 2015, p. 80). In this theory, economic factors cause the creation of "rings" of varying productivity around city-centers.

areas and the various production activities performed in these contexts, from food production to the production of industrial goods and later services.

When replying to the to critique that the removal of animals from the social sphere was largely due to improving knowledge in the realm of public health, readers should remember that the current literature largely argues that elements of pre-germ theory (such as the view that foul odors could transmit disease) helped increase the sanitary policing of urban spaces and culminated in the increased removal of non-human animals from urban spaces (Atkins 2012b; Stallybrass and White 1986). Even today, the topic of disease transmission between humans and animals is complex, dependent upon the specific species involved, and an area of knowledge that urban populations (including those involved in governing bodies) are often not well versed in (Bouvier 2012). Let's look at the example of the chicken to see how current knowledge of disease transmission largely does not play an important role in determining whether or not municipal ordinances. According to Bouvier (2012), the two diseases commonly raised as objections to the back-yard chickens are avian flu and salmonella. However, pubic health officials have found no evidence that the chance of incidences of these diseases in small flocks are a concern (Greger 2006; Pollock et al. 2012). Nonetheless, fear of these diseases play a role in whether or not ordinances are changed. Fairlie (2010) provides a similar analysis of the influence of bovine spongiform encephalopathy (Also known as BSE and mad cow disease) on ordinances regulating pig production, even though BSE does not infect pigs. My point here is not to downplay public health concerns or work in this area but to point out that a wide variety of factors, including current assumptions and fears, influence zoning-laws, municipal ordinances, and other regulating laws and mandates. Thus, the counter-argument does not hold.

9.11 Conclusion: Challenging Historical Barriers to Food Justice

Local food initiatives are steadily becoming a part of contemporary cities around the world and can take on many forms (Holt-Gimenez 2011; DeLind 2011; Martinez et al. 2010). While some of these initiatives are concerned with providing consumers with farm-fresh produce or improving urban population's access to foodstuffs in urban areas, a growing portion of movements are concerned with increasing the food sovereignty of marginalized urban communities. The purpose of this chapter was to provide a theoretical analysis of urban contexts with the aim of identifying conceptual barriers that may act as roadblocks to achieving food sovereignty in cities. Specifically, in this paper, I argued that (1) presupposed commitments created during the birth of the modern city could act as conceptual barriers for the implementation of food sovereignty programs and (2) that urban food activists and programs that challenge these barriers are helping to achieve the goal of restoring food sovereignty to local communities, no matter their reasons for doing so. At the very least, understanding the complexities of these barriers and how they operate helps to strengthen ties between urban food projects, provides these initiatives with ways to undermine common arguments used to support restrictive ordinances and policies, and illustrates the transformative potential of food sovereignty movements.

References

Angotti, T. 2015. Urban agriculture: Long-term strategy or impossible dream?: Lessons from prospect farm in brooklyn, New York. *Public Health* 129 (4): 336–341.

Atkins, P. 2012a. Introduction. In *Animal cities: Beastly urban histories*, ed. P. Atkins, 1–19. Burlington: Ashgate Publishing Inc.

Atkins, P. 2012b. Animal wastes and nuisances in nineteenth-century London. Introduction. In *Animal cities: Beastly urban histories*, ed. P. Atkins, 19–53. Burlington: Ashgate Publishing Inc.

Atkins, P. 2012c. The charmed circle. Introduction. In *Animal cities: Beastly urban histories*, ed. P. Atkins, 53–77. Burlington: Ashgate Publishing Inc.

Barnes, D. 1995. *The making of social disease: Tuberculosis in nineteenth-century France*. Berkeley: University of California Press.

Barthel, S., J. Parker, and H. Ernstson. 2015. Food and green space in cities: A resilience lens on gardens and urban environmental movements. *Urban Studies* 52 (7): 1321–1338.

Barthel, S., C. Folke, and J. Colding. 2010. Social-ecological memory in urban gardens: Retaining the capacity for management of ecosystem services. *Global Environmental Change* 20 (2): 255–265.

Bouvier, J. 2012. Illegal fowl: A survey of municipal laws relating to backyard poultry and a model ordinance for regulating city chickens. *Environmental Law Reporter* 42 (10888): 1–34.

Brantz, D. 2011. Domestication of empire: Human-animal relations at the intersection of civilization and acclimatization in the nineteenth century. In *A cultural history of animals in the age of empire*, ed. K. Kete, 42–59. New York: Bloomsbury Publishing.

Castillo, S.R., C.R. Winkle, S. Krauss, A. Turkewitz, C. Silva, and E.S. Heinemann. 2013. Regulatory and other barriers to urban and peri-urban agriculture: A case study of urban planners and urban farmers from the greater Chicago metropolitan area. *The Journal of Agricultural Food System Community Development* 3 (3): 155–166.

Coleman, W. 1982. *Death is a social disease: Public health and political economy in early industrial France*. Madison: University of Wisconsin Press.

Clutton-Brock, J. 2011. How domestic animals have shaped the development of human societies. In *A cultural history of animals in antiquity*, ed. L. Kalof, 75–95. New York: Bloomsbury Publishing.

DeLind, L. 2011. Are local food and the local food movement taking us where we want to go? Or are we hitching our wagons to the wrong stars? *Agriculture and Human Values* 28 (2): 273–283.

DeMello, M. 2011. The present and future of animal domestication. In *A cultural history of animals in the modern age*, ed. R. Malamud, 67–94. New York: Bloomsbury Publishing.

De Zeeuw, H., S. Guendel, and H. Waibel. 2000. The integration of agriculture in urban policies. *Urban Agriculture Magazine* 1 (1): 161–180.

Edwards, P. 2011. Domesticated animals in renaissance Europe. In *A cultural history of animals in the renaissance*, ed. B. Boehrer, 95–119. New York: Bloomsbury Publishing.

Fairlie, S. 2010. *Meat: A benign extravagance*. White River Junction: Chelsea Green Publishing Company.

Fenton, R. 2013. Hey, NYC—What's the pig deal? New York Post. http://nypost.com/2013/02/18/hey-nyc-whats-the-pig-deal/. Accessed June 12, 2016.

Flora, C.B. 2010. Schanbacer, William D: The politics of food: The global conflict between food security and food sovereignty. *Journal of Agricultural and Environmental Ethics* 24 (5): 545–547.

Food and nutrition security. 2015. FAO: Growing Greener Cities. http://www.fao.org/ag/agp/greenercities/en/whyuph/foodsecurity.html. Accessed June 09, 2016.

Gamber, W. 2005. Away from home: Middle-class boarders in the nineteenth-century city. *Journal of Urban History* 31: 289–305.

Greger, M. 2006. *Bird flu: A virus of our own hatching*. New York: Lantern Books.

Haraway, D.J. 1989. *Primate visions: Gender, race, and nature in the world of modern science*. New York: Routledge.

Harding, S. 1993. Rethinking standpoint epistemology: What is 'strong objectivity'? In *Feminist epistemologies*, ed. L. Alcoff, and E. Potter, 49–82. New York: Routledge.

Holt-Gimenez, E. (ed.). 2011. *Food movements unite!: Strategies to transform our food system*. Oakland: Food First Books.

Huang, D., and M. Drescher. 2015. Urban crops and livestock: The experiences, challenges, and opportunities of planning for urban agriculture in two Canadian provinces. *Land Use Policy* 43: 1–14.

Hunter park garden house. 2017. Allen Neighborhood Center. http://allenneighborhoodcenter.org/gardenhouse/. Accessed June 11, 2016.

Howell, P. 2012. Between the muzzle and the leash: Dog-walking, discipline, and the modern city". In *Animal cities: Beastly urban histories*, ed. P. Atkins, 221–243. Burlington: Ashgate Publishing Inc.

Inwagen, P. 2013. *Metaphysics: Third edition*, 3rd ed. Boulder: Westview Press.

Kalof, L. 2007. *Looking at animals in human history*. London: Reaktion Books.

Kimberton CSA. 2017. http://www.kimbertoncsa.org. Accessed June 11, 2016.

LeJava, J.P, and M.J. Goonan. 2012. Cultivating urban agriculture: Addressing land use barriers to gardening and farming in cities. *Real Estate Law Journal* 41: 216–245.

Light, A. 2003. Urban ecological citizenship. *Journal of Social Philosophy* 34 (1): 44–63.

Light, A. 2001. The urban blind spot in environmental ethics. *Environmental Politics* 10: 7–35.

Lyson, T. 2004. *Civic agriculture: Reconnecting farm, food, and community*. Medford, MA: Tufts University Press.

Martinez, S., et al. 2010. *Local food systems: Concepts, impacts, and issues*. Washington DC.: U. S. Department of Agriculture, Economic Research Service.

Mcgirr, H.K., and S.P.J. Batterbury. 2016. Food in the city: Urban food geographies and 'local' food sourcing in Melbourne and San Diego county. *Geographical Research* 54: 3–18.

McNeur, C. 2015. One of the first gentrification movements—the great piggery war. New York Post. http://nypost.com/2015/02/01/one-of-the-first-gentrification-movements-the-great-piggery-war/. Accessed June 11, 2016.

McNeur, C. 2014. *Taming Manhattan: Environmental battles in the antebellum city*. Cambridge, MA: Harvard University Press.

McNeur, C. 2011. The 'swinish multitude': Controversies over hogs in antebellum New York City. *Journal of Urban History* 37: 639–660.

Michelfelder, D. 2003. Valuing wildlife populations in urban environments. *Journal of Social Philosophy* 34 (1): 79–90.

Miller, J.R. 2005. Biodiversity conservation and the extinction of experience. *Trends in Ecology & Evolution* 20: 430–434.

Mitchell, D. 2003. Dead labor and the political economy of landscape—California living, California dying. In *Handbook of cultural geography*, ed. K. Anderson, 223–248. Sage, London; Thousand Oaks, CA.

Mizelle, B. 2011. *Pig*. London: Reaktion Books.

Noll, S. 2015. History lessons: What urban environmental ethics can learn from nineteenth century cities. *Journal of Agricultural and Environmental Ethics* 28 (1): 143–159.

Palmer, C. 2011. The moral relevance of the distinction between domesticated and wild animals. In *Animal cities: Beastly urban histories*, ed. T. Beauchamp, and R.G. Frey. Oxford: Oxford University Press.

Pascua, E. 2011. Domestication. In *A cultural history of animals in the medieval age*, ed. B. Resl, 81–102. New York: Bloomsbury Publishing.

Pimbert, M. 2008. *Towards food sovereignty: Reclaiming autonomous food systems*. London: The International Institute for Environment and Development.

Pollock, S.L., C. Stephen, N. Skuridina, and T. Kosatsky. 2012. Raising chickens in city backyards: The public health role. *Journal of Community Health* 37 (3): 734–742.

Pyle, R.M. 1978. The extinction of experience. *Horticulture* 56: 64–67.

Ricoeur, P., and K. Blamey. 1995. *Oneself as another*. Chicago: University of Chicago Press.

Roehr, D., and I. Kunigk. 2009. Metro Vancouver: Designing for urban food production. *Berkeley Planning Journal* 22 (1): 61–70.

Rose, M. 2003. Reembracing metaphysics. *Environment and Planning* 36: 461–468.

Schafft, K.A., E.B. Jensen, and C.C. Hinrichs. 2009. Food deserts and overweight schoolchildren: Evidence from Pennsylvania. *Rural Sociology* 74 (2): 153–177.

Schanbacher, W.D. 2010. *The politics of food: The global conflict between food security and food sovereignty*. Santa Barbara: Praeger International.

Stallybrass, P., and A. White. 1986. *The politics and poetics of transgression*. Ithaca: Cornell University Press.

Steinhoff, A. 2011. Nineteenth-Century urbanization as sacred process: Insights from German Strasbourg. *Journal of Urban History* 37: 828–841.

The school garden project. 2016. http://schoolgardenproject.org/. Accessed June 11, 2016.

Taylor, D., and K.J. Ard. 2015. Food availability and the food desert frame in Detroit: An overview of the city's food system. *Environmental Practice* 17: 102–133.

Thompson, P.B. 2015. *From field to fork: Food ethics for everyone*. New York: Oxford University Press.

Thompson, P.B. 1994. *The spirit of the soil: Agriculture and environmental ethics*. New York: Routledge.

Voigt, K.A. 2011. Pigs in the backyard or the earnyard: Removing zoning impediments to urban agriculture. *Boston College Environmental Affairs Law Review* 38 (2): 537–566.

Walker, G. 2012. *Environmental justice: Concepts, evidence, and politics*. New York: Routledge.

Werkheiser, I., and S. Noll. 2014. From food justice to a tool of the status quo: Three sub-movements within local food. *Journal of Agricultural and Environmental Ethics* 27 (2): 201–210.

Whittman, H., A. Desmarais, and N. Wiebe. 2011. *Food sovereignty: Reconnecting food, nature and community*. Oakland: Food First Books.

Wooten, H., and A. Ackerman. 2011. *Seeding the city: Land use policies to promote urban agriculture*. Oakland: Public Health Law & Policy.

Chapter 10
Race, Religion, and Justice: From Privilege to Solidarity in the Mid-South Food Movement

Emily A. Holmes and Christopher Peterson

Abstract This chapter examines the growth of the food justice movement in Memphis and the Mid-South in light of the problem of food insecurity experienced in this region. Although food insecurity disproportionately affects low-income people of color, the good food movement is primarily led by white middle-class individuals. Drawing on critical race theory and liberation theologies, this paper offers a critique of (race and class) privilege within organizations dedicated to food justice and suggests that humility and solidarity are the key virtues needed for those working in the food justice movement from a position of white privilege and economic security. Humility shapes our self-understanding and ability to listen to others, while solidarity promotes the work of justice as an expression of love. The chapter concludes by arguing for the cultivation of these virtues through religious narratives, communities shaped by a diverse range of experience, and concrete, on-going practices of shared commitment.

Keywords White privilege · Solidarity · Humility · Food justice · Memphis

Over the last decade, multiple strands of the "good food movement" (Allen 2013) have converged into a broad "polyculture" (Alkon and Agyeman 2011) dedicated to cultivating food justice.[1] In this polyculture, religious communities join health advocates, anti-hunger activists, foodies, environmentalists, and alternative farmers in their shared critique of the unjust effects of the industrial food system. Religious

---

[1]These strands include differing origins and priorities such as alternative local and organic food (Pollan 2006); environmental impact (Kingsolver 2007); labor and distribution (Gottlieb and Joshi 2010); food insecurity and food access (Sack 2001; *A Place at the Table* 2012); and sovereignty of food and seed (Shiva 2007).

E.A. Holmes (✉)
Christian Brothers University, Memphis, USA
e-mail: emily.holmes@cbu.edu

C. Peterson
Loch Holland Farm, Saulsbury, TN, USA
e-mail: petersochris@gmail.com

© Springer International Publishing AG 2017   113
I. Werkheiser and Z. Piso (eds.), *Food Justice in US and Global Contexts*,
The International Library of Environmental, Agricultural and Food Ethics 24,
DOI 10.1007/978-3-319-57174-4_10

communities are motivated in the practical dimensions of this work by the values of their traditions; these values are often embedded in scriptural narratives that provide an alternative vision of the possibilities and purposes of our food system (Sanford 2012; Ayres 2013; Wirzba 2011). In a just food system, small-scale farmers would be able to make a living growing food using environmentally sustainable methods and raising animals treated humanely. Food workers would earn a living wage. Farmers and eaters could connect face to face through local food hubs and markets. Food deserts would vanish through economic development in the poorest neighborhoods. School and community gardens would flourish. Everyone would have access to fresh, locally grown foods, and everyone would have enough to eat.

This idealized vision of a just, local, and sustainable food system provides a critical contrast with the current state of food in the U.S. and a motivation for religious communities to engage in the shared work of its transformation. Religious communities support the broader food justice movement while contributing their vision and values, taken from their narrative traditions, as well as the virtues embedded within those traditions. In this chapter, we argue for the importance of two particular Christian virtues in the practical work of achieving a vision of food justice at the local level: humility and solidarity.

Like all knowledge production, this chapter is written from a particular social and physical location. We are writing as a theologian (Emily) and a farmer (Chris), both active in our local food movement. We are white, middle class, over-educated, progressive Christians, educators and advocates for peace, justice, and reconciliation in Memphis and the Mid-South region. After years working in the local food movement, we have become attuned to the problems of white privilege, conflicts of race, class, and gender, and the systemic injustices that shape our city and region. We do not have all the answers, but we know that simply continuing the work while calling attention to these problems will not solve them. We are convinced that the work itself can be transformative, but in order effectively to do the work of building a more just, local, and sustainable food system, we need to cultivate particular virtues appropriate to this work, as it is undertaken in this particular place. We recognize the need for virtues that are grounded in our locality; shaped by our collective and individual narrative histories; and habituated by particular practices.

This chapter emerged from a series of on-going conversations across six years about the food system, racism, justice, and the history of Memphis, and what it means to do this work as a relatively privileged white person in this place. Because race and religion are, arguably, the two most influential forces that have shaped the history, culture, and agriculture of the south, we turn to critical race theory and religious narratives to rethink current practical food movement work. In what follows, inspired by critical race theory and liberation theologies, we suggest that humility and solidarity are the key virtues needed at this time, in this place, for those, like us, born to unearned privileges of whiteness and economic security. Humility shapes our self-understanding and ability to listen to others, while solidarity promotes the work of justice as an expression of love. We conclude by arguing for the cultivation of these virtues through religious narratives, communities shaped by a diverse range of experience, and concrete, on-going practices of shared commitment. But first, an introduction to Memphis, the place in which we write.

10.1   Food (in)Justice in Memphis

Memphis is an urban environment in the heart of an agricultural region in which too many citizens struggle to find enough to eat. The injustices of the industrial food system are keenly felt locally. In 2010, Memphis was ranked the hunger capital of the nation in a Gallup poll, with a shocking 26 percent of residents reporting an inability to afford sufficient food for their families at some point during the previous year (Conley 2010). Food insecurity, defined by the USDA as a household-level economic and social condition of limited or uncertain access to adequate food, is an on-going local and regional problem.[2] According to a 2014 report from Feeding America, 209,720 people were food insecure in Shelby County. 49% of those people had an income that was above the maximum to receive SNAP benefits, and yet still struggled with food insecurity (Feeding America 2014). Nearly 100,000 people rely on charitable organizations to supplement their food supply. These numbers correlate to Memphis' high poverty rate: according to the most recent U.S. census, 19% of the population lives in poverty, making Memphis the poorest city in America (Charlier 2011). A 2013 study again named Memphis as the nation's poorest metropolitan area with a population of at least 1 million, with a 19.9% poverty rate for the eight-county metropolitan area, and a 28.3% rate within the Memphis city limits (Charlier 2013). This poverty rate also masks a growing gap among racial groups: while the poverty rate for white residents declined for a third straight year, the 33.6% percent rate for African-Americans in the city limits has remained unchanged. Disturbingly, the poverty rate for African-Americans living in Memphis is higher than elsewhere in Tennessee and in the nation, while non-Hispanic white residents have a lower poverty rate than white state and national populations (Delavega 2015).

Many of the poor in Memphis live in de facto economically and racially seg-regated neighborhoods that coincide with so-called "food deserts." A Memphis newspaper describes how "In zip codes like 38126—the poorest in the city with a median income of $12,000—small corner stores and convenience stores abound, but a supermarket could be miles away. Stocked with little or no produce and priced out of range for many residents, convenience stores do not provide viable access to fresh foods" (Sayle 2010). The USDA maps food access through a research atlas

---

[2]United States Department of Agriculture Economic Research Service, "Food Security in the U.S.: Definitions of Food Security." 8 Sept. 2015. Web. 7 May 2016. See also Committee on National Statistics, Division of Behavioral and Social Sciences and Education, "Food Insecurity and Hunger in the United States: An Assessment of the Measure" (Washington, D.C.: The National Academies Press 2006).

that indicates both low-income and low access census tracks; these are used by researchers as indicators of food deserts. In an urban area, living in a food desert typically means living in a low-income census tract that is more than one mile from a supermarket or large grocery store.[3] Memphis is riddled with these food deserts, neighborhoods that lack the availability of fresh foods that are the foundation of a healthy diet and instead abound in cheap, processed and fast foods that are high in calories but low in nutrition and which contribute to obesity, diabetes, and other health problems.[4] The result, which is counterintuitive to people privileged not to live within these constraints, is often obesity coupled with malnutrition. While several studies have shown that simply introducing grocery stores into food deserts does little to improve the health or fresh produce consumption of people who continue to suffer the stress-related effects of poverty, giving people access to healthier and more affordable food options in their own neighborhoods remains an important local and national food justice goal.[5]

Memphis may be worse off than other large metropolitan areas for its high rate of poverty and food insecurity. But its citizens have a surplus of grit and determination to build a better community in the areas of education, healthcare, criminal justice, recreation, and sustainable living. We also have natural resources of abundant water, a long growing season, relatively good soil, and a rich, if complicated, agricultural history. For these reasons, in the area of food justice, Memphis also has recently seen innovative forms of urban agriculture; a growth in the number of regional small organic/naturally grown/biodynamic family farms; an abundance of farmers market and CSA options; a regional food hub; and numerous community, school, and church gardens. In these efforts by local citizens to increase access to local, just, and sustainable food, the food movement in Memphis and the Mid-South region reflects national and global trends. The past ten years in particular have seen a noticeable shift in awareness surrounding the ways in which food is grown, distributed, and consumed, as well as activist efforts to strengthen the regional food system and to use policy changes to increase access to "good food for all."[6]

---

[3]United States Department of Agriculture Economic Research Service, "Food Access Research Atlas," http://www.ers.usda.gov/data-products/food-access-research-atlas/documentation.aspx# definitions. See also The National Research Council, *The Public Health Effects of Food Deserts* (Washington D.C.: The National Academies Press 2009); and Mark Winne, *Closing the Food Gap: Resetting the Table in the Land of Plenty* (Boston, MA: Beacon Press 2008).

[4]For a vivid description of life in one of these neighborhoods, see Chris Peck, "What Obama Didn't see," *Memphis Commercial Appeal*, 30 September 2012; http://www.commercialappeal.com/news/2012/sep/30/chris-peck-what-obama-didnt-see/, accessed 13 May 2016.

[5]These studies are summarized by Heather Tirado Gilligan, "Food Deserts Aren't the Problem: Getting fresh fruits and vegetables into low-income neighborhoods doesn't make poor people healthier," *Slate* (10 Feb. 2014)

http://www.slate.com/articles/life/food/2014/02/food_deserts_and_fresh_food_access_aren_t_the_problem_poverty_not_obesity.2.html, accessed 13 May 2016.

[6]See Memphis and Shelby County Food Advisory Council Working Group, *Good Food for All: The Need for a Food Advisory Council for Memphis and Shelby County* (Memphis, TN 2012).

Increasingly, the work of the local food movement exemplifies many of the features of grassroots, democratic, and justice-driven social change: patient coalition building; empowerment of individuals and communities as both producers and consumers of food; and regional planning and policy changes developed from stakeholder input. But to fully realize the potential for food justice, Memphians have to contend with the history and impact of racism and class division in our city. Numerous well-intentioned efforts to make local foods available in impoverished areas have foundered on these rocks.

10.2   The Problem of Whiteness in a Majority Black City

Recent research in the fields of geography and cultural anthropology has demonstrated the degree to which the alternative food movement is marked by implicit, and sometimes explicit, notions of whiteness and class privilege (Guthman 2011a, b; Lambert-Pennington and Hicks 2016). Through her research into alternative food organizations, Julie Guthman, for instance, notes two distinct manifestations of whiteness that pervade and hinder the work of well-intentioned food organizations: "colorblindness" and an "assumption of universalism." Alternative food organizations have an inherent tendency towards the deliberate "absence of racial identifiers in language," or colorblindness, which simply "erases the privilege that whiteness creates" (Guthman 2011a, 267). In illustrating this idea, Guthman points to the fact that very few farmers markets (in her research area of California) are located in communities of color. Disputing the common reasoning that farmers markets tend to be located in wealthier, white areas because "farmers may make more money there," she points to the fact that even in neighborhoods with racially diverse populations, these markets still tend to be patronized primarily by the white residents of the given neighborhood. In ignoring the race patterning of these markets, the sponsoring organizations implicitly "naturalize inequality" (Guthman 2011a, 267–69).

Further, Guthman explores additional ways in which the assumption of universality in the values of the alternative food movement—healthy, clean, natural whole foods—actually reflect the privileged aesthetic values of white people. Those who hold these values assume that they are "normally and widely shared" (Guthman 2011a, 267), and when these supposedly universal values do not resonate in low-income communities and communities of color, it is assumed that this indicates the need for education and enculturation into these values rather than a flaw in the original assumption. In demonstrating this point, Guthman describes a survey of farmers market and CSA managers in which they were asked, "What do you think are some of the reasons that it is primarily European-American people who seem to participate in CSAs?" Her qualitative analysis revealed the common responses such as *"better education," "more concern about food quality," "more health consciousness,"* and even *"more time"* along with economic factors, all of which point towards a judgment that it is primarily a dearth of correct values that explains the lack of broad, non-white participation in alternative food. As her analysis demonstrates, this phenomenon is not limited to farmers markets and CSAs but is pervasive in different types of food movement work throughout the country (Guthman 2011a, 269–72).

Rather than accept these explanations, Guthman offers evidence that the lack of resonance of these supposedly universal values as well as overall limited participation by individuals of color is rooted in "exclusionary practices" including lack of "cultural competency" as well as "white privilege" as the primary reasons for a lack of diversity. In concluding her analysis, she points towards a need for the white leadership of this movement to thus abandon (or at least rethink) the dominant narrative of "bringing good food to others," and instead embrace a position of humility in the face of our lack of knowledge, ultimately opening up the movement to "allow for others to define the space and projects that will help spur the transformation to a more just and ecological way of providing food" (Guthman 2011a, 276).

The food movement in Memphis too often confirms Guthman's analysis of the dominance of whiteness in alternative food. While similar social scientific research on the effect of whiteness on alternative food in Memphis is currently underway (Lambert-Pennington and Hicks 2016), our experience bears out the relevance of her analysis to our local context. Many of the groups working towards food justice in Memphis include at minimum a rhetorical critique of power and privilege. However, privilege continues to manifest itself in counterproductive, even insidious ways. Here, we deal briefly with two prime manifestations of this phenomenon: (1) lack of leadership from within communities of color and (2) lack of diversity within community coalitions.

10.2.1 Leadership

One of the most glaring manifestations of white privilege within our local food movement is that, while issues like food insecurity most directly impact people of color in Memphis, very few of the organizations working in this sector are led by those individuals. In our experience this is true not only of paid leadership, but of volunteer boards of directors as well. Moreover, even organizations that take seriously the realities of privilege are not immune. For the sake of this section we will focus explicitly on trying to overcome this problem within an organization with which we have both been affiliated, GrowMemphis, whose mission is to help communities build gardens to improve access to locally grown food in their neighborhoods. In addition to supporting over forty gardens, its programming further includes a DoubleGreen$ program that doubles SNAP benefits spent at local farmers markets and convening the Food Advisory Council of Memphis and Shelby County to address local food policy concerns. GrowMemphis was originally launched as a program of the Mid-South Peace and Justice Center, an organization with a 35-year history in civil rights and grassroots movement building. Being rooted in an organization with a broad social justice platform, GrowMemphis progressed differently from many of its counterpart organizations in other urban areas. While the work of developing community gardens on vacant properties in low-income communities is similar to that done in other cities with similar levels of vacant property and low food access, GrowMemphis' approach to this work focuses explicitly on working directly with community members to empower *them* to create and sustain community gardens. GrowMemphis' approach to this work actively seeks to avoid the narrative that Guthman describes as "bringing good food to others" by championing a community driven, community owned approach.

During the course of our work with the organization, despite clearly recognizing the ways in which privilege operates (i.e. supporting community gardens *for* low-income communities), we still struggled regularly with the reality that, while individual projects were run by individuals of color within their own neighborhoods, and often within food insecure communities, the organization itself was staffed and overseen primarily by white people of privilege. In trying to address this problem through our board and employee recruitment, we were met with similar "colorblind" explanations for lack of qualified candidates of color. Hiring scenarios are emblematic of this issue. Though actively committed to diversity in hiring, we were often faced with a limited number of candidates of color, or more frequently, candidates of color who lacked the "right kind" of experience. That is, work experience such volunteer time on farms or in urban gardening programs (experience more common for white, college educated individuals) were often valued highly by hiring committees. Meanwhile, educational backgrounds, training, and job experiences of those who grew up in a low-income, food insecure community would not match the "type" of candidate being searched for. Rather than look inwards and realize that, in our hiring values, we had created an inherently white space, for expedience's sake we would often inadvertently "naturalize inequality."

While Guthman's research is helpful in calling out the implicit biases in this example, she does not address the systemic nature of problems such as leadership. Just as farmers markets can be faulted for creating exclusive spaces, our organization takes responsibility for combating privilege in developing leadership. However, there are also a number of systemic factors that must be taken into account. To keep with the example of hiring criteria, qualifications such as experience working in urban gardening or non-profit program management experience are not created in a vacuum by one organization alone, but are motivated in part by a non-profit system of funders that sets standards and expectations for our work. It is not our intention to excuse allowing privilege to operate within our organization, but rather to point out practical factors that slow progress even within progressive, anti-racist, and social justice focused organizations.

10.2.2   Inclusion and Ownership Within Coalitions

The example of privilege pervading hiring practices in many ways is symptomatic of a larger trend: valuing the knowledge of "experts" over the experiential knowledge of those most affected by the issues. This trend is particularly apparent in the development of community coalitions working against food injustice. Community coalitions are held up as ideal for tackling large social problems in Memphis as in many other cities. Not only does this allow, in theory, for resource and expertise sharing, but there is often an explicit desire to include members of the community with a diverse range of backgrounds and interests. While we do not want to denigrate any of the good work being done, the coalitions that design these projects are rarely an accurate representation of Memphis at large, much less of those communities that bear a disproportionate burden of diabetes, obesity, hunger, and blighted property. While we are not privy to the inner workings of every coalition, we can speak from our own experience on the struggle with the racial and economic privilege that permeates coalition work.

Coalitions are most often initiated from a position of power, such as an established non-profit organization. This privileged worldview deeply affects the construction of coalitions. As a cultural norm, most coalitions include at least a small number of people of color, most often African-Americans (who comprise just over 61% of the population of Memphis). However, those African-Americans sought out by coalition organizers tend to typify either what Cornell West calls "managerial leaders" (West 1994, 59–60)—that is, government or private sector leaders who have bought into the current power structure, and often seek to mute more radical voices—or one or two token community members. The latter token members are often disempowered in their isolation. They are asked to speak for their communities rather than from personal experience. This tokenism fails to capture the nuance and disagreement within the communities they are asked to represent. Further, it is not uncommon for coalitions to select those community members already in agreement with a proposed project, even while these individuals are ostensibly at the table to provide community critique.

We have primarily worked with organizations with a commitment to the inclusion and elevation of those voices traditionally not welcome at the table of power. In building the membership of our Food Advisory Council, for example, we codified the inclusion of non-affiliated community members with a stake in our food system (the food insecure, working poor not employed by non-profit organizations or government, farmers, etc.). While our aims in constructing our coalition were progressive, privilege can still serve to minimize these voices, and it is often the most benign form of privilege that excludes community participation. For instance, non-profit leaders operate within an accepted upper middle class meeting structure that is not accessible, much less welcoming for the types of voices we professed to value so highly. The meeting structure—whether timing, location, or format—for coalitions is rarely conducive to involvement by community members. Most of the voices we need to hear are working during our lunch meetings; our meetings are most often held in the board rooms of power for the sake of "central location" rather than in the affected communities; and, when these barriers are overcome, we rely on a technocratic vocabulary, full of jargon, which community members either need explained (thus emphasizing the divide) or which serves to provincialize their viewpoints. Most obviously, these structures create a system that is only accessible to a certain subset of the voices we seek, most often retirees. While their experience is important and appreciated, to allow their voices to suffice for their entire community is a mistake. This form of tokenism fails to appreciate the nuance and differences of opinion within communities. Thus, while our Food Advisory Council may be on the right track in understanding the need for community voices, our structure, borne out of privilege, still fails to become fully inclusive.

This privilege is also readily apparent in the ways "professional" voices are often treated as inherently more valuable than those voices speaking from experience. Whether doctors, economic development leaders, or academics, professionals in coalitions often act in their role with respect to community members in terms of analysis and interpretation rather than partnership. The role of experts is too often to take the raw material of community and package it into something that the entrenched powers can work with, thus objectifying the individuals and sanitizing their influence. This negligence fails to see the inherent value in experience. The divide between experience and detached expertise is a prime example of the dualistic thinking in western society that has been widely criticized by ecofeminists such as Vandana Shiva and Rosemary Radford Ruether (Shiva 1993, 2005; Ruether 1992). While experts' ability to map, analyze, criticize, and synthesize is a valuable and important skill set, the lived experience of those most affected by these issues is equally valuable and must be treated as such.

Recognizing how privilege operates in hidden and insidious ways, we need to continue to improve the way we do the practical work of food justice. We need to cultivate the sort of virtues that commit us not just to personal development, but to systemic change. In our particular context, religious virtues and narratives are particularly useful in articulating this approach. In what follows, we examine how the virtues of humility and solidarity, as situated in religious narratives, can help us to shift the way we think about and carry out our food justice work. While the responsibility for just engagement must be borne by everyone in the community, our discussion focuses on virtues that ought to be cultivated by those coming to the work from a position of privilege. In so doing, we are explicitly acknowledging our own privilege.

10.2.3 Narrative and Virtue

In her research on religion and agriculture in India, Whitney Sanford argues that those critiquing the unjust effects of industrial agriculture need better narratives to shape the work of the alternative food movement (Sanford 2012). Religious narratives, in particular, can inform and convey our relationship to food, to land, and, perhaps most importantly, to one another. Narratives express our values and they support our practices. Like the Vaishnava Hindu story of Balaram and the Yamuna River examined by Sanford, Christian narratives have shaped our Western attitudes toward land, food, and people. While early environmental ethics often blamed the Judeo-Christian story of dominion over the land (Gen 1–3) and spiritual otherworldliness for our current environmental problems (White 1974), more recent discussion demonstrates the potential for Christian narratives to link the way we grow, distribute, and consume our food to fundamental principles of embodiment, love, and justice.[7] Particularly in a place like Memphis, situated in the so-called "Bible Belt," Christian narratives can motivate communities to rethink the way they address problems such as food insecurity.[8] They can help us understand the meaning and symbolism of food beyond economic or nutritional value.

Because narratives provide the context for the development of virtues, they can help us to address problems of white privilege in the food movement, too. Virtues are those characteristics that allow us to succeed in practices valued in a particular community and shaped by a particular tradition (MacIntyre 1984). They are both character traits and habits, that is, they require practice, habituation, and community affirmation in order to become personal attributes. Two virtues in particular are necessary for white people working in the field of food justice in a place like ours: humility and solidarity. While they are both situated in the context of Christian narratives, cultivation of these two virtues can facilitate the work of the broader food justice movement.

Humility is needed for anyone doing social justice work from a position of privilege. This most Christian of virtues was dismissed by David Hume as a "monkish virtue," but it has been reclaimed by some philosophers in the field of environmental virtue ethics (Haught 2010; Hill 1983). Etymologically, it links us to the humus of the earth, keeping us grounded. As developed in his 6th-century *Rule* for the lives of monks, Benedict of Nursia describes humility as having the proper estimation of oneself: recognizing one's guilt for past wrong-doing and seeing oneself in fundamental relationship with others and dependent upon them (Benedict 1998). In a life of intentionally chosen community, this means recognizing dependence on one's fellow community member, across differences of age, rank, and ethnicity. Recognition of dependence encourages a sense of gratitude—for divine gifts, such as life and food, of course, but also gratitude for the gifts of others, their wisdom, skill, and life experiences. In Benedict's *Rule*, humility is also closely linked to obedience. While the concept of obedience does not sit easily with our modern distrust of hierarchy, in the *Rule*, to obey primarily means to listen ("Listen!" is its first word) to others and to put their needs first. Humility and its

---

[7]See for instance Norman Wirzba, *Food and Faith: A Theology of Eating* (Cambridge: Cambridge University Press 2011); Fred Bahnson and Norman Wirzba, *Making Peace with the Land: God's Call to Reconcile with Creation* (Madison, WI: InterVarsity Press 2012); Ellen F. Davis, *Scripture, Culture, and Agriculture: An Agrarian Reading of the Bible* (New York: Cambridge University Press 2009); Gary W. Fick, *Food, Farming, and Faith* (Albany, NY: State University of New York Press 2008); and Angel F. Mendez-Montoya, *The Theology of Food: Eating and the Eucharist* (Malden, MA: John Wiley & Sons 2012).

[8]See, for instance, the work of the Memphis Center for Food and Faith, whose mission is "to promote health of land and community through church-supported agriculture," and whose vision includes "Community partnerships cultivating a wealth of healthy soils and well-fed souls from urban core to rural edge."

sister obedience are, in the end, both the expression and the cultivation of love (Benedict 1998, 16–20).

For those beginning food justice work from a position of privilege, humility encourages us to listen to others, recognize our dependence, and cultivate gratitude and love. Humility is the personal recognition of our interdependence and shared dependence on the earth for our food. Like the pursuit of a more just food system, cultivating humility is an ongoing process. However, our experience with community coalitions testifies to its transformative potential. For example, after an internal review laid bare the lack of diversity within our Food Advisory Council, this virtue played a prominent role in the response of the leadership. Recognizing past fault for unintentional exclusionary practices, conveners of the council attempted to change the culture of our coalition in ways that subverted privilege and replaced power with humility through simple exercises like shifting meeting times to accommodate community members' schedules; increasing the number of non-affiliated community members in our meetings; and moving our meeting locations to community spaces in lower income neighborhoods. Additionally, we began the practice of opening every meeting by listening to an overview of the work being done by our host (generally a community organization within that particular neighborhood). By beginning our meetings by *listening* rather than dominating the agenda, we were better able to recognize the ways in which their priorities were interwoven with ours. Listening with humility can transform our work in unexpected ways. While allowing community voices to set the tone of meetings requires relinquishing control, it demonstrates an active commitment to the idea that we cannot achieve our food justice goals alone. This shift towards humility, trust, and dependence on the knowledge that comes from experience, rather than expertise, can also be extended beyond coalition work to influence organizations in matters such as hiring staff and recruiting board members and advisors from within those communities most affected by food insecurity.

In addition to humility, the virtue of solidarity is needed in the work of food justice. Solidarity is a particularly important virtue for Christians writing in the field of liberation theology.[9] Love of neighbor is the fundamental principle of Christian ethics. But both James Cone, writing as a Black theologian, and Ada Maria Isasi-Diaz, writing as a mujerista theologian, reject the traditional interpretation of this commandment as charity. Food charity such as soup kitchens and food pantries

---

[9]Liberation theology emerged as a mid-twentieth century movement in Latin America and interprets traditional theological concepts (e.g. sin, salvation) from the perspective of the poor and oppressed. Liberation theologians point out that in the biblical narratives, God repeatedly condemns unjust social structures and intervenes on behalf of the poor. In this view, salvation is not restricted to the afterlife, but includes liberation from all forms of oppression—economic, spiritual, political, and social—in this life and this world. Liberation theology was influenced by Marxism and emphasizes the role of praxis as the means by which God's liberation is accomplished; see, for example, Ada Maria Isasi-Diaz p. 90, "Our participation in the act of salvation is what we refer to as liberation. It consists of our work to transform the world." Liberation theology has influenced all contemporary forms of contextual, social, and political theologies, such as black theology, feminist theology, womanist theology, queer theology, theologies of (dis)ability, etc.

is necessary to fill the gaps in an unjust system, but it begins from a position of privilege and reinforces social division. Instead of charity, liberation theologians advocate solidarity with the oppressed as the expression of love of neighbor, which in society is manifested as justice. Whereas charity is a "doing for others," solidarity is a "being with others" (Isasi-Diaz 1996, 86). According to Cone, "Knowing God means being on the side of the oppressed, becoming *one* with them, and partici-pating in the goal of liberation" (Cone 2010, 69), and for Isasi-Diaz, "Solidarity has to do with understanding the interconnections that exist between oppression and privilege, between the rich and the poor, the oppressed and the oppressors" (Isasi-Diaz 1996, 89). The goal of solidarity is to participate in the ongoing process of liberation, the main obstacle to which is the alienation caused by unjust social structures, like a food system with unequal access, or a food movement that is blind to the reality of food insecurity, the way privilege operates, and the existence of movements by people of color for food sovereignty. The solution to alienation is what liberation theologians call the preferential option for the poor and the oppressed. In the narrative of the Hebrew and Christian scriptures, this option is exercised by God, who consistently sides with the marginalized, the poor, the oppressed, or the "least." For those Christians inspired by this narrative, they too are called to stand in solidarity with the oppressed.

Like humility, the pursuit of solidarity can be transformative for our work. Since its inception, GrowMemphis has built solidarity into its approach to supporting community gardening projects. The organization does not plant or operate gardens, but partners with community members to support their initiative. Partner garden leaders define their own mission and vision for their gardens. That is, it is the communities that desire and support those gardens that decide whether their garden will focus on educating youth, providing nutritious food to the community, eco-nomic development, community building and social cohesion, or any other goal they might have. GrowMemphis then stands in solidarity with these projects by providing funding, expertise, and other types of support where it is needed and requested. This form of solidarity can be challenging to the dominant corporate non-profit paradigm as it often requires altering institutional policies and practices to fit the needs of a particular community. For instance, to support a community garden organized and operated by people experiencing homelessness, our standard policies of grant reimbursement for tool purchases required different forms of support. Likewise, encouraging these groups to speak about their projects for themselves in public settings can be challenging in a structure where the privileged are prone to speaking *for* these groups.

Solidarity, however, extends beyond localized models of leadership and flexible policies and practices that prioritize support for the needs of particular communities. Solidarity is sometimes best seen in the shared embodied experience that comes from being physically present in the same space, discussing, laboring, and eating together. GrowMemphis' monthly garden leader meetings provide opportunities to build community and share knowledge across differences of race, class, gender, sexual orientation, gender identity, age, and religion, all around the shared love of growing and eating vegetables (Latta 2014). The location and topic of the meeting varies, but always includes shared experience, learning opportunities, and working together to solve problems. At one extreme, solidarity in a garden leader meeting might mean taking turns shoveling half-decomposed food waste from a compost bin, uncovering a nest of rats in the process. At the other, it might mean a garden tour on a hot and humid summer evening, the hospitality of lemonade, and the inspiration of new ideas shared with excitement, in a neighborhood most white Memphians have never set foot in. Sharing the same embodied experiences, in the same environment, is a powerful practice of solidarity that contributes to a broader and deeper vision of what food justice can be.

10.2.4 Humility and Solidarity in Theory and Practice

As virtues, both humility and solidarity exist at the intersection of theory and practice. As practice or strategy, Isasi-Diaz calls solidarity "a praxis of mutuality" or dialogue that begins with the consciousness and voices of the oppressed. The practice of humility requires that those who have historically benefited from oppression to be willing to listen and to be questioned by the oppressed in order to practice mutuality in solidarity with them. Cone, too, encourages whites who are serious about solidarity with the black community first to "keep silent and take instructions" (Cone 2010, 66). This does not mean that different groups cannot work together or build coalitions around common concerns. But it does mean that true dialogue does not begin from a position of power. Mutuality begins with the voices of the oppressed and can only proceed to dialogue when those who have benefited from privilege are willing to listen, in humility.

Such listening requires practice. But the practice of humility and solidarity opens us to a deeper theoretical understanding of "the commonality of interests that links humanity" (Isasi-Diaz 1996, 93). Oddly enough, acknowledging the painful history of our racialized differences, and the different forms of privilege that we bear in our bodies, is what allows us to recognize our deeper common humanity, for instance, the fact that everyone eats—it is utterly common, and it is precisely what we have in common (Miles 2007). This common fact enables us to proceed from humble listening, for instance in our garden leader meetings, to dialogue, to mutuality and solidarity. The practice of solidarity and humility urges both those who historically have been oppressors and those who have been oppressed together "to envision and work toward alternative nonoppressive systems" (Isasi-Diaz 1996, 98). Exercising these virtues, particularly by those who have benefited from power and privilege, is what allows different communities to work together for justice in our food system. Without them, just coalition work is impossible.

These practical virtues also assist us in building what Martin Luther King, Jr., called the "beloved community"—a community in which differences do not manifest as separation, but instead strengthen a community built on mutuality, solidarity, and a shared sense of interdependence. Humility and solidarity are virtues given meaning within Christian narratives of divine creation, human alienation, and redemptive liberation, but they are not, of course, limited to Christians. These virtues do, however, imply a worldview in which human beings are deeply inter-dependent with one another and meant to live and work together in a world in which everyone has enough to eat. They are also meant to be practiced; in other words, we also need a sense of humility about our failure to do better in our work to this point. We still have a long way to go in the work of food justice. We are too often impeded in this work by racial and political divisions within Christian communities themselves. But the narratives we use to shape the food justice movement, along with the virtues we admire, matter in the practical work we do. Cultivating virtues such as humility and solidarity is necessary for those, such as white people, who bring unearned privilege to this work. For some, particularly those working from a Christian background in places like Memphis, these virtues are usefully framed in Christian narratives of divine solidarity and love of neighbor to further the work of justice.

References

A Place at the Table. 2012. Directed by Kristi Jacobson. New York: Magnolia Pictures. DVD.
Allen, Will. 2013. *The good food revolution: Growing healthy food, people, and communities.* New York: Avery.
Alkon, Alison Hope, and Julian Agyeman 2011. *Cultivating Food Justice: Race, Class, and Sustainability.* Cambridge, MA: The Massachusetts Institute of Technology Press.
Ayres, Jennifer R. 2013. *Good food: Grounded practical theology.* Waco, TX: Baylor University Press.
Bahnson, Fred, and Norman Wirzba. 2012. *Making peace with the land: God's call to reconcile with creation.* Madison, WI: InterVarsity Press.
Benedict of Nursia. 1998. *Rule of Saint Benedict,* ed. O.S.B. Timothy Fry. New York: Vintage Spiritual Classics.
Charlier, Tom. 2011. Census calls city poorest in nation. *Memphis commercial appeal.* September 23. http://www.commercialappeal.com/news/2011/sep/23/census-calls-city-poorest-in-nation/. Accessed 3 April 2013.
Charlier, Tom. 2013. Memphis again nation's poorest large metro area as local poverty rates climb. *Memphis Commercial Appeal.* October 20. http://www.commercialappeal.com/news/memphis-again-nations-poorest-large-metro-area-as-local-poverty-rates-climb-ep-307335366-326361031.html. Accessed June 3, 2016.
Committee on National Statistics, Division of Behavioral and Social Sciences and Education. 2006. *Food insecurity and hunger in the United States: An assessment of the measure.* Washington, D.C.: The National Academies Press.
Cone, James H. 2010. *A Black Theology of Liberation,* 40th Anniversary ed. Maryknoll, NY: Orbis Books.
Conley, Chris. 2010. Going without: Memphis ranks No. 1 in national hunger poll. *Memphis Commercial Appeal.* March 30. http://www.commercialappeal.com/news/2010/mar/30/going-without/. Accessed April 3, 2013.
Davis, Ellen F. 2009. *Scripture, Culture, and Agriculture: An Agrarian Reading of the Bible.* New York: Cambridge University Press.
Delavega, Elena. 2015. 2015 Memphis Poverty Fact Sheet. The Mid-South Family & Community Empowerment Institute. Memphis, TN: The University of Memphis.
Fick, Gary W. 2008. *Food, Farming, and Faith.* Albany, NY: State University of New York Press.

Feeding America. 2014. Map the Meal Gap. http://map.feedingamerica.org/county/2014/overall/tennessee/county/shelby. Accessed 7 May 2016.

Gilligan, Heather Tirado. 2014. Food deserts aren't the problem: Getting fresh fruits and vegetables into low-income neighborhoods doesn't make poor people healthier. *Slate*. February 10. http://www.slate.com/articles/life/food/2014/02/food_deserts_and_fresh_food_access_aren_t_the_problem_poverty_not_obesity.2.html. Accessed 7 May 2016.

Gottlieb, Robert, and Anupama Joshi. 2010. *Food justice*. Cambridge, MA: The Massachusetts Institute of Technology Press.

Guthman, Julie. 2011a. 'If They Only Knew,' The unbearable whiteness of alternative food. In *Cultivating Food Justice: Race, Class, and Gender,* ed. Alison Hope Alkon and Julie Agyeman, 263–281. Boston: MIT Press.

Guthman, Julie. 2011b. *Weighing In: Obesity, Food Justice, and the Limits of Capitalism*. Berkeley, CA: University of California Press.

Haught, Paul. 2010. Hume's Knave and Nonanthropocentric Virtues. *Journal of Agricultural and Environmental Ethics* 23 (1–2): 129–143.

Hill, Thomas. 1983. Ideals of human excellences and preserving natural environments. *Environmental Ethics* 5 (3): 211–224.

Isasi-Diaz, Ada Maria. 1996. *Mujerista theology: A theology for the Twenty-First Century*. Maryknoll, NY: Orbis Books.

Kingsolver, Barbara. 2007. *Animal, vegetable, miracle: A year in food life*. New York: HarperCollins.

Lambert-Pennington, Katherine, and Kathryn Hicks. 2016. Class conscious, color-blind: Examining the dynamics of food access and the justice potential of farmers markets. *Culture, agriculture, food and environment* 38 (1): 57–66.

Latta, Kenneth S. 2014. A community of gardeners: Exploring GrowMemphis and the experiences of its community garden leaders. Master's Thesis. University of Memphis.

MacIntyre, Alasdair. 1984. *After virtue*, 2nd ed. Notre Dame, IN: University of Notre Dame Press.

Memphis and Shelby County Food Advisory Council Working Group. 2012. *Good food for all: The need for a food advisory council for Memphis and Shelby County*. TN: Memphis.

Mendez-Montoya, Angel F. 2012. *The theology of food: Eating and the Eucharist*. Malden, MA: John Wiley & Sons.

Miles, Sara. 2007. *Take this bread*. New York: Ballatine Books.

The National Research Council. 2009. *The public health effects of food deserts*. Washington D.C.: The National Academies Press.

Peck, Chris. 2012. What Obama Didn't See. *Memphis Commercial Appeal*. September 30. http://www.commercialappeal.com/opinion/chris-peck-what-obama-didnt-see-ep-384776846-329239171.html. Accessed 3 June 2016.

Pollan, Michael. 2006. *The Omnivore's Dilemma: A natural history of four meals*. New York: Penguin.

Ruether, Rosemary Radford. 1992. *Gaia and God: An ecofeminist theology of Earth Healing*. San Francisco: Harper Collins.

Sack, Daniel. 2001. *Whitebread protestants: Food and religion in American culture*. New York: Palgrave.

Sanford, Whitney A. 2012. *Growing stories from India: Religion and the fate of agriculture*. Lexington, KY: The University Press of Kentucky.

Sayle, Hannah. 2010. Unjust deserts: Finding quality fresh food is a daunting task for thousands of Memphians. *Memphis Flyer*. August 26. http://www.memphisflyer.com/memphis/unjust-deserts/Content?oid=2263963. Accessed 3 June 2016.

Shiva, Vandana. 2005. *Earth democracy: Justice, sustainability, and peace*. Boston: South End Press.

Shiva, Vandana (ed.). 2007. *Manifestos on the future of food and seed*. Brooklyn, NY: South End Press.

Shiva, Vandana. 1993. *Monocultures of the mind: Perspectives on biodiversity and biotechnology*. London: Zed Books.

United States Department of Agriculture Economic Research Service. 2015. Food security in the U.S.: Definitions of Food Security. http://www.ers.usda.gov/topics/food-nutrition-assistance/food-security-in-the-us/definitions-of-food-security.aspx. Accessed 7 May 2016.

United States Department of Agriculture Economic Research Service. 2015. Food Access Research Atlas. http://www.ers.usda.gov/data-products/food-access-research-atlas/documentation.aspx#definitions. Accessed 7 May 2016.

West, Cornel. 1994. *Race Matters*. New York: Vintage.

White, Jr., Lynn. 1974. The Historical Roots of our Ecologic Crisis [with discussion of St Francis; reprint, 1967]. *Ecology and Religion in History*. New York: Harper and Row.

Winne, Mark. 2008. *Closing the food gap: Resetting the table in the land of plenty*. Boston: Beacon Press.

Wirzba, Norman. 2011. *Food and faith: A theology of eating*. Cambridge: Cambridge University Press.

Chapter 11
Views from the Classroom: Teachers on Food in a Low-Income Urban School District

Sarah Riggs Stapleton, Person Cole, Melissa Washburn, Matt Jason and Tawny Alvarado

Abstract Schools have long been important sites for food justice work. However, rarely are formal educators (e.g. teachers, school administrators, school counselors) brought into the design of school-related food justice projects. This chapter argues that school-based food justice projects should substantively include formal educators because of the important insights that formal educators possess as insiders to schools. As evidence for this, this chapter reflects on a participatory action research project between an education researcher and four veteran teachers. The teachers in this collaboration teach in a low-income school district that serves a population that is approximately 80% free and reduced lunch eligible. Therefore, obtaining any food—and particularly healthy food—is of concern to the vast majority of students whom they serve. This chapter highlights the types of questions and explorations that can be done when food justice activists partner with formal educators as collaborators, making a case for why formal educators should be included in any school-based food justice work.

Keywords Schools · Participatory action research · Teachers

11.1 Schools, Formal Educators, and Food Justice

US public schools have long been important sites for food justice work (Gottlieb and Joshi 2013). Certainly the need has never been greater; for the first time in US history, the majority of US public school children are living in poverty (Southern Education Foundation 2015). As a result, schools provide up to three meals per day and some even supply weekend food assistance or host food pantries (USA Today 5 Apr 2015). Other initiatives, such as the rapid expansion of school gardens and

S.R. Stapleton (✉)
Department of Education Studies, University of Oregon, Eugene, OR, USA
e-mail: sstaplet@uoregon.edu

P. Cole · M. Washburn · M. Jason · T. Alvarado
Michigan Public School Teacher, Michigan, USA

© Springer International Publishing AG 2017
I. Werkheiser and Z. Piso (eds.), *Food Justice in US and Global Contexts*,
The International Library of Environmental, Agricultural and Food Ethics 24,
DOI 10.1007/978-3-319-57174-4_11

former First Lady Michelle Obama's campaign on childhood obesity, have focused on access to affordable, healthy food. Additionally, the battle to improve food that is served in schools has been fought by food activists for several decades (Gottlieb and Joshi 2013). Gottlieb and Joshi (2013) note by 2009, "school food changes had emerged as one of the most visible and promising results of the drive to make broader food system changes...and school food became the symbol of how food system change was necessary, and possible" (p. 91). In essence, there is no shortage of crucial food justice work being done in and around schools.

However, while food activists, politicians, and even celebrities are joining the school food bandwagon, attention to food from formal educators is noticeably less common. For example, my time as a participant in local food community activities within a small city in Michigan exposed a void in formal educator presence within local food justice work. This absence was particularly apparent in discussions that directly concerned schools, such as farm-to-school initiatives, school gardens, and healthier school lunches. At a local food justice conference, despite the existence of an educator and youth track and attempts to recruit local teachers, there were only a handful of formal educators out of nearly three hundred attendees. The need for formal educator voice in food conversations is not just a local phenomenon. From chef Alice Waters' Edible Schoolyard initiative to Jaime Oliver's Food Foundation, to FoodCorps, an Americorps-affiliate organization working to teach about healthy foods in schools and afterschool settings, school-based food work is primarily executed by people outside of formal education. In fact, in a conversation with a FoodCorps representative, I learned that, at that time, no formal educators were part of the leadership of the organization despite the fact that the organization had worked with over 82,000 youth across 15 states, primarily in classrooms and school lunchrooms (FoodCorps 2013). The impact of this missing perspective can be damaging. For example, Gottlieb and Joshi (2013) have noted that healthy menu changes in the Los Angeles Unified School District faced pushback because food activists did not educate and involve students and teachers.

Formal educators, including teachers, school administrators, and school counselors, have valuable perspectives to add to food conversations, such as their understanding about school curricula, school operations, and students' lives. If we are to more holistically address food justice issues in schools, we need to improve communication and collaboration between formal educators and outside activists/researchers. In this chapter, I argue that formal educators (those formally trained in the education sector and employed by schools such as teachers, school administrators, school counselors, etc.) have important insider-perspectives on food justice work within schools and should therefore be included in school-based food justice initiatives.

One way to involve formal educators is to include them in the *design* of projects, rather than just the implementation. Research methodologies created for collaboration can be useful for this endeavor. Action research is one such approach that involves practitioners, community members, or other non-researchers working to improve their communities, practice, and/or situations. Action research requires tangible, real-world action to occur as part of the research process. A tenet of action research holds that insiders have important, contextualized knowledge. Consequently, action research values practice as well as theory, aiming for theoretically-informed practice (McTaggart 1997). Attempting to categorize action research in a concise way is difficult because action research involves a large variety of perspectives and approaches and comes from a rich lineage (Reason and Bradbury 2008). While every project varies, the execution of action research typically includes planning, acting, observing, and evaluating the result of the action. Usually this process begins with a general idea that is agreed upon consensus by a group, and is refined throughout the steps (McTaggart 1997).

When working with groups whose perspectives are typically underrepresented or undervalued, a subtype of action research, called participatory action research, can be particularly helpful. Participatory action research (PAR) aims to address issues of power and is attentive to the *process* used toward knowledge generation, to whose voices are being heard and how (Fals Borda 2001; Greenwood and Levin 2007; Miller and Brewer 2003). PAR typically involves working with individuals from marginalized groups for emancipatory aims (Greenwood and Levin 2007). Miller and Brewer (2003) note that PAR, "aims to produce knowledge and action directly useful to people and also to empower people through the process of constructing and using their own knowledge" (2003, p. 225). In my work, I argue that teachers of marginalized youth often are themselves marginalized, so PAR can be a useful methodology for working with them.

11.2 The Project: Participatory Action Research with Teachers

To illustrate how formal educators can contribute in meaningful ways to school-based food justice work, I will describe a research project in which I was involved that included four veteran teachers in a low-income urban school district. I will briefly explain about how the project was conceived, executed, and the context of the work. Then I will summarize each of the teachers' projects to demonstrate how their perspectives and positions as school-insiders contribute to the ways in which they perceive food justice in their schools and their actions to address them.

Using participatory action research, I, an education researcher, partnered with the teachers to focus on food within their low-income urban school district. I determined the project's general focus on food in schools before recruiting teachers and limited my search to teachers in one urban school district because I felt that it would be important to focus on a place where the student population faced food security concerns. I then sought teachers in the city's school district who had interest in exploring food issues in schools and were willing to meet regularly. I wanted to find the right teachers for the project, so I spent several months searching for participants through networking. I met each of the four teachers at local food events or through personal contacts. I did not determine a set definition or scope for how we should view food justice; rather I wanted the teachers to be able to bring their own perspectives on food justice as it related to students in their district. As part of this, they identified school-based food issues and decided how they wanted to address them. For over a year and a half, we met monthly as a group to discuss and reflect on their observations and work. I also met regularly with each teacher, often for school visits. Though the teachers often asked for my input, the scope and focus of their projects were self-determined.

The place context that defines this research project is a small, post-industrial Midwestern city. The city reached its peak population in 1970, and has since declined. The city has seen considerable demographic changes in the last few decades. For example, white residents made up 86.4% of the population in 1970, but only 55.5% in 2010 (US Census Bureau). This upsurge in diversity is partly in response to an ever-growing refugee community; the city is in the top ten in refugee resettlement among "medium-sized metropolitan areas" in the US (Singer and Wilson 2006).

Compared to massive urban school districts in the US, this district is small, consisting of 26 elementary, middle, and high schools in the 2014–15 school year. The teacher co-researchers in this project work at three different schools in the district; all three schools possess extremely high levels of student poverty. The teachers estimate that on average, 80% of the students in the district are eligible for free and reduced lunch. In response to this high percentage of poverty, the district has opted to provide up to three free meals per day for all students.

11.3 The Teachers' Projects

Though the teachers' conceptions of food justice vary somewhat, there are several definitions that pertain to the work done across the projects. One relevant definition, coined by Gottlieb and Joshi (2013), outlines several aspects of food justice as "seeking to challenge and restructure the dominant food system and providing a core focus on equity and disparities and the struggles by those who are most vulnerable" (p. ix). Perhaps the most apt description of the food justice movement for this project is the following by Nancy Romer (2014):

> [The food justice movement] sees race, class, and gender as central to food oppression and leadership. It sees the food crisis as a result of corporate control over our land, water, agriculture, food processing and distribution with heavy assistance from neoliberal governments and the corporate media. It sees the necessity of sustainable food systems (agriculture, distribution, processing) to mitigate climate change: that means an end to factory farms. It sees feeding the world's people as dependent upon decentralization of the food system so people can build resilient, culturally-appropriate systems that meet their own needs (p. 5).

Each of the teachers conceives of food justice slightly differently, which is apparent in the way each conceptualizes and addresses her/his project. The focus of each teacher's work reflects her/his unique position within their school and points to the importance of formal educators from different backgrounds, schools, and teaching positions participating in the design of food justice projects. The teachers' projects—from focusing on food insecurity in school, to designing food curricula and school gardens—all speak to different needs within their particular schools.

11.3.1 Person Cole: Critically Examining Food Insecurity from an Insider's Perspective

Person has taught as a special education teacher in the district for 37 years and currently works at an alternative high school. She is extremely knowledgeable about the city and district contexts, likely a result of her many years working there and her involvement in the community. Person's own childhood was marked by hunger and she notes that this still impacts how she thinks about food security for herself and for her students. Her story is powerful: growing up in the infamously blighted urban center of Detroit, living in a food desert, having a teen mother, having to steal to eat. Her reflections give an important first-hand account of the direness of hunger for many children in the US. They also explain why Person is attentive to the details of food security for her students. She speaks of keeping snacks around to feed students—especially those who are pregnant—and worries about whether her students are getting enough food to eat. She also worries about their access to healthy food. As an adult, she is an avid gardener, partly because she sees it as a path to food security. Her complex background in relation to food makes her perspective particularly compelling and important.

Person has deep interest in food justice and food security and in her project she sought to explore and investigate how food insecurity was manifested at her school. She observed students' eating patterns within the school, including school lunches and the vending machine. She also worked closely with the school food coordinator, interviewing the coordinator about her systems and procedures for feeding students. Person monitored food being served on a daily basis, how many students were eating, and what students were saying about the food. She did this by listening to students during lunch, while they were picking out their lunch selections, and by chatting with them informally. She also observed the vending machine sales each week, noting how much of what items were being purchased. Person's project changed somewhat, primarily because the district changed their food service provider between the 2013–14 and 2014–15 school years. Initially, when we began working together, she noted that her school was itself a food desert because of the poor quality and quantity of food being provided to her students under the first provider, Aramark. Through her observations, she noticed considerable improvements in the quality and quantity of food being served to the students at her school when the food service provider changed for the district. She also witnessed students eating more food—a win, she recognized, for food-insecure students.

In recording the qualitative transformation in the food provided to her school across two food service companies, Person demonstrates the extent to which a school district's food service contract can impact the daily lives of students. Additionally, Person describes how the school food coordinator knew how much to order, how to make food stretch to feed afternoon students, how to "doctor" the food to make it more tasty, and how to encourage kids to eat fruits and vegetables. This seems to point to the fact that even individual food staff members can make a large difference in food security for students. Her project suggests that feeding food insecure kids in schools can be surprisingly precarious—both dependent on a company contract and on an individual staff member's prowess.

Person's interest in and frustration with the frequent use of the vending machine by students at her school reflects several concerns she has voiced. One concern is that students have and continue to eat junk food, despite the quality of food present in the cafeteria. Interestingly, her finding is corroborated by the literature, as Sandler (2011) notes, "Today, when children do not eat a school-staff provided breakfast and lunch, it is usually not because it is unavailable to them…but because they are eating food from school vending machines, snack bars, contracted private food providers, or supplementary programs" (p. 36). Sandler (2011) makes this point in reference to the increasing privatization of food available to students. In contrast, Person is concerned, first, that the vending machine food is nutritionally void and, second, that low-income students spend a large amount of money paying premium prices to purchase from the vending machine.

Through her project, Person has continually questioned *who* gets to make decisions around school food and *to what end* they are collecting data about who is given food (rather than about the rates at which food is consumed). Her sentiment is echoed in a comment by Robert and Weaver-Hightower (2011) that "asking who feeds whom, what, how, and for what purpose reveals a great deal about how children are cared for by a society" (p. 205). Person cares for her students deeply— she knows about their lives, their struggles, their successes—and she is clearly bothered by the lack of care they are shown by society through the low-quality food that they are given in school.

11.3.2   Melissa Washburn: Addressing Food Insecurity in School

Melissa Washburn ("Wash"), a high school health and social studies teacher, has been teaching for 31 years, with 24 of those years in this urban school district. Wash has done much to get food into the hands of kids. In past health classes, she made smoothies using healthy ingredients, such as fruit and yogurt. On a number of occasions, she has acquired local apples—both donated and bought with her own money—for her students. When I met her in the spring of 2014, she had recently begun a twice per week after-school activity program for the express purpose of getting district provided snacks into the hands of kids, since snacks were only provided for those participating in school-sanctioned activities.

Wash has also been involved in a number of district food-related committees. She has been part of the committee to rewrite the health and wellness plan for the district as well as a school-food action committee formed by community members, non-profit staff, and university extension staff to improve the quality of food in area schools. Wash has explained that she is motivated to improve her students' access to quality food for several reasons. First, she grew up eating fresh fruit and vegetables out of her parents' large garden and later, from their farm. As a result, she values fresh local food. Second, as a health teacher, she feels that she is finally "walking the walk" by doing something to help get food into the hands of food-insecure students.

As part of this project, Wash created and implemented a backpack feeding program for her high school. To set up the program, she had to create logistics within the school, raise funds, and find volunteers to staff it. Wash's backpack program ran weekly throughout the 2014–15 academic year, distributing food to approximately 15–25 students each week. To research this endeavor, she informally asked students about various aspects—whether they liked the food, what suggestions they have for the backpacks. She also created several surveys for the backpack recipients asking about their opinions of the program.

Wash and I were concerned that the food in the packs was not very nutritious. (It was primarily processed food as is typical for emergency assistance food). We decided that funding would be helpful to supplement the packs with local vegetables and fruits. Wash recommended that we approach a local non-profit organization that had created a system to connect local producers with small bulk-purchase buyers. For the 2015–16 school year, Wash procured a local grant, used the funds to purchase local fruits and vegetables through the non-profit hub, and had the food delivered to the school using a local bike delivery service. It was a wonderful combination of healthy and local foods that were made freely available to some of her school's most food insecure families.

Wash started the first feeding program at a high school in the district, while most of the elementary schools in the district already have such programs. It is worth noting that food assistance organizations target elementary schools *before* high schools. Perhaps the sentiment is that high school students are able to earn their own money (though asking a high school student to feed themselves through a part time job is undoubtedly problematic). Or, perhaps they have assumed that high school students would be too proud to take the assistance. An important reason to give older students weekend food assistance is brought up by Rodgers and Milewska (2007) who found that a number of middle school and high school students receiving food assistance were also responsible for feeding younger children in their house. This fact lends considerable support to the need for distributing backpack food assistance to high school students.

Wash's work on the backpack feeding program has answered a tangible question for the local food assistance organizations about whether the program could function in a local high school. Rodgers and Milewska (2007) have reported that though backpack feeding programs originated in 1995, there have been few studies about them. Wash's project may help to bring more awareness about these important supplemental food programs. Her work toward getting local and healthy food added onto the backpacks could also be an important contribution. Rodgers and Milewska (2007) report that a complaint about backpack programs is the processed nature of the food given. They note that some schools requested healthier foods, like fresh produce and meat. But they reported that there are logistical challenges in delivering perishable foods efficiently to a large number of schools. This speaks to the potential of Wash's model, which involves a dedicated school coordinator to make the connections to healthy food for a given school. Her model was also particularly innovative because she partnered with a local producers group to get deliveries that were not only healthy, but local.

Wash has shared stories about the ways in which food has helped to build community between her and students. She believes that students see teachers as caring when they bring in food for them. This speaks to the idea that when teachers attend to student needs beyond the classroom, students notice. Her reflections are supported by literature; Rodgers and Milewska (2007) shared observations of school personnel that students participating in a backpack program showed more trusting, better relationships with people at the school as a result of the program.

Wash's determination to help alleviate food insecurity for her students shows that teachers can take on roles serving their students in non-academic ways. Wash feels she is finally doing something to address significant life issues faced by students in her school.

11.3.3 Matt Jason: Bringing Food, Food Systems, and Food Culture into the Classroom

Matt has been teaching in the city's school district for 15 years, one of which was spent in the urban center's sister city in Japan. As a social studies educator, he has taught grades seven through twelve, on subjects such as geography, history, and economics. As someone who brews beer and raises chickens, food has been an important part of Matt's personal life. It has also played a role in his professional life, as he describes, "Throughout my career, and across all these subjects, even teaching English in Japan, food has been a connecting point." Prior to this project, Matt had already pursued a number of projects related to food in schools. He had attempted to obtain a grant to work on school food waste through worm bins. He had also tried to get a hoop house for early season gardening built for his high school. Matt had taught about child labor and trafficking within the chocolate industry. As part of this exploration, he showed the documentary *The Dark Side of*

Chocolate and asked students to follow up by calling chocolate companies to inquire about their labor practices. I would describe Matt's interest in food justice primarily in terms of fair and just food production systems. Additionally, Matt sees the importance of culturally relevant food for sustaining students' familial identities. Matt's project supposes that *just* eating includes access to food that is culturally sustaining.

As a social studies teacher, Matt is interested in students' food stories as a way to understand more about student identities and cultures. Matt's aim in his project was to further incorporate food and students' familial cultures into his curriculum, and reflect on this process to share with and encourage other teachers to explore food in their curricula. As part of our work, Matt created and taught a unit about food in the fall of 2014 for a 9th grade study skills course, a new course that he was tasked with designing. The unit lasted for 12–14 class periods and was based around the guiding question: *"If we are what we eat, then what are we?"* The unit included viewing and discussing the documentary *Food, Inc.* and creating infographics about the food production system, investigating ingredients in packaged foods, reflecting on the USDA's MyPlate recommendations, creating menus based on MyPlate guidelines, making connections between food and culture, and writing "Culture and Food" essays.

Often, when discussing food-insecure students, the focus rests (often necessarily) on the lack of food in students' homes. This lack of food may, in turn, be interpreted as a deficit of family food cultures or traditions. Though the vast majority of students in his school are eligible for free or reduced-price lunches and many are food insecure, Matt appreciates that they nonetheless have important family food cultures. Despite being food-insecure, particular dishes—ethnic or otherwise—are significant and meaningful to their families. By focusing on the food cultures of students despite their socioeconomic situation, Matt's work adds an authentic and meaningful way to incorporate students' cultures into curricula and provides a cultural focus often missing in food justice work.

In addition, assumptions are often made that those with low incomes simply do not have the capacity to care about issues such as factory farming. In other words, the ability to pay is sometimes conflated with the ability to care. Matt's experience with students suggests otherwise, as he recognized that 40% of his students expressed anger, disgust, or frustration in their writing about the poor treatment of animals within factory farming. This observation suggests that students from low-income households *can* care about how animals are treated within food production.

Overall, Matt has observed that a curricular focus on food has positively impacted student engagement. He notes that food is "a common language that all people speak...that can be used by teachers to reach their students in ways that are distinct from other common classroom themes and strategies." Matt thinks about food broadly and considers it from many aspects—social, economic, and environmental. His work therefore serves as an important example of the myriad ways through which to look at food and incorporate it into varied curricula as well as the positive impacts of doing so with youth in low-income urban schools and/or youth with diverse cultural backgrounds.

11.3.4 Tawny Alvarado: Creating an Afterschool Garden Club

Tawny has been teaching at the upper elementary and middle school level in the district for 23 years. She currently works as a STEM specialist teacher at a district elementary school. In addition to spending her teaching career in the district, Tawny grew up as a student in the city, attending the districts' schools. Her lifelong history within this school district allows Tawny to be able to reflect on how much the district food situation has changed over time. When she was young, students went home for lunch; no meals (to her knowledge) were served through the schools. She noted that the demographics had changed substantially since then.

As a teacher, she has observed the district food situation decline steadily. At first, the middle school where she taught served a hot breakfast and the teachers had access to a soup and salad bar at lunch. Starting four or five years ago, the hot breakfast ended and teachers began serving breakfasts in their classrooms. The items served were cold, pre-packaged foods such as cheese sticks, milk, juice, Chex mix, cereal with milk, or Pop-tarts. That the students were served Pop-tarts frustrated her because of their high sugar and low nutritional content. She recalled that in the breakfasts, mini pancakes in plastic wrappers were occasionally served, but that they were sometimes delivered to students partially frozen. When this occurred, she heated them in a microwave she had personally bought for her classroom. She admitted that if there was food remaining after an in-class breakfast, she would covertly collect the extra food and put it away instead of returning it as she was told. She did this so that she could serve food to students whom she tutored after school. In fact, she even kept a refrigerator in her room so that she could provide snacks and milk to students, and shared that other teachers kept refrigerators in their rooms so that they could feed kids as well.

In terms of food justice, Tawny is concerned about the quality of food to which her students have access. As part of this project, Tawny started an after-school garden club with 4th–6th graders at her school in part because she views the school garden as a way to encourage healthier eating. She has discussed having taste tests of various vegetables with the students to increase excitement about what they might like to plant, since it was clear from their conversations that the students had not tried many of the vegetables on the lists of potential plants for the garden. Her first homework assignment for her garden-club students was for them to mark which vegetables from a list they had tried as well as which ones they knew they liked. She then planned to bring in samples of vegetables that they did not know so that they could taste them. That students are not familiar with many vegetables—especially those in gardens in northern regions—has been reported in the literature; Blair (2009) has noted, "Gardening in America's northern regions during the school year requires elongating the growing seasons in both spring and fall, thus stretching children's knowledge and taste for cool-season vegetables, particularly for dark leafy greens. Because of our supermarkets' global reach and constant supply of heat-loving vegetables, many cool-season crops remain unfamiliar" (p. 18). Tawny has also expressed concern that students do not know much about food production and she aims to help students understand more about the larger food industry.

11.4 Concluding Thoughts

Given the need for school-based food justice initiatives, this project attests to the importance of substantively including formal educators in school-based food justice projects. In terms of methodology, action research—and participatory action research more specifically—can provide structure and direction through which researchers and/or food activists can engage formal educators. As a result of our work together, the teachers created projects that were meaningful to them and needed in their schools. A few insights from the teachers' projects include food-focused curricula as engaging to students, possibilities and challenges of implementing food assistance programs at the high school level, the deep impact of district food service providers in addressing food insecurity for students, and the potential for student ownership of garden projects. As these findings demonstrate, teachers can contribute much in terms of knowledge about students' individual stories and backgrounds, and awareness of school culture and climate. Moreover, because place-context is crucial in food issues, formal educators who have many years of experience in a given place and school have valuable knowledge about how the context and community has changed over time. Formal educators can also benefit from partnerships with food activists and researchers because these collaborations move food from the periphery into the core of their work. Working with those outside schools can also help teachers, in particular, gain needed recognition and/or support beyond their schools for the multi-faceted work that they do.

References

Bello, M. 2015. Schools becoming the 'last frontier' for hungry kids. *USA Today,* April 5, 2015.

Blair, D. 2009. The child in the garden: An evaluative review of the benefits of school gardening. *The Journal of Environmental Education.* 40 (2): 15–38.

Children's Defense Fund. Children in the states: Michigan. 2015. http://www.childrensdefense. org/library/data/state-data-repository/cits/2014/2014-michigan-children-in-the-states.pdf. Accessed 16 Aug 2015.

Fals Borda, O. 2001. Participatory (action) research in social theory: origins and challenges. In P. Reason & H. Bradbury (Eds), Handbook of action research: participative inquiry & practice. London: Sage. pp. 27–37.

FoodCorps. Our impact. 2013 (September), Retrieved from https://foodcorps.org/about/our-impact.

Gottlieb, R., and A. Joshi. 2013. *Food justice.* Cambridge, MA: MIT Press.

Greenwood, D. and M. Levin. 2007. Introduction to Action Research: social research for social change. London: Sage.

McTaggart, R. 1997. Guiding principles for participatory action research. In R. McTaggart, (Ed.) Participatory action research: international contexts and consequences. Albany, NY: SUNY Press. pp. 25–43.

Miller, R. and J. Brewer. 2003. The A to Z of social action research: A dictionary of key social science research concepts. London: Sage.

Reason, P. and H. Bradbury. 2008. Introduction. In P. Reason & H. Bradbury (Eds), The SAGE Handbook of action research: participative inquiry & practice. 2nd ed. London: Sage. pp. 1–10.

Robert, S., and M. Weaver-Hightower. 2011. Coda: Healthier horizons. In *School food politics*, ed. S. Robert, and M. Weaver-Hightower, 201–208. New York: Peter Lang.

Rodgers, Y., and M. Milewska. 2007. Food assistance through the school system: Evaluation of the *Food for Kids* program. *Journal of Children and Poverty* 13 (1): 75–95.

Romer, N. 2014. The radical potential of the FJ movement. *Radical Teacher* 98: 5–14.

Sandler, J. 2011. Reframing the politics of urban feeding in the U.S. public schools: Parents, programs, activists, and the state. In *School food politics*, ed. S. Robert, and M. Weaver-Hightower, 25–45. New York: Peter Lang.

Singer, A., and J. Wilson. 2006. From 'There' to 'Here': Refugee resettlement in metropolitan America. The Brookings Institution. Metropolitan policy program. Accessed from http://www.brookings.edu/~/media/research/files/reports/2006/9/demographics%20singer/20060925_singer.pdf.

Southern Education Foundation. 2015. A new majority research bulletin: Low income students now a majority in the nation's public schools. Accessed from http://www.southerneducation.org/Our-Strategies/Research-and-Publications/New-Majority-Diverse-Majority-Report-Series/A-New-Majority-2015-Update-Low-Income-Students-Now.

Chapter 12
Healthy Food on Wheels: An Exploration of Mobile Produce Markets Through a Food Justice Lens

Elisabeth (Elle) Mari

Abstract The built environment significantly influences the accessibility of healthy affordable foods. However, examination of built environments alone is insufficient to understanding why some neighborhoods in the United States are bountiful with fresh produce while others experience severe scarcity. This research investigates mobile produce markets (mobile markets) in order to understand how and to what extent they improve food availability and how they address structural inequities associated with poor food access environments. Race, class, and gender dimensions of food insecurity are reviewed, followed by an explanation of the emergence of mobile markets as a response to food insecurity. Eight geographically diverse U.S. mobile markets are studied using critical discourse analysis and interviews. A food justice framework is applied to the sample to understand the obstacles and opportunities to improving food availability and addressing structural inequities. Analysis reveals most mobile markets are not positioned to address root causes of inadequate access to healthy foods and it is unreasonable to build the case that mobile markets have transformational power as a food access intervention. Mobile markets are better situated to support and promote food availability goals on a short-term basis. This research contributes to understanding the efficacy of mobile markets and offers the recommendation of employing a food justice framework in the development of food access interventions to effectively address structural inequities associated with poor food access environments.

Keywords Mobile markets · Food justice · Food security · Food access · Alternative food

E. (Elle) Mari (✉)
University of California, San Diego, USA
e-mail: emari@ucsd.edu

© Springer International Publishing AG 2017   141
I. Werkheiser and Z. Piso (eds.), *Food Justice in US and Global Contexts*,
The International Library of Environmental, Agricultural and Food Ethics 24,
DOI 10.1007/978-3-319-57174-4_12

12.1 Race, Class, and Gender Dimensions of Food Insecurity

The examination of built environments only take us so far in understanding why some neighborhoods in the United States are bountiful with fresh produce while others experience severe scarcity. The importance of race, class, and gender dimensions of food insecurity become apparent when we investigate correlations between the food landscape and the people who live there. Low-income neighborhoods across the United States experience significant disparities in access to healthy affordable food as compared to most middle and high-income neighborhoods (Algert et al. 2006). Capital divestment and the devaluation of land in numerous pockets of once thriving urban areas have led to an influx of junk food swamps and barren food landscapes (McClintock 2011). 23.5 million people live in low-income neighborhoods located more than one mile from a supermarket or large grocery store. Of this particular population, nearly half are low-income. Low-income people living in poor neighborhoods are also less likely to own a vehicle, which creates an additional barrier to accessing affordable healthy food (Ver Ploeg et al. 2012). Healthy food access issues are especially apparent in predominately African American neighborhoods across the United States, regardless of income level (Baker et al. 2006; Williams et al. 2008), and are also palpable in predominately white high-poverty neighborhoods (Baker et al. 2006). A study on the grocery gap in Los Angeles revealed there are 2.3 times as many supermarkets per household in low-poverty areas compared to high-poverty areas and predominantly white neighborhoods have 3.2 times as many supermarkets as predominately African American areas and 1.7 times as many as in Latino neighborhoods (Shaffer 2002).

Coinciding with inadequate access to healthy food, African Americans, Latinos, and Native Americans experience higher rates of obesity, diabetes, and other diet-related diseases as compared to whites and have fewer local options available to make healthy food choices for a nutritious balanced diet (Baker et al. 2006; Larson et al. 2009). Unmistakably then, race and class are social determinants of health that affect one's ability to conveniently locate affordable healthy foods. Race and class also affect how well people are environmentally situated to lead healthy lifestyles (Guthman 2008).

There is a similarly daunting narrative to be told on the relationship between food insecurity and gender, as a disproportionate number of women experience food insecurity and greater consequences of hunger. While nearly 50 million Americans lived in food insecure households in 2013, single mothers experienced food insecurity more than any other demographic (USDA ERS 2014). Food insecure women are also more likely to be overweight or obese than men and experience significantly more economic hardship in procuring food for their families (Dinour et al. 2007; Carney 2012; Larson et al. 2009).

12.2   Mobile Produce Markets as a Response to Food Insecurity

Given the challenging food access disparities in the United States, localized alternative food initiatives such as farmers' markets and healthy corner stores have emerged to alleviate inequities and close the food security gap. Alternative food initiatives in low-income communities, especially those located in neighborhoods of color, are designed to increase opportunities for affordable healthy food purchasing, organize and activate residents to take action on improving local food landscapes, and promote food security and/or food justice. Mobile produce markets (hereafter mobile markets) have emerged as an alternative practical solution to the problem of inadequate access to healthy affordable foods in underserved neighborhoods. Mobile markets recreate a small grocery market out of a vehicle, typically a repurposed utility truck or bus, to sell fresh produce at affordable prices in neighborhoods known to experience scarcity.

Mobile markets as a food access improvement strategy have increased in popularity over the past decade. A few examples of mobile markets include: Mobile Market, started and operated by the People's Grocery in West Oakland in 2003, Fresh Moves started and operated by Food Desert Action in Chicago in 2006, and the Baltimore-based Real Food Farm Mobile Farmers' Market, started and operated by Civic Works in 2011. Mobile markets continue to gain support and interest from food policy activists, as exemplified by the Los Angeles Food Policy Council's work on street food and healthy mobile vending policy recommendations (Los Angeles Food Policy Council 2014). Public health officials have embraced mobile markets too, such as New York City's establishment of the Green Carts Program (Lucan et al. 2011; New York City Department of Health and Mental Hygiene 2014).

No known research has examined how mobile markets contribute to long-term food accessibility changes for the people and neighborhoods they serve. And, perhaps most importantly, no research has examined how mobile markets contribute to more equitable social conditions whereby historically marginalized people are positioned as active leaders in shaping a localized fair food system. What do we know about mobile markets? We know geographic modeling of mobile market distribution patterns is a useful strategy to ensure these markets on wheels reach their intended shoppers effectively and thereby have the potential to increase food availability in a city or region (Widener et al. 2012). We also know mobile markets may have a positive effective on promoting healthy food choices amongst its shopper base (Zepeda and Reznickova 2013). However, when considering the viability of mobile markets as a food access solution, we know, at least among farmworker populations, that they are only an, "interim solution and not economically sustainable" (California Institute for Rural Studies 2012, 21). This research investigates mobile markets in order to understand how and to what extent they improve food availability and how they address structural inequities associated with poor food access environments so that those working to improve conditions may apply new knowledge to future policy work and practice in the food system.

12.3 Framing Mobile Produce Markets and Engaging a Food Justice Framework

Can gains be made towards a more equitable food system if the measure of success for food access initiatives is situated only within the parameters of the physical presence or absence of food? If a new supermarket is built, freshly stocked with affordable healthy fresh food, in a neighborhood that has gone without this resource, does that automatically indicate success? What happens if local residents don't shop there? What if that particular new supermarket only has one aisle dedicated to what they deem "ethnic foods" and those particular foods are priced at a premium? Some food access initiatives focus entirely on improving food availability and thereby sometimes neglect the societal and cultural factors that affect the end consumer intended to be fed and enriched with new food options (Caruso 2014).

Contrary to working solely to increase food availability, food access initiatives situated in the context of food justice assume structural social inequities related to race, class, and gender are key contributing factors to the existence of an inequitable local food landscape. Thus, food justice activists attempt to provide solutions driven by improving social conditions for people of color, low-income working class communities, and women (though not always addressing each of those groups equally or explicitly). For example, in 2003 the first known U.S. mobile market rolled onto the streets of West Oakland in the form of an old postal truck painted in bright colors, retrofitted as a supermarket aisle stocked with fresh produce, and staffed by young activists employed by Oakland-based non-profit, People's Grocery (Suutari 2006). People's Grocery started the Mobile Market to support their mission to, "improve the health and local economy of West Oakland through investing in the local food system" (People's Grocery 2014). People's Grocery founders believed access to healthy fresh foods was an immense obstacle to healthy living for residents and identified food as a key area for community involvement to serve as a catalyst for social change. People's Grocery operated the market for four years within the fold of the West Oakland community in order to, "ensure community self-determination plays a primary role in revitalizing underinvested neighborhoods" (People's Grocery 2014). Self-determination plays a critical role in doing the work of food justice and self-determination of a community is described as the long-term goal of food justice activists (Sbicca 2012, 462).

Not all mobile markets, however, nor all food access initiatives for that matter, are so unambiguously concerned with a community's ability to thrive in a fair food system. Several bodies of research have detailed the failures of alternative food initiatives to stimulate long-term positive social impact in the food system for historically marginalized communities. Namely, food access initiatives that identify neighborhoods as "food deserts" and attempt to deliver consumer-based solutions are perhaps most subject to criticism. Consumer-based solutions, such as farmers' markets, attempt to improve food access conditions with a singular focus on creating new food retail environments but do not address the economic constraints people experience in affording healthy food (Alkon et al. 2013). Scholars have panned alternative food initiatives that problematize food insecurity as an endemic consequence of a deficient environment, and therefore respond to such environments with supply solutions (Guthman 2013). Similarly, other scholars have criticized alternative solutions to food insecurity for relying upon consumer-driven mechanisms for change instead of addressing poverty as a root cause of food insecurity from an oppositional standpoint (Allen et al. 2003; Alkon et al. 2013). The validity of this critique is illustrated in Roots of Change's (2013) report on the collective efforts and contributions of the California Market Match Consortium, "The Market Match program has been critical to both the sustainability of several new and existing farmers' markets and to the growth of a loyal customer base for many new markets" (Roots of Change 2013, 18). In other words, it is the economic investment in low-income people through expanding food entitlements to boost purchasing power, such as California's Market Match Program (The Ecology Center 2015) and Michigan's Double Up Food Bucks (Fair Food Network 2015) that helps sustain an alternative food initiative like a farmers' market, instead of the mere presence of the farmers' market and fresh foods alone contributing to more equitable food conditions.

Alternative food initiatives have also come under fire for neglecting to address racial determinants to poor food access and subsequently poor health, and instead unwittingly reproducing themes of white privilege (Slocum 2006; Alkon and McCullen 2010). Guthman (2008) contends farmers' markets and CSAs attempt to bring the "good food movement" to communities who perhaps didn't ask for a movement in the first place. Copious analyses on the transformative potential of farmers' markets in low-income communities reveal common themes that "farmers' markets are for white people" or "farmers' markets are for rich people", and still perhaps most poignant to the topic of food accessibility, "farmers' market prices are too expensive" (Gottlieb and Joshi 2010). These themes further illustrate the disconnect between many alternative food initiatives and the communities in which they serve, as well as shed light on the challenging nature of utilizing consumer-based solutions to improve food conditions.

Given the failure of several existing alternative initiatives to address structural inequities in the food system and provide opportunities for systemic change, is it reasonable to assert mobile markets are promising food access strategies, equipped to return long-term benefit to people and communities writ large? What if these mobile markets were constructed at inception with a food justice approach in mind? A food justice approach to mobile markets can potentially address the described critiques on localized alternative food solutions. A food justice approach positions both the problem of inadequate access to healthy food and the solutions to improve access in a deeper discourse on the systemic issues that affect one's ability to access healthy affordable foods and goes beyond attempting to merely fix a deficient food landscape.

12.4   Food Justice and Mobile Produce Markets

Gottlieb and Joshi (2010) identify food justice as a "powerful idea" that has the ability to "bring about community change and a different kind of food system" by connecting advocates from many sectors. Now is a critical point in time to connect advocates involved in healthy food access work and explore the transformative potential of mobile markets. Are these markets merely charity vehicles taking fresh foods into neighborhoods on a temporary basis or are they able to offer a deeper sense of community engagement whereby neighborhood residents are positioned to take control of their local food system by shaping solutions?

Understanding how organizations frame the problem of inadequate access to healthy foods and then craft their solutions to the ensuing problem definitions can inform the efficacy of mobile markets. An analysis of mobile markets' ability to meet food justice goals offers critical insights to those working in the food movement; especially to those engaged at a policy level with the power to recommended and fund intervention strategies to promote a fair food system.

12.4.1   Getting to Know Mobile Markets

I asked the following questions in order to get to know mobile markets and how they relate (or do not relate) to food justice. How and to what extent do the mission statements, goals, and program activities of organizations leading mobile markets promote food justice implicitly or explicitly? How do organizations leading mobile markets define the problem of inadequate access to healthy foods in the communities in which they operate? How are mobile markets differently enabled and constrained in addressing food justice versus food availability?

I chose a sample of eight geographically diverse U.S.-based mobile markets that primarily serve urban and peri-urban clientele. A convenience sample was also constructed based on the availability of online content and mobile market leaders' ability to participate. I used critical discourse analysis of online content in the following categories: mission statements, goals and definitions of success of the mobile market, identified target audiences the market serves, description of program activities, listing of funders, and testimonials of shoppers of the mobile market. Mobile market sample:

1. Beans & Greens, Kansas City, KS
2. Fresh Moves, Chicago, IL
3. Freshest Cargo Mobile Market, Concord, CA
4. Lowcountry Street Grocery, Charleston, SC
5. Mobile Fresh, Moreno Valley, CA
6. My Street Grocery, Portland, OR

7. Real Food Farm Mobile Farmers' Market, Baltimore, MD
8. Ujamaa Freedom Market, Asheville, NC.

Data were qualitatively analyzed by theme to better understand how and to what extent these organizations explicitly or implicitly promote food justice through their mobile markets. Themes include: institutional racism; classism; sexism; ancillary community engagement programs related to food (e.g., economic development, leadership development, small business incubators, youth development, control and power of food; culturally relevant food, and a right to food).

Telephone interviews were conducted with leaders of four different geographically diverse mobile markets. Participants were asked questions about their organizations' motivations to start a mobile market and how they understand food access problems in the communities their mobile markets serve. The interview sample was based on the following criteria: geographic diversity within the United States, urban or peri-urban community service area, market leaders' availability, and a focus on serving low-income people. Leaders from the following mobile markets were interviewed:

1. Beans & Greens, Kansas City, KS
2. Lowcountry Street Grocery, Charleston, SD
3. Mobile Fresh, Moreno Valley, CA
4. Real Food Farm Mobile Farmers' Market, Baltimore, MD.

Interviews were transcribed and analyzed to determine how leaders' answers to questions fit or do not fit the characteristics of a food justice framework. Data were also organized to reflect themes in short-term food access solutions and long-term food access solutions. Finally, data were organized into themes reflecting discussion related to race, class, and gender inequities.

12.4.2 Identifying Food Justice

The food justice framework utilized here is based upon (Gottlieb and Joshi's 2010) broad identification of food justice as a concept which ensures, "the benefits and risks of where, what, and how food is grown and produced, transported and distributed, and accessed and eaten are shared fairly" (Gottlieb and Joshi 2010, 6). The food justice framework I employ is also grounded in the work of Alkon and Norgaard (2009), who define food justice as a concept that situates access to affordable, healthy, and culturally appropriate food within the contexts of historical institutional racism. This framework applies the work of Joshua Sbicca (2012), who provides further credence to the idea that food justice is a concept that, "pursues a liberatory principle focusing on the right of historically disenfranchised communities to have healthy, culturally appropriate food, which is also justly and sustainably grown" (456). The work of Cadieux and Slocum (2015) clarifies the definition of food justice by providing analysis of food justice literature; the authors suggest, "food justice practices would seek ways to intervene against structural inequities". They categorize four "food justice organizing nodes and practices" as: trauma and inequity, exchange, land, and labor (14). Cadieux and Slocum contend that activists must first address the "inequities that need to be undone" in order to build and observe transformation in the food system (15).

Given the definitions of food justice provided, I have situated an organization as being food justice-oriented if the organization meets the following criteria: attention to institutional and historical food system oppressions (e.g., racism, sexism, and classism), attention to community engagement, attention to economic justice for disenfranchised communities, attention to culturally relevant food, attention to environmentally sustainable production and distribution of food, and attention to the right to healthy affordable food for all people.

Identifying food justice at work and a food justice-oriented organization is challenging because there is no singular, unified definition of what constitutes food justice. According to Gottlieb and Joshi (2010), "even as the language of justice is embraced by a growing number of food groups, exactly what constitutes a food justice approach still remains a moving target" (6). Likewise, Sbicca (2012) contends there is a lack of united definition and understanding of food justice and food justice-oriented solutions, therefore often conflicting and complex ideas and messages emerge. Nonetheless, despite the challenge, scholars continue attempts at identifying shared food system priorities, characteristics, and discourses to construct a useful understanding of the food justice movement that will contribute to deeper and cohesive engagement in activist work to build social change.

12.5 Promoting Food Justice

Almost all organizations leading mobile markets appear to be more concerned with increasing food availability and promoting healthy eating habits, than they are about the social injustices that led to poor food access conditions in the first place. To be fair, activists working on food access are expected (by funders and boards of directors, typically) to take action and demonstrate tangible outcomes based on their efforts, usually on a relatively short timeline. All mobile market organizations included in this study are certainly taking action to improve conditions, however the possibility for long-term sustainability of their efforts to build or shape transformative change in food access is debatable. Both Beans and Greens and Fresh Moves are mobile markets that have come and gone in the last five years. Why? The easy answer is insufficient funds to maintain operations. The more complex answer involves probing further to consider what we, as scholars and activists, ought to expect food access interventions, like mobile markets, to realistically do for the people they serve (Table 12.1).

Most mobile market organizations do not operate from a food justice perspective, but instead, operate from a food availability perspective. As such, these

Table 12.1 Mobile Produce Market Organizations & Food Justice-Orientation Criteria

	Oppressions-race, class, gender	Community engagement	Economic justice	Culturally relevant food	Sustainable food-environment	Right to food
Beans & Greens, Kansas City, KS		X	X			
Fresh Moves, Chicago, IL	X	X	X	X		X
Freshest Cargo Mobile Market, Concord, CA		X	X		X	
Lowcountry Street Grocery, Charleston, SC		X	X		X	
Mobile Fresh, Moreno Valley, CA		X	X			
My Street Grocery, Portland, OR		X		X	X	X
Real Food Farm Baltimore, MD		X	X		X	X
Ujamaa Freedom Market, Asheville, NC	X	X	X	X	X	X

markets are better situated to deal with short-term food availability issues than they are at making broader systemic shifts that allow historically marginalized people and communities to determine how they get their food. A food justice perspective might shift the way mobile markets do business in that they would involve the community in all aspects of the market, not just at the point of purchase stage. For example, in a video on Ujamaa's website, we see one of the leaders engage in community-based participatory research by asking local children to vote on the design of their market logo and explain what they like or don't like about it. This is a small action but it demonstrates and encourages a lasting dialogue between the community being served and the leaders of the mobile market as partners in creating change, and not merely one group of people offering food for sale to another group of people. Aligned with this idea of engagement, only three of eight organizations mentioned or emphasized culturally relevant foods as part of their model.

12.6   Defining Food Access Problems

Critical discourse analysis of the available online content of eight mobile markets is a starting point to determining if they're situated to address and promote food justice, however it is a limited method for providing a comprehensive understanding on why activists select mobile markets as a solution to food access problems. Through interviews with leaders or founders of four of the eight mobile markets, I explain how organizations leading these markets define the problem of inadequate access to healthy foods in the communities in which they operate and how that definition influenced or led to the inception of their respective market.

At the time of interview in 2014, Beans & Greens was funded and operated by Menorah Legacy Foundation, which is a Jewish health foundation. This changed as of 2015 and Beans & Greens is now operated by Cultivate Kansas City, which focuses on urban sustainable agriculture. Beans & Greens is an organization that primarily runs a robust SNAP incentive program at farmers' markets (e.g., Market Match, Double Bucks) at 16 farmers' markets in and around Kansas City. The former program manager of Beans & Greens, a white female, was interviewed. According to the interviewee, Beans & Greens' now out-of-business mobile market was started in order to meet the needs of customers who lived in communities without farmers' markets and transportation. When asked about how healthy food access is a problem or challenge for the communities the mobile market served, the program manager said, "Good food, fresh food, is not easily affordable in inner-city neighborhoods. There are environmental deficiencies". When probed further about why the food landscape is deficient, the leader said she believed people in poor communities experience generational problems with poor eating habits. In summary, she said poor people are surrounded by bad food choices (e.g., fast food and liquor stores) and do not have enough choices available to them because most people in those communities are not interested in eating healthfully. She also contended that poor people, referring to primarily African American people in the urban Kansas City area, do not know how to cook with fresh and healthy ingredients as previous generations did. These comments seem to indicate inadequate access to healthy foods is attributable to a lack of demand for fresh healthy foods and lack of education on the part of community residents. This reasoning suggests a mobile market would sufficiently address this problem by increasing food availability and including educational outreach materials and activities, however this reasoning completely neglects a discussion on root causes of inequitable access to good food.

When asked about why the mobile market closed, the manager noted challenges related to a staff person who ran the market. She mentioned the employee was someone, "who grew up on food assistance and was a product of poverty. He could relate to customers and was a great guy but had a lot of personal challenges. If we had hired a white person from a middle-class background the market may have looked and worked differently." The interviewee further explained the employee's economic background—"he had lots of money troubles". Interestingly, this same topic of conversation also led the interviewee to state she believed sufficient infrastructure was not in place to support and train the employee for success. The employee was responsible for all operations of the market including: marketing, managing produce suppliers, managing distribution sites, driving, and maintaining the market vehicle. In an ordinary retail market business, those tasks would be distributed among several teams of employees, though the interviewee did not mention the involvement of any other employees to support the mobile market.

Lowcountry Street Grocery is a new organization that launched a mobile market in 2015. The founder and director, a white male, was interviewed. When asked what motivated him to start a mobile market, he said he was motivated by a concern for fairness and justice in the food system. He mentioned that Charleston, South Carolina is, "at the convergence of a rich food culture that is adjacent to food deserts". When asked more specifically about how or why food access is a challenge for some communities but not others, this leader again mentioned food deserts and also indicated that in regard to using the mobile market model to reach customers experiencing poor food access, he stated, "We don't tell the community what we think they need. We don't want to get involved in a toxic charity model. We're huge on empowerment. Charity often creates a disconnect". This comment seems to reveal a desire to support the self-determination of communities. Lowcountry stressed economic justice as a part of their business model. They try to create systemic change by using a pricing structure that allocates revenue generated from higher-income customers at stops in affluent neighborhoods to support lower prices and nutrition education outreach in poor neighborhoods. Lowcountry hopes to provide a 60% price markdown for low-income customers and are looking for a corporate partner to fund a SNAP incentive program to further increase purchasing power for low-income customers.

Mobile markets seem to require considerably less capital investment and political maneuvering than a traditional brick and mortar market. Lowcountry's director believed a mobile operation would better meet the needs of more people because it will serve a geographically and economically diverse consumer-base, and like

Beans and Greens, spoke to the significant reduction in barriers to entry as a new business compared to that of a traditional grocery retailer. Trends have shown retail industry giants and their investors are not willing to put their money in neighborhoods perceived to be unprofitable, thus continuing the grocery gap. For instance, U.K. retail giant Tesco entered the U.S. market in recent years with much fanfare for establishing a new smaller store chain called Fresh & Easy Neighborhood Market in Southern California neighborhoods labeled as food deserts. However, critics argue Tesco has received more praise than deserved and has not, in fact, greatly impacted food access for communities in need (Gottlieb and Joshi 2010, 50–51). The Fresh & Easy is my own neighborhood in San Diego closed last year and was vacant for nearly two years before a local small grocery chain called Baron's opened in the revamped space.

Mobile Fresh, California's Inland Empire mobile market, is in its third year of operation. The program manager, a Latino male, was interviewed. The manager affirmed Family Services Association (FSA), Mobile Fresh's parent non-profit, was motivated to start the market because of the high food insecurity rate in Riverside and San Bernardino Counties. The manager also mentioned that he believes it is, "no longer proving advantageous for grocery stores to open in our area, especially in rural or even semi-rural neighborhoods". He further explained the disastrous effects the economic recession and housing crisis had on the development of grocery stores in California's Inland Empire, "when the recession hit, new subdivisions with track homes out in the high desert were never completed. So then the families who were already there, the ones who didn't lose their homes in foreclosures, didn't have grocery stores or even gas stations to go to because they were never built as originally planned".

The manager noted the mobile market is founded on the knowledge that food access is a problem because of a lack of infrastructure in rural areas, an abundance of liquor stores and fast food outlets with poor food choices, and a lack of transportation for poor consumers, particularly for seniors. Mobile Fresh seems to be very practical with their approach and continues to add more stops to their weekly schedule based on shopper demand, which has continued to grow beyond 20 unique stops per week, according to the manager. This market seems to be working from a food security perspective, and does not conclusively appear to be pursuing food justice goals that might permanently fix the systemic issues that contribute to poor residents in rural areas, particularly Latinos, to go without adequate access to convenient affordable healthy food options.

One has to wonder, what will happen if FSA is no longer able to fund Mobile Fresh's operations? Mobile Fresh, like the other markets interviewed (with the exception of Lowcountry), relies on grant funding to operate. Beans & Greens lost money for three straight years before closing. According to the manager, Mobile Fresh is positioned to break even on the cost of produce inventory and sales revenue, however it is not financially sustainable with regards to labor, vehicle maintenance, gas, and liabilities, and all other operating costs.

Finally, Real Food Farm's Mobile Market's program coordinator, a white female, was interviewed. Real Food Farm started their Mobile Market because, "one important element of food access is transportation so we decided to bring the point of sale as close to people as possible". Real Food Farm builds change in the food system by engaging in and promoting urban agriculture to both grow the food supply and create new distribution points, like the mobile market, to increase access for consumers. The program coordinator said she believes, "The mobile market is a reaction to the current food landscape in Baltimore city. Ideally, we wouldn't need the market anymore and then we'll know we've improved food access conditions." This leader believes a long-term solution to food access problems is situated in the expansion of urban agriculture. Given this definition, it seems Real Food Farm's mobile market is situated to support the local food economy from a producer's perspective, however it is not apparent they are situated to promote or pursue food justice goals.

12.7 Pursuing Food Justice for Long-Term Healthy Food Access Solutions

Given most organizations do not appear to be food justice-oriented and are not well positioned to address root causes of inadequate access to healthy foods, it would be unreasonable to build the case that mobile markets have transformational power as a food access intervention. Mobile markets are better situated to continue to support and promote food availability goals—that is, they can effectively increase the presence of healthy affordable food in underserved communities and provide complementary education and promotional activities.

While first embarking on this research process, I, perhaps naively, anticipated uncovering more mobile markets pursuing food justice goals, and specifically, more markets that discussed food access problems in relationship to race, class, and gender oppressions. Interestingly, no organizations, through critical discourse analysis or interviews, demonstrated attention to gender or race. Even Fresh Moves in Chicago and Ujamaa in Asheville, despite defining food access problems and solutions as related to class inequality, did not explicitly talk about the racial and gender dimensions involved in who has access to healthy foods and who goes without. Perhaps these discourses are implicit within organizations working within communities of color and engaging with predominately women shoppers at market stops. That said, I am curious as to how mobile markets, and food access interventions writ large, can demonstrate efficacy in building or influencing sustainable social change in the food system if open discourse on social inequities is not standard practice and procedure?

Mobile markets can better interact with the food justice movement and promote food justice if they existed within a comprehensive and strategic approach to improving food access conditions. For example, one mobile market seemed to be operating in isolation from other programs and community services offered by the parent non-profit organization. When asked if they engaged in leadership development work, the mobile market manager replied, "no." However, a quick online investigation revealed the organization operates a youth workforce development program geared towards training young leaders in useful job skills. The integration of programs like this with a mobile market can help lead to valuable economic opportunities for community members.

12.7.1 Where and How Can Mobile Markets Best Effect Change in the Food System?

Despite the interest and good intentions of food activists from several sectors including, but not limited, to sustainable agriculture, public health, and social food entrepreneurs, my work suggests mobile markets are not suited to improve food access conditions in the long run for those who are most vulnerable to food system inequities. Furthermore, while it continues to be important to implement innovative solutions to meet immediate needs, improving the geographic availability of food does not necessarily improve or lead to improvements in the social conditions that make food access a problem in the first place. However, food availability and food justice are not competing concepts. They are not a binary. Pursuing food justice goals can happen in concert with working to improve food availability in under-resourced neighborhoods. Each mobile market leader I interviewed spoke with compassion about the neighborhoods and people they serve and recognized the general unfairness of a food system that permits such pervasive disparities in accessibility to healthy affordable food.

Further study on mobile produce markets should continue to consider what differences are observable in how mobile market organizations define the problem of food access and how those definitions influence the way they attempt to go about improving food access conditions. It may also be worthwhile to compare and contrast cooperative, worker-owned mobile markets, such as Ujamaa Freedom Market to cooperative, worker-owned retail stores such as Mandela Market Place in Oakland, CA or Our Table in Portland, OR to determine if differences exist in long-term potential to pursue food justice goals. How much of a factor does the mobility of the food, or conversely, the investment in a brick-and-mortar market, play in creating lasting change?

Mobile markets and other retail projects focused on increasing healthy food access should devote considerable resources to engaging community residents in economic development activities. The opportunity for training, employment, and ownership by community residents seems to be missing from most models of community-oriented food enterprises in the local food movement. An intentional focus on economic justice is important to realizing food justice. The ability for residents to participate in shaping food access solutions and owning both the means and product of their labor is a key contributing factor to ensure we're making an impact on the systemic barriers to healthy food access.

Finally, activists should continue advocacy efforts to support public policies that ease the cost of entry to operating a mobile food enterprise. Similarly, activists can work towards appealing prohibitive local laws that ban mobile vendors (including banning vendors focused on selling fresh fruits and vegetables) from operating on municipal property. For instance, the Los Angeles Food Policy Council has organized multiple stakeholders to develop street food vending policy recommendations for the City of Los Angeles. The Council is working to enact a supportive mobile policy similar to New York City's Green Carts Initiative to support both increased food availability in underserved neighborhoods and existing and would-be business owners who are currently marginalized by a prohibitive law and, "occupy a socioeconomically and legally precarious position" (Los Angeles Food Policy Council 2015). Mobile markets can be important tools used in a comprehensive, justice-oriented approach to improving food access conditions.

Appendix

Interview Questions

1. What motivated you and/or your organization to develop a mobile produce market?
2. How does your organization try to create change in the food system?
3. How is healthy food access a problem or challenge for the community/communities you serve?
4. Is your organization involved in leadership development work related to your mobile product market? If yes, please explain.
5. Is your organization involved in economic development work related to your mobile produce market? If yes, please explain.
6. How does your organization engage community members with your mobile produce market?
7. Does your organization work on other healthy food access initiatives in addition to your mobile produce market? If yes, please explain.
8. How is your mobile produce market funded?
9. If you are no longer operating a mobile produce market, why did it cease operation?
10. How do you see your mobile produce market being able to improve healthy food access conditions in the community/communities you work in the short-term and long-term?
11. How do you see your mobile produce market improving social inequities in the community it serves? Short-term and long-term ideas?
12. What are the biggest challenges or barriers you face in trying to improve social inequities through your mobile produce market?

References

Algert, S.J., A. Agrawal, and D.S. Lewis. 2006. Disparities in access to fresh produce in low-income neighborhoods in los angeles. *American Journal of Preventive Medicine* 30 (5): 365–370. doi:10.1016/j.amepre.2006.01.009.

Alkon, A.H., D. Block, K. Moore, C. Gillis, N. DiNuccio, and N. Chavez. 2013. Foodways of the urban poor. *Geoforum* 48: 126–135. doi:10.1016/j.geoforum.2013.04.021.

Alkon, A.H., and C.G. McCullen. 2010. Whiteness and farmers markets: Performances, perpetuations...contestations? *Antipode* 43 (4): 937–959.

Alkon, A.H., and N. Norgaard. 2009. Breaking the food chains: An investigation of food justice activism. *Sociological Inquiry* 79 (3): 289–305.

Allen, P., M. FitzSimmons, M. Goodman, and K. Warner. 2003. Shifting plates in the agrifood landscape: The tectonics of alternative agrifood initiatives in California. *Journal of Rural Studies* 19: 61–75.

Baker, E.A., M. Schootman, E. Barnidge, and C. Kelly. 2006. The role of race and poverty in access to foods that enable individuals to adhere to dietary guidelines. Preventing Chronic Disease [serial online] 2006 Jul. Retrieved from http://www.cdc.gov/pcd/issues/2006/jul/05_0217.htm.

Cadieux, K.L., and R. Slocum. 2015. What does it meant to *do* food justice? *Journal of Political Ecology* 22.

California Institute for Rural Studies. 2012. Farmworker mobile market feasibility study. California Institute for Rural Studies.

Carney, M. 2012. Compounding crises of economic recession and food insecurity: A comparative study of three low-income communities in santa barbara county. *Agriculture and Human Values* 29 (2): 185–201. doi:10.1007/s10460-011-9333-y.

Caruso, Christine C. 2014. Searching for Food (Justice): Understanding Access in an Under-served Food Environment in New York City. *Journal of Critical Thought and Praxis* 3, no. 1, Article 8. Available at http://lib.dr.iastate.edu/jctp/vol3/iss1/8.

Dinour, L.M., D. Bergen, and M.-C. Yeh. 2007. The food insecurity–obesity paradox: A review of the literature and the role food stamps may play. *Journal of the Academy of Nutrition and Dietetics* 107, no. 11: 1952–1961. Retrieved from http://www.andjrnl.org/article/S0002-8223 (07)01616-1/pdf.

Ecology Center. 2015. Retrieved February 22, 2015, from http://ecologycenter.org/marketmatch/.

Fair Food Network. 2015. Retrieved February 22, 2015, from http://www.fairfoodnetwork.org/what-we-do/projects/double-up-food-bucks.

Gottlieb, R., and A. Joshi. 2010. *Food justice*. Cambridge, Mass: MIT Press.

Guthman, J. 2008. Bringing good food to others: Investigating the subjects of alternative food practice. *Cultural Geographies* 15 (4): 431–447.

Guthman, J. 2013. Too much food and too little sidewalk? Problematizing the obesogenic environment thesis. *Environment and Planning* 45: 142–158.

Industrious Productions. (n.d.). Exploring foods + ujamaa freedom market (video file). Retrieved from http://vimeo.com/110076855.

Larson, N., M. Story, and M. Nelson. 2009. Neighborhood Environments Disparities in Access to Healthy Foods in the U.S. *American Journal of Preventative Medicine* 36 (1): 74–81.

Los Angeles Food Policy Council. 2014. Retrieved October 14, 2014, from http://goodfoodla.org/policymaking/working-groups-2/.

Los Angeles Food Policy Council. 2015. Retrieved March 3, 2015, from http://goodfoodla.org/resources/local-good-food-organizations/street-food/.

Lucan, S., A. Maroko, R. Shanker, and W. Jordan. 2011. Green carts (mobile produce vendors) in the bronx-optimally positioned to meet neighborhood fruit-and-vegetable needs? *Journal of Urban Health* 88 (5): 997. doi:10.1007/s11524-011-9593-2.

McClintock, N. 2011. *From industrial garden to food desert: Demarcated devaluation in the flatlands of Oakland, California*, 104–135. Cultivating Food Justice: Race, Class, and Sustainability. MIT Press.

NYC Department of Health and Mental Hygiene. 2014. Retrieved October 14, 2014, from http://www.nyc.gov/html/doh/html/living/greencarts.shtml.

People's Grocery. 2014. Retrieved October 14, 2014, from http://www.peoplesgrocery.org/.

Roots of Change. 2013. *2013 market match: Snaping up benefits for farmers and shoppers*. Public Health Institute.

Sbicca, J. 2012. Growing food justice by planting an anti-oppression foundation: Opportunities and obstacles for a budding social movement. *Agriculture and Human Values* 29 (4): 455–466.

Shaffer, A. 2002. The persistence of L.A.'s grocery gap: The need for a new food policy and approach to market development. UEP Faculty & UEPI Staff Scholarship. Retrieved from http://scholar.oxy.edu/uep_faculty/16.

Slocum, R. 2006. Anti-racist practice and the work of community food organizations. *Antipode* 38 (2): 327–349. doi:10.1111/j.1467-8330.2006.00582.

Suutari, A. 2006. The people's grocery: Bringing healthy food to low-income neighborhoods. *The Ecotipping Pointing Project*. Retrieved from http://www.ecotippingpoints.org/our-stories/indepth/usa-california-oakland-people-grocery-healthy-food.html.

United States Department of Agriculture, Economic Research Service. 2014. Community food security. http://www.ers.usda.gov/topics/food-nutrition-assistance/food-security-in-the-us/community-food-security.aspx#.U4Ou4ZRdWDh.

Ver Ploeg, M., V. Breneman, P. Dutko, R. Williams, S. Snyder, C. Dicken, and P. Kaufman. 2012. Access to affordable and nutritious food: Updated estimates of distance to supermarkets using 2010 data. Economic Research Report No. (ERR-143).

Widener, M., S. Metcalf, and Y. Bar-Yam. 2012. Developing a mobile produce distribution system for low-income urban residents in food deserts. *Journal of Urban Health* 89 (5): 733–745. doi:10.1007/s11524-012-9677-7.

Williams, E.M., B.O. Tayo, B. McLean, E. Smit, C.T. Sempos, and C.J. Crespo. 2008. Where's the Kale? Environmental Availability of Fruits and Vegetables in Two Racially Dissimilar Communities. *Environmental Justice* 1 (1): 35–43. doi:10.1089/env.2008.0503.

Zepeda, L., and A. Reznickova. 2013. Measuring Effects of Mobile Markets on Healthy Food Choices. Madison, University of Wisconsin. Retrieved from http://dx.doi.org/10.9752/MS142.11-2013.

Part III
Food Justice and Governance

Chapter 13
A Vignette from Our Table Cooperative

Mallory Cochrane

Abstract Mallory Cochrane illustrates the fertile possibilities for food justice outside of state institutions by introducing us to a group which tries to implement radically egalitarian governance structures internally, while promoting those values to reform food markets and other institutions.

Keywords Food justice · Governance · Multi-stakeholder cooperatives · Egalitarianism · Public policy

From an outside perspective it looks like many other farms across the U.S.—rows of budding crops, roaming chickens, red tractors carefully navigating the fields—a beautiful scene with a picturesque mountain view backdrop. While Our Table Cooperative may look like many other farms from the surface, the real beauty of this farm is not found solely within the food in the ground nor bees in the air; it's the radical ownership structure that sets it apart from the rest. As a truly community owned food system, this farm and business is a profound shift from the traditional and dominant ownership structures we see across the American food system. It all began when a community started asking critical questions about who controls our food, how consequences of the decisions made are distributed, and why we need to create models that redefine the relationships between producers and consumers. What emerged was a new paradigm that equally prioritizes ecological soundness, social justice, and communal economic prosperity.

Our Table Cooperative (OTC) is using the multi-stakeholder cooperative model to build an integrated and sustainable food system in the Pacific Northwest. Located 15 miles south of Portland, OR, OTC is uniquely positioned to meet the diverse needs of the nearby rural, suburban and urban populations. The cooperative, which

M. Cochrane (✉)
Our Table Cooperative, Sherwood, USA
e-mail: mallory@ourtable.us

© Springer International Publishing AG 2017
I. Werkheiser and Z. Piso (eds.), *Food Justice in US and Global Contexts*,
The International Library of Environmental, Agricultural and Food Ethics 24,
DOI 10.1007/978-3-319-57174-4_13

brings workers, producers, and consumers together through the multi-stakeholder model, officially incorporated in 2013 and now operates as one of only a dozen multi-stakeholder food and farming cooperatives in the U.S. and is the only one owned by these three particular stakeholder groups. OTC is a "new paradigm for a more localized food system based on a new form of agriculture that blends the wisdom of the past with the science of the present" (Our Table n.d.). This type of cooperative model provides the foundation for OTC's governance and operational principles, which prioritize social relationships and communal benefit above profit-only motives.

As a cooperative value chain, OTC is a Certified Organic farm, as well as an aggregator, wholesaler, distributor, and with the recent opening of an on-farm grocery and deli, a retailer. The co-op's farm, coordinated by the worker members, operates a CSA, participates in farmers' markets, and distributes wholesale products to local restaurants and retailers. In late 2014, regional producer and consumer members were added to round out the multi-stakeholder cooperative model in its entirety. Twelve independent Oregon businesses have joined as co-owners of OTC, supplying the cooperative with a variety of Oregon grown and produced products that showcase the best of sustainable Oregon agriculture. The ownership structure of OTC allows the producer members to own and control their product from field to customer, without the costs and work associated with direct marketing. This represents a stark difference from both conventional and alternative models that tend to distance and muddle relationships as the supply chain gets longer. While OTC sells and distributes food to a wide community, the co-op's 200+ consumer members have committed to the long-term success of the business by becoming co-owners and supporters beyond simply purchasing the food that the OTC workers and producers sell. With all aspects of a local food system represented in a single business, trust and interdependency are created and sustained.

The unique ownership structure of OTC creates a network of participants by making a space in their local food system for the sharing of resources, skills and knowledge among a community of workers, producers, and consumers. The cooperative model unites and gives equal emphasis to each part of the food system, and by doing so, OTC is facilitating a shared investment in the health and vitality of the community food system (Our Table n.d). The foundation of the cooperative is built on the relationships between stakeholders, which are based on principles of interdependence, transparency and trust. Through all of its enterprises and activities, OTC builds a loyal consumer-member base for the direct sale of local foods and developing a community-scale aggregation and distribution system that combines and strengthens the efforts of independent regional producer members, and supports paying living wages and provides ownership opportunities to worker-members on the farm.

As OTC grows and matures, the benefits to all stakeholders of the multi-stakeholder cooperative emerge and expand. This is only possible, however, if integrated and committed relationships are established in order to strengthen the whole. OTC is not unlike any other small food and farming business, however; the worker and producer members still face weather and market challenges (among many others) and consumers must make intentional decisions about how and where they access their food, often foregoing convenience for the sake of quality and connection. It is through the commitment of this diverse group of people and enterprises that sets OTC apart from other businesses. When both challenges and successes arise, the members of the co-op frequently come back to its unofficial motto "resiliency through diversity", demonstrating that collective strength often overcomes any individual aim.

Chapter 14
Introduction to Food Justice and Governance

Paul B. Thompson

Abstract As Paul Thompson points out in this introduction, when discussing governance and justice there is a tendency to emphasize government and de-emphasize non-state governance structures, even though these structures have profound, daily impact on our lives in many different ways.

Keywords Food justice · Governance · Multi-stakeholder cooperatives · Biopower · Egalitarianism · Public policy

14.1 Food Justice and Governance

Contributions to this section emphasize governance, a concept that should be distinguished from government. Starting with Thomas Hobbes' *Leviathan* (1651) and especially after John Locke's *Second Treatise on Government* (1689), English language discussions of justice have been focused almost exclusively on state action. Governance, in contrast, is a term that is intended to include a wider array of mechanisms for social control. Voluntary associations, private standards (such as fair trade), business practices and even the informal expectations that E.P. Thompson refers to as "moral economy" function to structure and control human activity without resort to the state's monopoly on the use of violence (see Thompson 1991). The relative lack of state involvement in such institutions should not imply that questions of justice are irrelevant to them. And at the same time, we should not suppose that government itself is irrelevant, either.

Mallory Cochrane opens this section with a description of the Our Table Cooperative (OTC), a multi-stakeholder co-op outside of Portland, Oregon.

P.B. Thompson (✉)
W. K. Kellogg Chair in Agricultural Food, and Community Ethics, Michigan State University, East Lansing, MI, USA
e-mail: thom@anr.msu.edu

© Springer International Publishing AG 2017
I. Werkheiser and Z. Piso (eds.), *Food Justice in US and Global Contexts*,
The International Library of Environmental, Agricultural and Food Ethics 24,
DOI 10.1007/978-3-319-57174-4_14

As Cochrane explains, activists working to address injustice and oppression in the food system have experimented with a number of new institutional forms for structuring the production, distribution and consumption of food. Farmer co-ops were already forming in the 19th century and became common in some sectors by the mid 20th century as farmers organized to resist power of railroads and processors (see Gray 2014). Similarly, local consumer co-ops are a staple of the alternative food movement in the United States, where people organize to source a wide array of food and other products from outside the supply chains of major retail stores. OTC is a further development of the cooperative movement that links farmers and consumers as well as others who work in food to create a community of people whose relational ties are much richer than those of a transactional food economy.

Additional attempts to discipline the economist's picture of the cash transaction are explored through the essays in this section. John Stuart Mill's portrayal of "economic man" (and perhaps the gendered term here was not insignificant) portrayed the value-for-value trade as the epitome of rational behavior. Economic agents traded one good for another solely on the basis of their respective expectations for realizing satisfaction from the possession or utilization of said goods. Given the highly unrealistic assumptions of neo-classical economic theory, unfettered exchanges of this sort appear to fulfill the demands of the utilitarian maxim. One achieves the greatest good for the greatest number by allowing people to make any and all trades that they perceive to be in their own best interest (Mill 1836). During the twentieth century, the notion that virtually all human behavior could be modeled as the buying and selling of goods has been incorporated into libertarian and neo-liberal governance philosophies that recommend leaving virtually everything to "the market."

As I have argued in *The Agrarian Vision,* even the most optimistic advocates of this model must admit that the market will only optimize according to an existing distribution of resources. If the starting point is unfair or oppressive, leaving things to a market will hardly ever make things better. Policy models that reinforce existing property rights and encourage us to think that trade-based efficiencies should be the governing norms of a food economy thus run contrary to more robust conceptualizations of food justice (Thompson 2010, 71–77). Various cooperative models like OTC emerged as a response to the power of wealthy or well-capitalized actors within market economies, actors whose very size inures them to the plight of weaker parties who would be forced to accept unfair trades, exploitative wages or even defective products simply because they have no other alternative. There is a palpable sense in which such "exchanges" are coercive, yet governments have been reluctant to interfere on the grounds that doing so would defeat the "efficiencies" of unregulated markets. Co-operatives where individuals or smaller firms cooperate to attain a more equitable bargaining position with respect to large corporate players represent an alternative model that still plays by the rules of a capitalist economy.

Yet however much those of us who root for the underdog are dismayed by it, rules that were originally intended to constrain the powerful players in capitalist economies can frequently be applied in ways that disadvantage weaker parties. Anti-trust legislation, for example, can in some cases prevent cooperative activities that allow the parties to a cooperative endeavor to obtain higher prices than they otherwise would. The entry by Amber Leasure-Earnhardt, Carrie A. Scrufari, and Rebecca Valentine takes us deep into the weeds of legal liabilities that have constrained another practice that has traditionally allowed those at the margins to escape some of the most extreme oppressive tendencies of market-based food economies: gleaning. As depicted in the famous 1857 painting by Jean-François Millet, gleaning was an activity performed by the poorest of the poor who would follow along behind the primary harvest picking kernels of grain up with their hands. The gleaners of Millet's painting would not necessarily have been allowed to simply keep the proceeds of their labor, however. Liana Vardi has debunked the notion that peasant women's opportunity to obtain this residual grain was widely accepted right in the moral economy of pre-capitalist society (Vardi 1993).

Nevertheless, the thought that leavings in the field are the natural property of the poor has functioned as a compensating brake on the ruthless machinations of the market, at least in the expectations of many people today. It has in fact been extended to include the leftovers and scraps from restaurants and institutional kitchens as well as products that have gone beyond their "sell-by" dates in contemporary grocery stores. These locations on the margins of the industrial food system become sites where marginalized people may escape the powerful net of monetized exchange, and where they enjoy the protection of a widely-shared presumption that they are morally entitled to do so. Peter Singer and Jim Mason endorse the ethics of "freegan" dumpster diving as form of resistance to profit-seeking in the food system (Singer and Mason 2006). Yet as Leasure-Earnhardt, Scrufari and Valentine write, gleaners and those who aid them do not, in fact, elude the powers of government entirely. Soup kitchens and food pantries are not only subject to regulatory oversight for the purposes of sanitation and food safety, they are also possible targets for legal action when foodborne illness inadvertently occurs. Even food retrieved from dumpsters could trigger a lawsuit or criminal charges. Their article discusses the National Gleaning Project, which aims to more fully integrate the practice of scavenging and more fully utilizing potentially wasted food into the institutional governance structure of the current food system. This includes not only protections against liability, but also incentivizing contributions to food banks and similar initiatives that enable gleaning.

In contrast to Leasure-Earnhardt, Scrufari and Valentine's policy-dense chapter on gleaning, Melanie Bowman undertakes a more philosophical discussion of the food justice issues associated with wild rice. The role that harvesting, preparation and consumption of wild rice plays in the tribal food culture of Anishinaabe people has become a signature case study for work in food justice. In 2007, Native American activist Winona Laduke published an article in *Orion Magazine* arguing describing rice harvesting by Ojibwa elders and criticizing the University of Minnesota for a genome mapping project that would colonize this tribal food (Laduke 2007). Kyle Whyte has also discussed wild rice harvesting, describing the tribal governance practices that determine who will be allowed to participate in the harvest, as well as measures to conserve threatened populations of the rice from the pressures of economic growth and industrial pollution (Whyte 2016). Bowman adds her voice to this literature with a chapter that also explores tensions between tribal attitudes toward this culturally central food and its exploitation by plant breeders and geneticists.

Bowman emphasizes the epistemological dimension of this controversy. She argues that the academic environment of the modern university has conceded to a philosophy which views knowledge as a commodity good—something that can be traded very much along the lines of the value-for-value exchange that Mill identified in his writings on economic man. In contrast to this view, Bowman stresses the relational character of knowledge. She references Robin Kimmerer's suggestion that knowledge should be viewed within the context of a gift economy and cites Shawn Wilson to the effect that "… a common feature of Indigenous worldviews is an emphasis on relationship and the centrality of interconnection." Bowman argues that when knowledge is viewed in terms of its ability to generate monetary income, these relational aspects of knowing are suppressed, distorted, and ultimately stifled. Although Bowman is not hopeful about the prospects for more respectful and relationally based attitudes toward knowledge production on the part of university scientists, her analysis underlines the sense in which market-oriented governance models creep into virtually every dimension of the human condition.

Avi Brisman and Nigel South's chapter, "Food, Crime, Justice, and Security: (Food) Security for Whom?" focuses on yet another aspect of policy and governance: international regimes of securitization against war and crime. While this may at first seem only distantly related to food issues, the authors show both practical and theoretical connections between the two. Practically, securing food independence or reliable access to food systems is a serious security issue for many countries, increasingly so as our climate changes. The Arab Spring, which was at least in part sparked off by protests around food insecurity, was taken as a sober warning by many nations.

More theoretically, the word "security" is used in international discourses around war and crime as a reasonable-sounding cover for policies to which citizens might otherwise object. The authors suggest that this background should raise our suspicions that the equally reasonable-sounding goals of "food security" espoused by powerful countries and intergovernmental groups may be less lofty than they originally sound, and have more to do with pacification and maintaining the status quo. The authors push us to ask important questions about what it is nations are securing, for whom, and from whom. Given the ever more interconnected global world in which these states and other powerful institutions (as well as activists) operate, these are important questions.

The final chapter to be introduced in this section discusses how elements of the food system can be extremely effective venues for experiential education efforts. Elissa Johnson's essay includes several vignettes from her experience working with the Vermont Food Systems Summer Study Tour in 2015, the Higher Education Consortium for Urban Affairs (HECUA) in St. Paul, Minnesota and the Pine Island Community Farm, a partnership with the Vermont Land Trust. Johnson emphasizes the integrative impact of educational experiences that give students an opportunity to participate either in primary food production (often called "farming") or in activities intended to address structural inequalities in the food system. Not only do these activities stimulate learning, they result in a greater awareness of how seemingly disparate activities occurring at different sites along the chain that runs from farm to fork actually impact one another. Johnson's essay indicates why obtaining a more complete picture of the interconnections and networks that hold the food system in place will be a critical part of reform activities that are intended to address the systemic problems of injustice that plague domestic food systems within the United States.

In conclusion, the chapters in the section address the interconnection of food justice and governance by taking a look at specific cases where issues arise. Both gleaning and the harvesting of foods from common pool resources such as forests and fisheries (or, in this case, areas estuaries where rice grows without the aid of human cultivation) occur within loci where people have obtained food outside the parameters of monetized exchange and seemingly beyond the reach of state power. When considering food justice from a global perspective, it would be important to recognize that millions of subsistence producers still grow a significant portion of what they eat today. These practices are coming under pressure not only from the physical expansion of commercial production and its environmental spillovers, but through colonization of our very conception of governance itself. The presumption that all human relations can and should be optimized through mechanisms of competitive markets and commodity exchange has become a pervasive theme in contemporary thinking on governance. Like the multi-stakeholder cooperative described by Mallory Cochrane, the experiential education cover by Elissa Johnson can be viewed as an attempt to innovate new forms of cooperation that can resist the governance modes that the 20th century built on Mill's abstraction, the economic man.

It is worth noting that Mill himself would probably have been horrified by what has happened in contemporary food systems. In *The Birth of Biopower* Michel Foucault traced the way in which the "free market" ideal has been transformed from Mill's time, when *laissez-faire* literally meant that the government should just stay out of the economy. In contrast to 19th century *laissez-faire* (which Mill did *not* fully advocate) the architects of 20th century economic markets understood that government must be involved in creating the structures that would drive people into competitive markets, seeking to maximize their returns through exchange relations. Thus far from being a form of freedom, free markets are, in Foucault's rendering a governing hand, invisible only in the sense that it its rendering of power is concealed from view (Foucault 2008). There is an overarching philosophical question we should not overlook: Do the examples of innovation in food system governance that we have before us in this section represent resistance to what Foucault called biopower, or are they simply another form of capitulation to it?

References

Foucault, Michel. 2008. *The Birth Of Biopolitics: Lectures At The Collège De France, 1978–1979.* New York: Palgrave-Macmillan.

Gray, Thomas. 2014. Agricultural cooperatives. In *The Encyclopedia of Food and Agricultural Ethics,* ed. P.B. Thompson, and D. Kaplan, 45–54. Dordrecht, NL: Springer.

Laduke, Winona. 2007. Ricekeepers. *Orion Magazine.* https://orionmagazine.org/article/ricekeepers/. Accessed 15 Sep 2016.

Mill, John Stuart. 1836 [1877]. On the definition of political economy and the method of investigation proper to it. *Some Unsettled Questions of Political Economy.* 3rd ed, 120–164, London: Longmans Green & Co.

Our Table. n.d. Retrieved from http://www.ourtable.us.

Singer, Peter. 1976. *Animal Liberation.* New York: Ballantine Books for the New York Review of Books.

Singer, Peter, and Jim Mason. 2006. *The Ethics of What We Eat: Why Our Food Choices Matter.* Emmaus, PA: Rodale Press.

Thompson, E.P. 1991. *Customs in Common.* London: Merlin Press.

Thompson, Paul B. 2010. *The Agrarian Vision: Sustainability and Environmental Ethics.* Lexington: University Press of Kentucky.

Vardi, Liana. 1993. Construing the Harvest: Gleaners Farmers, and Officials in Early Modern France. *The American Historical Review* 98: 1424–1447.

Whyte, Kyle. 2016. Food justice and collective food relations, *122–134.* In *Food, Ethics and Society: An Introductory Text with Readings,* ed. A. Barnhill, M. Budolfson, and T. Doggett. New York: Oxford University Press.

Chapter 15
The National Gleaning Project: The Importance of Gleaning and Fresh Food Recovery in a Sustainable and Just Food System

Amber Leasure-Earnhardt, Carrie A. Scrufari and Rebecca Valentine

Abstract Although agricultural gleaning is an ancient practice, there are numerous legal issues that need to be addressed at the state and federal level in order for gleaning operations and the farms they work with to thrive in our modern food system. Today, gleaning is part of a larger movement intended to confront issues of food waste and food insecurity. Nonprofit organizations now engage in gleaning and fresh food recovery in a wide variety of settings including farms, backyard fruit trees, farmer's markets, and grocery stores. Gleaning needs support from many sectors, particularly the legal sector, as food donors and charitable organizations seek assistance with liability, tax, infrastructure, and labor issues. The National Gleaning Project (NGP—an initiative of the Center for Agriculture and Food Systems at Vermont Law School, sponsored by the National Agricultural Library, a part of the United States Department of Agriculture's Agricultural Research Service) is a multi-year endeavor that seeks to address gleaning issues and provide resources to farmers, advocates, organizations, and individuals involved in gleaning or fresh food recovery, or those seeking to develop such a program. To accomplish these goals, the NGP has developed an online gleaning resources hub, the first of its kind in the nation. By creating a resource clearinghouse and elevating the gleaning model of food waste reduction, community development, and social justice for the food insecure, the NGP will assist in the growth of the gleaning movement by reducing the information and resource barriers to organizing and implementing gleaning programs.

Keywords Gleaning · Food recovery · Liability · Food waste · Food insecurity

A. Leasure-Earnhardt (✉) · C.A. Scrufari · R. Valentine
Center for Agriculture and Food Systems, Vermont Law School, Royalton, VT, USA
e-mail: a.leasure.earnhardt@gmail.com

C.A. Scrufari
e-mail: cscrufari@vermontlaw.edu

© Springer International Publishing AG 2017   171
I. Werkheiser and Z. Piso (eds.), *Food Justice in US and Global Contexts*,
The International Library of Environmental, Agricultural and Food Ethics 24,
DOI 10.1007/978-3-319-57174-4_15

15.1 Introduction: Examining Gleaning and Food Waste in America

Several 20, 30, and 40-somethings stand bent over rows of acorn squash, wearing an assortment of blue jeans, boots, and hooded sweatshirts. The morning air carries a touch of autumn chill although the sun is warm, quickly drying the last remaining dew drops glistening on the farmer's fields. The team of volunteers work under an azure blue sky, sorting squash into various bins—those that will carry the food to market, those that will make their way to the compost heap, and those that will arrive later in the day at the food bank. Last year, thanks to the volunteers' efforts, the food bank collected 300,000 pounds of fresh, local produce and redistributed it to more than 225 food shelves and meal sites throughout the state (Vermont Foodbank 2016).

The same scene unfolds across many towns and states throughout the country. These volunteers are engaging in an ancient practice known as "gleaning." Black's Law Dictionary defines gleaning as, "[t]he gathering of grain after reapers, or of grain left ungathered by reapers" under Hebrew common law. The practice of gleaning reaches back to biblical times, illustrated in 2 Ruth 2:23, as a means of providing for the poor and disenfranchised members of society (see also Lev. 19:9, 23:22 and Deut. 24:19). Farmers allowed orphans, widows, and travelers to glean food from their fields at the end of the season after harvest (Farmer 2011). Often, farmers would also refrain from harvesting the edges of their fields, leaving them for the poor to glean (Farmer 2011). As will be discussed further, the practice of gleaning has also been expanded to include the terms "food recovery" or "food rescue" to reflect the practice of preventing food waste created by for-profit wholesalers, retailers, manufacturers, and other businesses (Gleaning for Hunger Relief 2002, 102.11).

Today, gleaning is part of a larger movement meant to address the troubling fact that the U.S. wastes 30–40% of the food it produces while 6.4 million households are unable to meet the nutritional needs of their families (Horton 2014). Rather than individual orphans, widows, and travelers gleaning from farmers' fields, today, nonprofit organizations engage in gleaning related activities in a wide variety of settings—on farms after the commercial harvest and from restaurants, grocery stores, and farmer's markets after the business day has ended (Gleaning for Hunger Relief 2002, 102.11).[1] These organizations then redistribute the food they have gleaned to food banks, food shelves, and soup kitchens, feeding hundreds or potentially thousands of Americans (Gleaning for Hunger Relief 2002, 102.11).

---

[1]Many advocates argue that gleaning and food recovery, while related, are separate and distinct. Specifically, both practices consist of the collection of surplus crops or food, but for some advocates, gleaning involves a relationship with the farmer who has grown or produced the food whereas food recovery does not. Regardless of the specific definition of gleaning one chooses to adopt, the main intent of the practice is to recover surplus food for redistribution to food insecure populations, encompassing a charitable dimension to the act.

Some organizations estimate that they have 35,000–40,000 volunteers assisting with gleaning (Gleaning for Hunger Relief 2002, 102.11). These efforts result in the diversion of tens of millions of pounds of fresh food from landfills to 40 million hungry Americans living below the poverty line (Society of St. Andrew 2016).

The United States Department of Agriculture (USDA) defines gleaning as "simply the act of collecting excess fresh foods from farms, gardens, farmer's markets, grocers, restaurants, state/county fairs, or any other sources in order to provide it to those in need" (USDA 2010, 2). In practice, all of these venues can donate fresh foods, and resources are available in communities across the country to allow farmers, farmer's market organizations, local grocers, or even the backyard gardener to donate excess fresh foods to people in need. Although a gleaning network is not currently established in every community, the resources available to create or tap into existing resources are rapidly increasing as gleaning awareness grows. Essentially, gleaned food—as used in this article—is a collected second harvest from farms and other venues, comprised of fresh food that would otherwise become waste, that is captured for redistribution to food insecure populations.

Gleaned fruits and vegetables can supplement the diet of many low-income families who otherwise could not afford—or would not have access to—nutrient-dense, high quality food (USDA 2010, 2). As such, it is widely recognized that the practice of gleaning can contribute significantly to a sustainable and just food system. Congress defined sustainable agriculture for the first time in the *Food, Agriculture, Conservation, and Trade Act of 1990* (the 1990 Farm Bill), explaining that a sustainable food system is one that integrates plant and animal production practices to satisfy human food needs, enhances environmental quality, and efficiently uses natural resources, while allowing farming to remain an economically viable vocation that enhances farmers' quality of life [7 U.S.C. § 3103 2014].

Gleaning can help shape the future of sustainable agriculture and promote the ideals of food justice by serving those traditionally marginalized by society. Since there is a significant portion of food wasted everyday on farms, at the store, and in the home, this wasted food can be part of the solution to alleviate hunger in local communities without growing more food, if the food can be quickly collected and redistributed to those in need. There is a crucial need for the development of laws and policies to incentivize gleaning and to promote this important practice as an integral part of the food system. Despite the value of gleaning as a means of reducing food waste and combatting hunger, farmers and volunteer organizations participating in gleaning have legal concerns (such as liability) and a host of practical challenges (including lack of infrastructure and capacity). While gleaning can be an important part of a sustainable and just food system, the practice needs more legal, financial, and structural support to expand and sustain the movement across the U.S. The National Gleaning Project (NGP) of the Center for Agriculture and Food Systems at Vermont Law School is an attempt to raise awareness and provide some of the support gleaning needs to continue succeeding as a sustainable practice.

15.2 The National Gleaning Project (NGP)

Nationally, gleaning and fresh food recovery organizations are growing in number and strength, providing their communities with healthy and sustainable foods. The NGP[2] strives to capture these organizations' innovative and collaborative processes and share them with the gleaning community at large so they can be replicated. The NGP was created in response to the need for a national network connecting modern agricultural gleaning and fresh food recovery organizations across the United States. The NGP is intended to connect these organizations with each other and to legal resources that can help grow their work. Through the use of detailed case studies, the NGP provides models for how gleaning organizations can work in tandem with existing aid services such as Food Banks and Food Shelves to complement other hunger relief efforts and expand access to fresh, local, whole foods.

Specifically, the NGP is a multi-year project seeking to address the challenges to gleaning while providing resources to food donors, advocates, organizations, and individuals involved in gleaning or seeking to develop a program. In broad terms, the NGP aims to work with organizations across the U.S. to reduce food waste, incentivize and/or remunerate farmers, restaurants, and other establishments to increase food donations to charitable organizations, address issues of food insecurity, and support local economies. The NGP specifically focuses on addressing legal barriers to enhancing gleaning and fresh food recovery in the United States.

As part of its initial phase, the NGP developed an online gleaning resources hub (National Gleaning Project 2016). The hub includes a catalogue of state and national laws and regulations pertaining to gleaning and fresh food donation, a comprehensive directory and map of gleaning and food recovery organizations across the U.S., and additional web-based resources on a variety of legal and non-legal issues associated with gleaning and food recovery practices. In addition, the NGP involves interviewing dozens of local, regional, and national gleaning organizations across the U.S. about their specific legal and non-legal challenges. These organizations also provided details regarding their creative models for success. In later phases of the project, the NGP will highlight these successful models and collaborative concepts through detailed case studies. The gleaning resources hub is a living resource intended to be periodically updated and to evolve with the needs of the gleaning movement.

Through preliminary research and interviews, the NGP learned that there is a need for a national network to connect the organizations and individuals engaged in gleaning and fresh food recovery across the United States to share resources, advocate for their work, and provide a support system. These organizations vary in size, structure, mission, etc. and may engage solely in gleaning or glean as part of a

---

[2]The National Gleaning Project is an initiative of the Center for Agriculture and Food Systems (CAFS) at Vermont Law School, sponsored by the National Agricultural Library, a part of the United States Department of Agriculture Agricultural Research Service. For more information, please visit: http://nationalgleaningproject.org.

larger food recovery program. The NGP aims to create a centrally located resource that elevates the gleaning model of food waste reduction, community development, and social justice for the food insecure. In doing so, the NGP will assist in the growth of the gleaning movement by reducing the information and resource barriers to organizing and implementing gleaning programs across the U.S. The NGP would not be possible without the efforts of gleaning and fresh food recovery organizations that already invest time and resources towards creating a more just and sustainable food system every day.

15.3 Types of Gleaning

After preliminary conversations with gleaning organizations, the NGP created a comprehensive list of gleaning organizations and resources. This list allowed for the NGP to understand the general climate of gleaning and food recovery activities across the country. The list will also help expand and formalize the national gleaning conversation and connect organizations to each other. The information provided in this section stems from the outline of the types of gleaning and fresh food recovery activities happening in both urban and rural communities across the U.S. The types of gleaning and fresh food recovery organizations described within this section can now be easily located on a map housed within the online resources hub of the NGP (National Gleaning Project 2016).[3]

The U.S. Environmental Protection Agency (EPA) recognizes four types of food recovery that illustrate how the reach of gleaning activities has been broadened: field gleaning, wholesale produce salvage, perishable and prepared food rescue, and non-perishable food donations (EPA 1999, 11). These methods of food recovery (with the exception of non-perishable food items and prepared foods, which are not addressed in this article) address a vast array of situations in which fresh food can be re-routed into the emergency food system. The emergency food system is that sector of the overall food system that reroutes food through "food banks, food pantries, soup kitchens or other sources for people seeking food who would otherwise not get enough to eat or not eat well" (Johns Hopkins Center for a Livable Future 2016).

Further, traditional field gleaning can effectively and efficiently gather excess produce. This type of gleaning also creates unique opportunities for volunteers to develop community relationships with growers and agencies that receive food donations. Field gleaning success is evident by the scope and work of many organizations across the U.S. that glean hundreds of thousands of pounds of produce annually. However, field gleaning no longer solely applies to a large agricultural field in a predominately rural area. Urban settings are now popular and

---

[3]See NGP online map of gleaning and fresh food recovery organizations, *available at:* http://forms.vermontlaw.edu/farmgleaning/GleaningOrgs.cfm.

valuable places to glean—from urban community gardens, farms, and public orchards, to private gardens, backyard fruit trees, and orchards. Also, retailers and distributors now provide excess fresh fruits and vegetables to the charitable food system that would otherwise be discarded, a process defined as fresh food recovery in this article (Feeding America 2016).

The transportation of gleaned food is rapidly evolving. Gleaning organizations contribute the most effort to picking up gleaned fresh foods, but farmers also play a vital role in closing the gap between farm and field by dropping off fresh foods to gleaning organizations and food banks. Some organizations concentrate solely on transporting fresh foods from large grocery store chains. Other organizations facilitate and support retail contribution and transportation, but do not engage in these activities themselves. The rise of urban gleaning is also resulting in creative ways to move perishable produce quickly through densely populated areas, such as through the use of bicycles. Volunteers supplied with pick-up and drop-off information use bike baskets or trailers to move small amounts of fresh food quickly, keeping small donations from slipping through the cracks, while maintaining potentially low infrastructure costs (Dansky et al. 2013).

Farmer's markets are also growing in popularity as a venue for fresh food recovery. Depending on the size of the market and the variety of its vendors, farmer's markets can literally be a "one-stop-shop" where fresh, perishable, and prepared foods can be recovered all at once. Most farmer's markets are open air and/or seasonal and therefore are not a consistent source of fresh food. However, the U.S. Department of Agriculture's Economic Research Service (USDA-ERS) conducted a study which found that in select areas of the country, particularly in the northeast and on the west coast, there may be anywhere from 22 to 68 farmer's markets of varying size within 30 miles of a gleaning organization (Price and Harris 2000, 6). As such, farmer's markets present an opportunity not only to recover fresh produce, but also to meet farmers and vendors who may be interested in working with gleaning organizations and open to education on the topic of fresh food recovery.[4] Further, organizations can diversify their gleaning operations to cover backyard fruit tree harvests, wholesale, and farmer's markets to effectively combat food waste and food insecurity in their communities (Food Forward 2016).

Some consider the role of fresh food recovery from grocery stores, wholesalers, and other food vendors as the best option for relieving systemic hunger in the United States. Yet field gleaning remains significant in its unique ability to address stressors on the modern food system that can impact the flow of fresh food donations. For example, downward economic trends force suppliers to restrict food order sizes and individuals to buy only what they will use, thereby minimizing the flow of surplus into the emergency food system. Gleaning can build relationships with

---

[4]For the purposes of this article, "fresh food recovery" refers to perishable foods such as fresh produce and does not include food items such as canned vegetables, boxed foods (i.e. cereals or rice), etc.

farmers and sustainable sources of fresh product that aids organizations in meeting hunger needs during times of declining donations from other sources (Price and Harris 2000, 2). Further, field gleaning, both rural and urban, can strengthen community relationships and offer vulnerable individuals ownership of their community and food source. For example, the Oregon-based Linn Benton Food Share partners with fourteen gleaning groups whose members participate in gleaning efforts. Therefore, those receiving the benefits are actively involved in the solution (Linn Benton Food Share 2016). Communities are best served by assessing the various options of fresh food recovery to meet their area's specific demands of food insecurity, food waste, and community health. Yet, at the state and national level, there are legal and logistical issues that must be addressed if the practice of gleaning is to realize its full potential towards alleviating hunger and food waste.

15.4   Good Samaritan Laws

15.4.1   State Good Samaritan Laws

Many individuals, producers, growers, and organizations working within the emergency food system desire increased donations of fresh produce, but continue to face uncertainties regarding the legal implications of gleaning-related activities. Concerns about liability deter many individuals, corporations, and farmers from participating in gleaning or donating fresh produce from their farms, gardens, markets, and stores. All fifty states and the District of Columbia have enacted various forms of a "Good Samaritan" law that limit the liability of food donors [H.R. Rep. No. 104-661 1996]. These laws were enacted to encourage food donations because all states typically hold food producers, distributors, or retailers strictly liable[5] if their product causes injury to the end user [H.R. Rep. No. 104-661 1996]. Therefore, the food producer, distributor, or retailer is liable regardless of whether an act of negligence occurred.[6] Although the state laws are well-intentioned, the inconsistency among them creates another barrier for food donation, especially for multistate corporations (Van Zuiden 2012, 243).

---

[5]Strict liability "exists when a defendant is in legal jeopardy by virtue of an wrongful act, without any accompanying intent or mental state." Cornell University Law School Legal Information Institute. 2016. Wex Legal Dictionary. https://www.law.cornell.edu/wex/strict_liability. Accessed 13 April 2016.

[6]"Negligence is the omission to do something which a reasonable [person] guided by those considerations which ordinarily regulate the conduct of human affairs, would do or doing something which a prudent and reasonable [person] would not do. It must be determined in all cases by reference to the situation and knowledge of the parties and all the attendant circumstances," Black's Law Dictionary. 2016. http://thelawdictionary.org/negligence/. Accessed 13 April 2016.

15.4.2 Bill Emerson Good Samaritan Act

All state Good Samaritan laws were passed prior to the passage of a federal version of the law. State variations of the law created a need for legal uniformity to raise donor confidence and guarantee a base level of protection (Van Zuiden 2012, 243). Thus, Congress passed the Bill Emerson Good Samaritan Food Donation Act (Bill Emerson Act) in 1996 [42 U.S.C.A. §1791 1996]. The Bill Emerson Act partially preempts[7] state Good Samaritan laws by creating a federal baseline level of protection for food donors that states must meet. States can choose to pass stronger laws that build on the federal minimum and provide greater protection for donors and gleaners than the federal law (Food Recovery Committee 2007, 9).

Specifically, the Bill Emerson Act protects a "good faith" donor (a person, gleaner, or nonprofit organization distributing donated products who acts with good, honest intention or belief).[8] These donors are not liable for an injury or death that results from the condition of the donated product, except in cases of gross negligence or intentional misconduct, as defined by the Bill Emerson Act [42 U.S. C.A. §1791 (b)(7-8) 1996]. The Bill Emerson Act also protects landowners from civil and criminal liability in the event gleaners collecting food for donation to charity are injured on the landowners' property, except in cases of gross negligence or intentional misconduct by the landowner [42 U.S.C.A. §1791(c)(1)(d) 1996].

15.5 Legal Challenges to Gleaning and Fresh Food Recovery

After initial conversations with gleaning organizations, the NGP discovered some of the legal and non-legal challenges that are outlined within this section. The NGP also researched the applicable laws, outlined in Sect. 1.4, and other government materials pertaining to gleaning and food recovery in order to better disseminate information to organizations engaged in these activities. Therefore, the NGP attempts to bridge the practical and legal variations in terminology and practice with regards to the different types of gleaning and fresh food recovery activities. The NGP intends to make recommendations to government agencies and gleaning organizations, including (1) identifying solutions to gleaning challenges, based on

---

[7]Preemption is the right to "set up a prior claim, or the right to gain an advantage before anybody else," Black's Law Dictionary. 2016. http://thelawdictionary.org/preemption/. Accessed 13 April 2016.

[8]The term "good faith" is not defined within the definitions section of the Bill Emerson Act. See 42 U.S.C.A. §1791(b). Different sources in different areas of the law define "good faith" in various ways. Generally, good faith is the "honest intention to abstain from taking any unconscientious advantage of another, even through the forms and technicalities of law, together with an absence of all information or benefit of facts which would render the transactions unconscientious" (Bouvier's Law Dictionary, P. 1359 (3rd Ed., West, 1914).

organization feedback from personal interviews and a general survey to gleaning organizations; (2) building awareness on how to better legitimize and support gleaning organizations, including the development of a national or regional gleaning network where organizations can effectively connect with one another; (3) providing options to the types of donor incentives organizations or the government could offer (e.g., offsetting costs and/or tax incentives); and (4) explaining gaps in the laws to increase protection for all entities involved in gleaning.

15.5.1 Liability

Despite the existence of the Bill Emerson Act and state laws that shield volunteers and gleaning organizations, liability remains a concern for both donors and organizations. Part of this uncertainty stems from the lack of legal precedent, meaning there are no examples of lawsuits that demonstrate how these laws would be interpreted in court (Haley 2013). Liability concerns range from volunteer labor to food safety. For example, some organizations, farmers, and landowners have expressed concerns regarding: (1) the boundaries of volunteer vs. employee labor in gleaning operations, (2) how to apportion liability if a volunteer sustains injuries during a gleaning event, and (3) the legal remedies available if gleaners harm the landowner's property.

In addition, the Bill Emerson Act does not negate local and state health regulations, so violating these laws may create instances of gross negligence or intentional misconduct, which fall outside the protection of the Act itself (Food Recovery Committee 2007, 10). Therefore, gleaning organizations and food donors have lingering concerns about food safety and quality beyond conforming to the Bill Emerson Act. Gleaning organizations need to know what kinds of training and action are needed to comply with food safety standards (such as safe handling and storage of fresh donated food) in order to increase food donations and partnerships with growers and businesses (Food Recovery Committee 2007, 9). For those organizations interested in pursuing light value-added processing of food products to extend shelf life (i.e. canning or freezing), there are additional concerns related to food safety liability and the extension of liability coverage to organizations involved in processing.

Specifically, many gleaning organizations are also concerned with adhering to the new federal Food Safety Modernization Act (FSMA) rules and determining how these rules could impact their current operations [21 U.S.C. § 301 et seq. 2011]. Although the impacts of the new rules are unclear, gleaning organizations have expressed concern over whether or not gleaners must follow the same training, farm hygiene, and produce standards the law requires of food producers. Organizations are also worried about the impact these standards will have on the perception and willingness of farmers to allow gleaners on their land. Several gleaning organizations also expressed the importance of groups implementing efficient tracking and traceability methods into their everyday work in order to stay ahead of the food safety laws. These methods take time, training, and money, creating additional challenges for many resource-strapped gleaning organizations.

15.5.2 Donation Incentives

Because the process of fresh food donation at the farm or small vendor level is onerous, popular policy methods for increasing food crop donation from farms and other small food vendors involve offsetting the costs of donation through incentives. The Bill Emerson Act requires that food and grocery items donated in good faith be "apparently wholesome" and meet "all quality and labeling standards imposed by Federal, State, and local laws and regulations even though the food may not be readily marketable due to appearance, age, freshness, grade, size, surplus, or other conditions" [42 U.S.C.A. §1791(b)(1)(2) 1996]. To comply with this provision, farmers and other food donors must consider the methods by which they will separate products intended for the market from products destined for donation, including how that product will be packaged and whether the product needs to be stored before pick-up. These are costly considerations requiring significant effort on the part of the donor. Until recently, federal tax law incentivizing charitable donations provided an enhanced tax deduction for charitable food donations. However, this deduction was only available to C-corporations and did not account for S-corporations or other small business models typical of many farms. This situation left many would-be donors unable to recoup the costs and time spent complying with the Bill Emerson Act's requirements, as well as any other local or state donation standards. Many farmers who decided to donate had to absorb the cost of harvesting, packing, and transporting those donations.

In 2016, Congress permanently extended the deduction for charitable food donations by making the deduction available to all business models. The expanded deduction allows for a donor to take the lesser of two deduction routes: either twice the basis of the donation [§170 (B) (ii) (2016)] or 25% of the fair market value of the product [§170 (C) (iv) (II)]. Either option is capped at 15% of the donor's net income during that taxable year [§170 (c) (i-ii) (2016)], though donors can carry forward excess deductions for up to five years [§170 (D)(1)(A)(ii)].[9] The expanded federal tax deduction is the first amendment to this section of the IRS code in recent history and is a significant step forward towards encouraging charitable food donations directly from farmers and other vendors.

Before the recent amendment to the federal tax code, some states began offering their own tax incentives for charitable food donations. Colorado House Bill 14-1119, effective January 1, 2015, gives taxpayers a credit of "either twenty-five

---

[9]See also, The National Gleaning Project. April 2016. The Federal Enhanced Tax Deduction for Food Donations: A Guide for Farmers and Gleaning Organizations. Center for Agriculture and Food Systems, Vermont Law School. http://forms.vermontlaw.edu/farmgleaning/EnhancedTaxDeductionGuideForFarmers.pdf. Accessed 25 April 2016.

percent, not to exceed five-thousand dollars, of the wholesale market price or...the most recent sale price of the food contribution" [C.R.C.P. § 39-22-536]. The bill also restricts taxpayers claiming a corporate credit for food donations from claiming the additional 25% credit [C.R.C.P. §39-22-536(4)(a-b)]. A similar Oregon law provides a 10% credit to an "individual or corporation that is a grower of a crop and that makes a qualified donation" at the "computed wholesale price" [ORS §315.156 (1)(a-b)]. There is debate as to whether state-lead initiatives would be more useful than blanket federal tax incentives, as state initiatives may better address local needs by catering to regional commodities and/or climate. Other state-lead options include state-appropriated funds for farmer incentives or ways for gleaning organizations to secure grant funding with which to pay farmers to harvest and pack fresh produce (Powers and Snow 2014, 10). In some cases, such as the Florida Association of Food Banks' program Farmers Feeding Florida (Florida Association of Food Banks 2016), measures of this kind have been successfully implemented. However, such programs are not common. It is unclear if and how the recent and significant changes to the federal tax incentive for charitable food donations will impact individual state incentives.

15.5.3 Infrastructure and Capacity

The lack of donation incentives for most farmers and small food vendors leaves gleaning and hunger relief organizations to absorb the costs of harvesting, sorting, packing, storing, and transporting fresh foods. These costs vary depending on the size of the organization, use of and availability of volunteers, donation size, freshness, the length of time the food needs to be stored, whether the food requires processing, and the distance food must travel. Lack of adequate infrastructure is an epidemic among gleaning and hunger relief organizations. From large food banks to small community pantries, all need access to walk-in coolers, refrigerated trucks, large-scale processing kitchens, and storage and packing facilities at inconsistent and often unpredictable times. One of these organizations may have reliable transport, but no kitchen access, while another has shared kitchen access but limited cold storage. This is especially problematic when considering that gleaning fresh produce can create massive amounts of one kind of food crop that may not be suited to light processing and may be difficult to store, such as leafy greens.

Hunger relief organizations note that the inability to meet these infrastructure needs at critical times "disrupts the flow of food," often leading to food waste (Powers and Snow 2014, 57, Citing Rotary First Harvest, https://www.firstharvest. org). A compounding factor is the reality that "many charitable/emergency food providers and food assistance programs work within certain political boundaries, often defined by zip code, town, county, or state lines" (Powers and Snow 2014, 65). It is often not the case that organizations share their infrastructure resources. Advocates note that "with sufficient administrative support, transportation, receiving, and distribution procedures," facilities could capture massive amounts of surplus fresh food (Powers and Snow 2014, 11). This capture could be achieved through the pooling of local, regional, and state resources with a focus on a properly functioning supply chain (Powers and Snow 2014, 65). Regional gleaning networks and statewide gleaning collectives continue growing, despite the challenges of limited resources. Many organizations achieve success through collaboration, sharing resources to encourage gleaning and distribution efforts on a larger scale. For example, the Vermont Gleaning Collective provides technical assistance, staff training, and operational support to its member groups to support gleaning efforts statewide (Salvation Farms 2016).

Food hubs offer services that could fill these supply chain and infrastructure challenges. Hubs help manage the local supply chain by aggregating locally and regionally produced food, coordinating distribution, and offering permanent cold storage and processing facilities (Azzarello et al. 2012, 7–10). Hubs also have the capacity to incorporate gleaning into their aggregation activities and charitable distributions into their distribution methods (Azzarello et al. 2012, 7). While not all gleaning and hunger relief organizations have access to a food hub facility, they can still use the food hub model. Throughout Washington State, gleaning networks connect through Rotary First Harvest (RFH) to collect, transport, and distribute gleaned and recovered foods to food banks and other agencies, primarily through existing, underused infrastructure. RFH solicits donations of surplus produce all across the state and then contacts trucking companies to donate or significantly discount the extra space available between deliveries to transport millions of pounds of fresh, healthy produce to partner hunger agencies every year, referencing themselves as a "non-profit produce broker" (Rotary First Harvest 2016). While not a food hub in the traditional sense, RFH demonstrates that slight tweaks to the model can grow success and use existing resources to advance the work of hunger relief efforts.[10]

Finally, lack of program funding is a significant barrier to gleaning organizations throughout the country. The presence of dedicated full-time staff able to oversee gleaning efforts and coordinate between organizations, farmers, and volunteers is essential to a successful program. Long-term staff develop and sustain relationships with community members and offer a reliable presence. For many organizations, the Corporation for National and Community Service is a resource for placing AmeriCorps volunteers into various positions (Corporation for National and Community Service 2016). AmeriCorps is a significant and valuable source of affordable workers, but AmeriCorps positions are usually only one or two year placements. The AmeriCorps system thereby causes a high rate of turn over that requires organizations to retrain new personnel on a regular basis.

---

[10]Another example of an innovative food hub model for donated produce is the *Hub and Spoke Program* in Maryland. For more information see Southern Maryland Agricultural Development Commission. 2016. Hub and Spoke Program. http://www.smadc.com/food_farms_nutr/hubspoke. html. Accessed 13 April 2016.

15.6   Conclusion

"Food is a basic human need...its production, distribution, and consumption present issues of deeply important social concern" (Mitchell 2011, 7). On a daily basis in the U.S., individuals and communities address food justice issues: food insecurity, food waste, nutrition education, and providing respect and dignity to those who grow and prepare our food. Gleaning has the potential to play a significant role in relieving hunger and enhancing quality of life for the most vulnerable members of our society (USDA 2010). It can also address the problem of food waste rapidly accumulating in landfills as food insecurity paradoxically grows.[11]

The movement requires legal, financial, and structural supports in order to continue expanding and filling gaps in the food system. The NGP aims to provide some of these supports and solutions in order to grow the movement. Prominent agricultural law professor and writer Neil D. Hamilton notes that, "[t]he laws we write, the goals we pursue, and the choices we make help determine the health of the system we create. Sustainability is not something somebody else does for us—it is something we choose in the decisions we make and the foods we eat" (Hamilton 2011, 145). With the proper support and thoughtful legal action, gleaning can reduce food insecurity and waste, thereby creating a more just and sustainable food system.

References

42 U.S.C.A. §1791. 1996. The Bill Emerson Good Samaritan Food Donation Act. http://www.gpo.gov/fdsys/pkg/USCODE-2010-title42/pdf/USCODE-2010-title42-chap13A-sec1791.pdf. Accessed April 13, 2016.

Azzarello, M., Capece, A., Cassidy, M., Camberlain, L., Faust, B., Franklin, S., ... J. Jensen. 2012. *Community food hubs: Community food security and economic development.* Edward J. Bloustein School of Planning and Public Policy, Rutgers University. http://rwv.rutgers.edu/wp-content/uploads/2013/07/FoodHubFinalReport.pdf. Accessed April 13, 2016.

Corporation for National and Community Service. 2016. http://www.nationalservice.gov. Accessed April 13, 2016.

Dansky, H., H. Katich, B. Higbee, and C. Phillips. 2013. *Create a food rescue program in your community.* Boulder Food Rescue. https://docs.google.com/file/d/0B0HUVA0kzTpUWG NVT2FZVU82RkE/edit. Accessed April 13, 2016.

EPA. 1999. Waste not, want not: Feeding the hungry and reducing solid waste through food recovery. http://nepis.epa.gov. Accessed April 8, 2016.

---

[11]The federal government defines food insecure households as those households that "do not at all times have 'enough food for an active, healthy life.' Garcia, Liza Guerra. 2015. "Free the Land": A Call for Local Governments to Address Climate-Induced Food Insecurity in Environmental Justice Communities. *William Mitchell Law Review* 41: 572, 594, citing USDA. 2015. Overview: Food Security. http://www.ers.usda.gov/topics/food-nutrition-assistance/food-security-in-the-us.aspx. Accessed 13 April 2016.

Farmer, Blake. 2011. Gleaning a harvest for the needy by fighting food waste. N.P.R. http://www.npr.org/2011/01/20/133059889/gleaning-a-harvest-for-the-needy-by-fighting-waste?ft=1&f=1053. Accessed April 13, 2016.

Feeding America. 2016. Securing and distributing meals. http://www.feedingamerica.org/about-us/how-we-work/securing-meals/. Accessed April 13, 2016.

Florida Association of Food Banks. 2016. Farmers feeding Florida. http://www.fafb.org/programs/farmers-feeding-florida. Accessed April 13, 2016.

Food Forward. 2016. https://foodforward.org. Accessed April 13, 2016.

Food Recovery Committee. 2007. Comprehensive guidelines for food recovery programs. http://www.foodprotect.org/media/guide/Food-Recovery-Final2007.pdf. Accessed April 13, 2016.

Gleaning for Hunger Relief. 2002. *Journal of the American Dietetic Association* 102.11. Expanded Academic ASAP.

Haley, James. 2013. The legal guide to the Bill Emerson Food Samaritan Food Donation Act, *Arkansas Law Notes*. August 8. http://media.law.uark.edu/arklawnotes/2013/08/08/the-legal-guide-to-the-bill-emerson-good-samaritan-food-donation-act/. Accessed April 13, 2016.

Hamilton, Neil D. 2011. Moving toward food democracy: Better food, new farmers, and the myth of feeding the world. *Drake Journal of Agricultural Law* 16: 117.

Horton, Emily. 2014. The gleaners: Advocates against hunger and waste are reviving an ancient practice. *Vegetarian Times: Agriculture Collection*, September 1.

H.R. Rep. No. 104-661. 1996. http://www.gpo.gov/fdsys/pkg/CRPT-104hrpt661/html/CRPT-104hrpt661.htm. Accessed April 13, 2016.

Johns Hopkins Center for a Livable Future. Hunger and food security vocabulary definitions. http://www.jhsph.edu/research/centers-and-institutes/teaching-the-food-system/curriculum/_pdf/Hunger_and_Food_Security-Vocabulary.pdf. Accessed April 13, 2016.

Linn Benton Food Share. 2016. http://communityservices.us/nutrition/detail/category/gleaning-programs/. Accessed April 13, 2016.

Mitchell, Jay A. 2011. Getting into the field. *Journal of Food Law and Policy* 7: 69.

National Gleaning Project. 2016. Vermont law school's center for agriculture and food systems. http://nationalgleaningproject.org. Accessed April 13, 2016.

Powers, Jessica and Theresa Snow. 2014. Beyond bread: Health food sourcing in emergency food programs. WhyHunger. http://www.whyhunger.org/uploads/beyondbread/0596-WH%20Book_BEYOND%20BREAD_Single.pdf. Accessed April 13, 2016.

Price, C., and M. Harris. 2000. Increasing food recovery from farmers' markets: A preliminary analysis. USDA ERS. http://www.ers.usda.gov/publications/fanrr-food-assistance-nutrition-research-program/fanrr-4.aspx. Accessed April 13, 2016.

Rotary First Harvest. 2016. http://firstharvest.org/about-rfh/core-work/. Accessed April 13, 2016.

Salvation Farms. 2016. http://salvationfarms.org/programs.html#gleaning. Accessed April 13, 2016.

Society of St. Andrew. 2016. End hunger in America. http://endhunger.org/hunger-in-america/. Accessed April 13, 2016.

USDA. 2010. Let's glean! united we serve toolkit. http://www.usda.gov/documents/usda_gleaning_toolkit.pdf. Accessed April 13, 2016.

Van Zuiden, H. Stacey. 2012. The good food fight for good samaritans: The history of alleviating liability and equalizing tax incentives for food donors. *Drake Journal of Agricultural Law* 17: 237.

Vermont Foodbank. 2016. Gleaning program. http://www.vtfoodbank.org/OurPrograms/FreshFoodInitiatives/GleaningProgram.aspx. Accessed April 8, 2016.

Chapter 16
Food, Crime, Justice and Security: (Food) Security for Whom?

Avi Brisman and Nigel South

Abstract This paper explores food and food crime in the context(s) of the increasingly powerful discourse(s) of security and securitization. Because food has long been tied to conflict, we recognize it as a material need that frequently contributes to or drives conflict. In the post-9/11 world, however, food is taking on new discursive and material roles and dynamics in global conflicts. Critical security scholars have noted the ways that political and cultural calls for increased security lead to problematic processes of securitization that often serve to enlarge the power of elite state and corporate actors. As climate change continues to impact agriculture and access to food, calls for increased food security from within "northern" nation-states have become louder and more common, and are likely to be answered by the exercise of both military power and intensified "commercial colonialism" in "southern" and non-western contexts, as with examples of "land grabbing" in various parts of the world. Importantly, it should not be forgotten that although the context of accelerating climate change is new, colonial food expropriation and land appropriation are not. Demands for food security require that we ask important questions: What exactly is being secured? *From* whom/what is it being secured? *For* whom is it being secured? This paper draws on green criminology and security studies to pose these questions and provide possible insight to their answers. It begins with a discussion of different conceptions of "food security" and "food insecurity" and then considers relationships between food, crime, justice and security.

Keywords Security · Inequity · Colonialism · Famine

A. Brisman (✉)
Eastern Kentucky University, Richmond, USA
e-mail: avi.brisman@eku.edu

A. Brisman · N. South
Queensland University of Technology, Brisbane, Australia
e-mail: n.south@essex.ac.uk

A. Brisman · N. South
University of Essex, Colchester, UK

© Springer International Publishing AG 2017   185
I. Werkheiser and Z. Piso (eds.), *Food Justice in US and Global Contexts*,
The International Library of Environmental, Agricultural and Food Ethics 24,
DOI 10.1007/978-3-319-57174-4_16

16.1 Relating to Food and Security: Introduction and Definitions

Food and water are central to the viability of social organisation. Their availability has been the reason for the rise and fall of civilisations; the human relationship with them has been interpreted and explored in innumerable ways, for example through art, music, psychoanalysis, and, of course, different styles of eating and drinking. Despite their everyday 'ordinariness', they are also—alongside air—the most precious things on the planet. They are, in short, central to survival. Historically, this has made them subject to property claims, ownership conflicts and absolutely essential for individual and state security. Focusing on food, Coveney (2014: 2, 13, 30) asserts: 'food plays a central role in virtually all aspects of our lives. From providing us with sustenance, to lubricating social engagements, to enriching our cultural practices, to creating big business, to destroying local economies and endangering lifestyles, food is it… [F]ood makes us what we *are*, in the sense that we *are* cultural, social and sentient beings, as well as physical shapes'. This chapter explores several ways of thinking about and defining our relationship to food, primarily through different understandings of the idea of 'security'.

The nature of access to food—its abundance, distribution, scarcity, value and so on—and hence our understandings or interpretations of 'food security'—are shaped by causal forces and contingent factors, both local and global. The chapter outlines some of these, particularly as they affect large emergent economies, but also shows how food insecurity is (of course) also a problem in the economies of 'the west' and how other security pre-occupations of these countries, such as the quest for energy security, have contributed to food insecurity. 'Security' can provide a message of reassurance—all are provided for and well-being is assured—but it can also be a signal of distinction and discrimination, drawing a line that divides those who 'have' from those who 'have not'. In the section on food, crime and security, we describe some ways in which this line is contested by some but also supports profiteering by others. At the heart of tensions around food security are questions of justice and power. Historically, food has been one of the ultimate tools used by those in charge of the administration of power: having control over the supply of food means having control over the power of life and death (see generally Brisman 2008). We provide a reminder of this historical lesson about the relationship between food and the pacification of peoples in different contexts and at different times. The chapter concludes with a series of open questions.

16.2 Food Security

Without food, the result is malnutrition and starvation. By contrast, the public health history of nations with a historically improving supply and diet of food shows improved skeletal, muscle and mental development. Food is necessary for basic survival but more than this—it is also a supporter of and catalyst for human dignity and human development (Apodoca 2014: 349). Despite—or perhaps *because of*—food's vital and multifaceted contribution to individual and collective life, the availability and absence of food, as well as its nutritional adequacy and safety, present a range of micro- and macro-level challenges as reflected in the concept of 'food security'.

The term 'food security' signifies different things to different people and entities and takes on different meanings in various contexts. Slater et al. (2012: 152) provide a useful history of the phases of approaches to food security and it is important to understand how and why the definition of 'food security' has changed over recent decades. Thus, in the 1960s and 1970s, 'Malthusian fears that the rapid growth of the world population would outstrip food production and lead to widespread famine meant that food production and *availability*, rather than access, were the key concerns' (emphasis in original). By contrast, during the 1980s, it was increasingly recognised that economical, social and cultural factors were important in terms of influencing '*access* to food'. In particular, this followed from the work of Amartya Sen (1981) on *Poverty and Famines*, which challenged the conventional belief that famines stemmed from food shortages. While Sen acknowledged that droughts and harvest failures could be contributing factors, he argued that *absolute scarcity* was relatively rare: "If one person in eight starves regularly in the world, this is... [really] the result of his inability to establish *entitlement* to enough food; the question of the *physical availability* of... food is not directly involved" (emphasis added). Thus, central to Sen's thesis was the idea of 'entitlement' to food: the basis on which this entitlement might be met and the various ways in which it might be undermined. Entitlement is related to production, trade, labour and gifts. In other words, it is not necessarily the sum of food available for national consumption that leads to famine, however. Distortions in the social system of distribution can contribute alongside limitations placed on the access afforded to individuals and groups regarding availability of food. Correspondingly, Devereux (2012: 177) maintains, 'Even when a drought causes crop failure, only some groups suffer "entitlement failure" and face starvation—wealthy families and urban residents are rarely affected'. We expand below upon this history of the concept of food security as well as the paradox that those close to the land—the peasants and farmers—can also be the populations most vulnerable to famine. This may occur because, as Sen suggested, their 'entitlement' to food is overly dependent on rain-fed agriculture and hence drought is particularly devastating. In some cases, however, this hardship is exacerbated, as we describe below, by the ways in which food entitlement and distribution are organised. In such circumstances forms of 'food grabbing'—diversion and export—may leave workers of the land facing starvation while consumers in urban centres, within the country and overseas, are the beneficiaries of the harvest.

The Food and Agriculture Organization of the United Nations (FAO) (1996, 2002) defines 'food security' as a situation where 'all people, at all times, have physical, social and economic access to sufficient, safe and nutritious food to meet their dietary needs and food preferences for an active and healthy life' (van Uffelen

2013: 58; see also Hall 2013: 37, 2015: 46–49; Wutich and Brewis 2014: 445). For Coveney (2014: 73, 91), 'food secure' means 'that the population has available a food supply that is safe, affordable and nutritious' (Coveney 2014: 73), while 'food security' indicates 'that individuals, communities and populations have control over what foods they have access to and under what terms and conditions'. Stack et al. (2013: 161) suggest that in order to achieve 'global food security', food must be '(1) available, (2) affordable, (3) safe, (4) nutritious, and (5) in a culturally appropriate form of choice within each society'. The United States Department of Agriculture (USDA) has adapted a definition proposed by the Life Sciences Research Office, wherein 'food security' refers to '"access at all times to enough food for an actively healthy life"'. It includes, at a minimum, '"the ready availability of nutritionally adequate and safe foods"' and '"an assured ability to acquire acceptable foods in socially acceptable ways (e.g., without resorting to emergency food supplies, scavenging, stealing, or other coping strategies)"' [Wilde 2013: 174 (quoting National Research Council 2006)].

Slater et al. (2012: 152) add a third dimension to understanding of food security: the importance of the '*utilization* of food for nourishment' with health playing 'a major role'. There is therefore an important distinction to be drawn between 'food security' and 'nutrition security' (World Bank 2006: 66). This turns on the idea that food security is fundamentally concerned with physical and economic *access* to food, made available in appropriate quantities and ways (culturally and socially). More broadly, nutrition security correlates with positive measures of individual health, environmental health and care.

In some respects therefore, 'nutrition security' is a more capacious concept than 'food security'. The former exists when food security is combined with a sanitary environment, adequate health services, and proper care and feeding practices to ensure a healthy life for all household members. Thus, a family (or country) may be food secure, yet have many individuals who are nutritionally insecure. 'Food security' is therefore often a necessary but not sufficient condition for 'nutrition security'.

van Uffelen (2013: 58) argues that '[a] precise operational definition of food security does not exist'. Rather, he contends, 'food security has four dimensions relevant to food security programming: production of food, access to food (either via informal or formal networks such as markets), the utilisation of food, and the stability of the food system' (van Uffelen 2013: 58). As van Uffelen explains, analyses of food security and explanations of vulnerability to famine often fall into two camps or 'clusters', a 'physical ecology cluster', which 'focus[es] on population growth, declining soil fertility, and drought', and a 'political economy cluster', which 'blame[s] government policies, weak markets, and institutional failure' [2013: 62 (citing Devereux 2000)]. In his study of Ethiopia's key famine periods (1972–73, 1984–85) and the food crises of 1999–2000, 2003–03, 2008, and 2011, van Uffelen (2013: 63) finds merit for both the 'physical ecology' and 'political economy' interpretative lenses, but cautions against ascribing sufficient explanatory power to one to the exclusion of the other. The point to take away from van Uffelen's (2013) work and others (e.g., Stack et al. 2013; Walters 2006, 2011) is that food insecurity is quite often *not* caused by food shortages, but by a combination of factors that include inequity, poverty, the concentration of food production, and ecological degradation. In other words, starvation and hunger are often less reflective of food production problems than issues of distribution and economy (Sen 1981, discussed above; see also Lee 2009: 139).

Furthermore, ecological degradation and the political economy of growth (both economic and in population numbers) are frequently interlocked. The ongoing, ecodical impacts of global 'business as usual', serving corporate and state interests, creating and responding to the power of everyday consumption, all occur despite well-known evidence of consequences (Brisman and South 2014). This pattern reflects our simultaneous dependence on growth economics and domination of nature to meet its demands. In the face of this, Klein (2014: 2–3) suggests that humans are engaging in cognitive dissonance on a planetary scale but the signs of planetary distress are increasingly evident and one of the first arenas in which strains are showing is food supply. As Juniper (2013: 37) points out, 'It is estimated that about 1 billion people today live in regions experiencing land degradation and declining [food] productivity. Much of the worst damage is in China, Africa south of the equator and parts of South East Asia. In other words, in places undergoing rapid population increase' alongside growing economies. One response from nations with a food production challenge, rising consumer demand and the wealth to invest elsewhere is to farm on land in other countries. Acquisition of agricultural land and 'natural capital' (Kareiva et al. 2011) in developing countries by larger emerging economies, such as China, Brazil and India and countries such as the Gulf States with large revenues to invest, has become a significant trend since around 2008 (Geary 2012; Heinimann and Messerli 2014). Such large-scale investments have become, broadly termed, 'land grabbing'. This kind of 'colonisation' usually involves an agreement between national governments as owners of land and major transnational corporations and investors. The traditional small-scale farmers that will often have had use-rights to farm these lands find these 'rights' are withdrawn and their means of subsistence have been removed. The negative label of 'land-grab' is applied when such transactions are in violation of human rights, deny traditional land users a role in negotiation, proceed without their prior and informed consent, and dilute, circumvent or ignore social and environmental impact assessments (see, e.g., Geary 2012; Heinimann and Messerli 2014; see also Bennett et al. 2015; Dauvergne and Neville 2010; Green et al. 2015). The anti-poverty and famine charity Oxfam (Geary 2012: 15) has argued that 'if the rising tide of interest in farmland investment is to have any positive impact on the food security of local communities and the livelihoods of small-scale producers, it is crucial that land governance and investments in land do not undermine food security by facilitating the transfer of land rights away from people living in poverty'. But this is exactly what is happening.

It should come as little surprise that 'food insecurity' has also been conceptualized in a number of ways. Carney (2014: 2) defines 'food insecurity' rather succinctly as 'a situation in which people lack enough food to meet basic needs'. For Berkowitz et al. (2014: 304), 'food security' refers to '"limited or uncertain availability of nutritionally adequate and safe foods, or limited or uncertain ability to acquire acceptable foods in socially acceptable ways"' (quoting Bickel et al. 2000), and they find that food insecurity is strongly associated with cost-related medication under-use. As with 'food security', the USDA adapted a definition of 'food insecurity' proposed by the Life Sciences Research Office: 'food insecurity' exists '"whenever the availability of nutritionally adequate and safe foods or the ability to acquire acceptable foods in socially acceptable ways is limited or uncertain"' [Wilde 2013: 174 (quoting National Research Council 2006)].

'Food insecurity' is not confined to developing countries. According to the U.S. Department of Agriculture, 17.5 million families—or 1 in 7—were 'food insecure' in 2013. As Fessler (2014) reports, '[this] means that at some point during the year, the household had trouble feeding all of its members. In 2012, the number was 17.6 million. The number of households experiencing what the government calls "very low food security"—which means people actually miss meals or cut back their intake because they don't have enough money for food—was also essentially unchanged last year at 6.8 million households' (Apparently, there has been a drop in unemployment, but higher food prices and inflation have offset the benefits of a brighter job market). Safo (2014) also reports that '[t]he number of American households suffering from food insecurity is down from its peak in 2011... The decline was a modest 2.7%—down to 17.5 million households where access to enough food for healthy and active living (how the USDA defines food security) was inconsistent or not dependable. The report also said the number of households with severe food insecurity, including one member of the household who is going hungry, remained unchanged at almost seven million'. Safo (2014) explains that droughts are to blame for higher food prices in certain categories, as well as inflation in certain categories, like proteins and dairy.

While some sections of advanced economies face the problem of food insecurity, those same countries also pursue policies that have been criticised for contributing to food insecurity elsewhere. Ironically, in some cases, these policies and scientific innovations are argued to be environmentally necessary. Hence, as Hulme (2008: 12) notes, '[i]n 2008 food insecurity was felt across the world. Aside from adverse weather conditions globally and disease affecting agricultural productivity, a shift to biofuels was partly to blame. Criticism was levelled at the policies of the European Union and the United States for inadvertently encouraging forceful land grabs and deforestation'. The development of biofuels has encouraged countries seeking to be less reliant on oil to reduce land in cultivation for food crops (Friedman 2008: 41; see also Dauvergne and Neville 2010; Gillis 2014; Sayer et al. 2012). In fact, the relationship between food production and energy production is easily overlooked but crucial because when fuel prices rise this increases the costs of farming (including the cost of petrochemical fertilizers used by many farmers) and hence increases food prices.

16.3 Food, Crime and Security

As suggested above, '[w]e need food to develop human cultural and social patterns. We need food to establish and maintain our identities. But the most important aspect in our lives is the need for food to survive' (Coveney 2014: 88). It follows, then, that the absence of food or threats to food security can lead to or provoke a range of legal and illegal responses. For example, research has demonstrated negative consequences of food insecurity for children's health and developmental outcomes, including cognitive development and school achievement, socio-emotional development and overall health (Wight et al. 2014: 1; see generally Maupin and Brewis 2014)—all of which have been found in varying ways and in differing degrees to increase the likelihood of crime (see Cullen and Agnew 2011). At the household-level, food insecurity may increase the risk of domestic violence (Wight et al. 2014: 13n.2). But apart from the contribution of poor nutrition to individual development and life chances—or food availability to domestic household economies and related inequalities and risks—food and food insecurity also influence or underpin larger scale symptoms of social malaise.

For example, food insecurity, when it is manifested in rising food prices, may find expression in food riots (Bohstedt 2010; Bradsher 2008; Coveney 2014: 88–89; Hall 2013: 38, 40, 146, 2015: 48; Lee 2009: 50; MacFarquhar 2010; Scheffran and Cannaday 2013: 268; Trauzzi 2014; White 2011: 44; Wilde 2013: 64). But, in turn, price rises can have complex causes. In 2010, food protests in Mozambique were a reflection of interactions between changing weather patterns in different agricultural regions and speculation in the global commodity markets. The riots and the robbery of grocery stores in the towns of Mozambique followed a period of price increases for bread as well as for water and electricity supply. However the rising bread prices had followed a local inability to meet demand—partly due to inadequate levels of domestic crop production and partly because purchase from abroad became more difficult when wildfires destroyed parts of the Russian grain belt during the hottest summer heat-wave for a century, leading to export bans from that country (South 2012). This situation encouraged international commodity speculators to engage in 'gambling on hunger in financial markets' (Patel 2010: 31). This scenario combines street crimes, predatory financial deals, and climate change (which many see as the result of 'crimes against the planet'). As Hall (2013: 40; see also 2015: 47–48) elaborates, 'those who protest or even riot as a consequence of a lack of food, or high food prices, are arguably only doing so as a result of the harms... visited upon them'. As with all scarce commodities, limited availability can both increase prices and also encourage illegal markets. In turn these may involve violence and exploitation and connect to other facets of informal economies and illegal enterprises.

Clearly, given such a situation, there is (also) a concern that the poorest people will turn to illegal food markets. As well as being criminal in themselves, such illegal markets will undoubtedly be run by those who will be willing to use threats or actual violence to ensure that they get the prices they want for the goods they sell, and who may also be involved in other allied trades (such as the supply of weapons and drugs and the control of prostitution) which can only prompt still further victimizations.

While crime and violence (in the form of riots and/or illegal markets) may stem from food insecurity, the direction of causation may be reversed. Wight et al. (2014: 13n.2) have noted that intimate partner violence may increase the risk of experiencing food insecurity. At a more macro level, Kushkush (2014) reports that ethnic fighting between South Sudan's two largest groups, the Dinka and the Nuer, has displaced thousands of people, disrupting the planting season, causing livestock to be lost or abandoned, and preventing fishermen from fishing the rivers. While 'humanitarian corridors' have been created to try to bring food to starving people, the rainy season renders South Sudan's mostly dirt roads impassable, cutting off large populations. Kushkush (2014) notes that United Nations' food warehouses have been looted, further complicating relief efforts.

Looking ahead, Wutich and Brewis (2014: 444) predict that 'insufficient food and water are... two of the greatest natural resource—and social justice—challenges that many [regions] will encounter in the current century' (citations omitted). A related challenge that will impact food, water and justice is climate change. As Revkin (2007) reports, 'Scientists say it has become increasingly clear that worldwide precipitation is shifting away from the equator and toward the poles. That will nourish crops in warming regions like Canada and Siberia while parching countries—like Malawi in sub-Saharan Africa—which are already prone to drought'. Climate change(s) will thus impact food availability and 'food security' in some fairly obvious and some less obvious ways (Agnew 2012a, b; Butts and Bankus 2013; Hall 2013; Kramer and Michalowski 2012; Mares 2010; Maas et al. 2013; McNall 2011; Scheffran and Cannaday 2013; Stack et al. 2013; White 2011, 2012). For example, access to food resources can be complex where traditional diet, cultural symbolism and 'a way of life' are bound together but opposed or undermined by interest groups and legal interventions (e.g., Duffy 2010; Hauck 2007; Walters 2005, 2011; see also Brown 2014; Cave 2009; Ervine 2011: 67 [citing McCarthy and Prudham 2004: 277]; Onishi 2011; Weeks 2012). This can lead to cultural conflict and the extinction of a way of life with tragic consequences (Samson 2003; Brook 1998). The hunting of seals and polar bears and other nonhuman animals to provide food and other resources illuminates certain tensions (e.g., Kaufman 2011; Mooallem 2013). For example, O'Keeffe (2010) writes of the struggles and accommodations that Arctic Inuits face as their environment has changed:

> To understand the significance of food security for the Inuit, we must recognise that food security isn't simply reliable access to nutritious food. It is linked to climate change, wildlife management, pollution and economic vulnerability—and to cultural security. ... In May 2008, the US Fish and Wildlife Service announced its decision to list the polar bear as a threatened species. ... This decision was the outcome of a process which had seen environmental groups square off against the Inuit. ... The Americans' move has been celebrated by environmental groups as a positive step in their campaign to pressure the

nation's government into changing its position on climate change. For the Inuit, this represents a further erosion of their capacity to manage the resources that have sustained them through the centuries.

Polar bears and seals are endangered and must be protected. But human culture, traditions and ways of life may also become endangered, and thus merit consideration. Thus, efforts to avoid ecocidal threats to wildlife can generate expressions of prohibition regarding traditional ways of life and supply of food that can be seen as a variety of "cultural genocide."

Brown (2012: 57, 71) emphasises another threat to food security—the non-substitutable, unique nature of water and the reliance of food production on a resource facing increasing demand while supply is increasingly depleted. Brown is particularly concerned with the ways in which water and food insecurity might impact fragile states. As he describes, in Yemen, where population growth is rising but water availability and the grain harvest have been falling, 'nearly 60% of its children' are 'physically stunted and chronically undernourished'. Brown's (2012: 62) prognosis is that 'The likely result of the depletion of Yemen's aquifers, which will lead to further shrinking of its harvest and spreading hunger, is social collapse... For the international community, the risk is that Yemen's internal conflicts could spill over its lengthy, unguarded border with Saudi Arabia'.

The conditions in Yemen and elsewhere appear dire and the geo-political implications of food and water deficits should not be overlooked in an era that has tended to focus on oil supply and security (Wenar 2016). The slow but steady expansion of the Sahara through Mali, which is killing crops and leaving farmers starving (see Lee 2009: 105), may have been a contributing force in the jihadist uprising in that African country in 2012. Since then, Al Qaeda in the Islamic Maghreb has seized control of the northern part of Mali and remains in conflict with the Malian government (Davenport 2014). Western governments may fear a similar situation unfolding in the Yemen and Saudi Arabia. Indeed, as Stack et al. (2013: 169) warn, '[f]or any nation experiencing political and/or social instability, food insecurity can be the proverbial last straw'. That said, it is best to be cautious rather than reflexively interpret the threat of food and water insecurity in developing countries as a *national* security concern for the United States and other Western countries.

16.4 Security, Control and Pacification

Arguably it is advisable to exercise restraint in the application of the words 'security' and 'insecurity' to situations involving food. This has less to do with differences in the ways that 'food security' and 'food insecurity' have been defined and employed to describe both individual/household/micro-level and macro-level phenomena than to the enlargement of the concept of 'security'. As Warner (2013: 78–79) observes, the fall of the Berlin Wall was accompanied by 'a widening of the security domain beyond its traditional concern with territory. Fossil fuels, oil, drugs, economic competition, the environment and terrorism could all be elevated to national security concerns and, in so doing, legitimize special measures, such as the deployment of the military and the suspension of civil rights'.

While remaining sensitive to global needs for food and water, and to the ways in which various phenomena (such as global climate change) may contribute to conditions of scarcity and suffering, we would argue for close inspection of official claims about food insecurity and efforts undertaken by states in the name of ensuring food security to make sure that ensuing measures are not veiled attempts at increasing state power—or at 'mak[ing] bourgeois all that is inherently communal' (Neocleous and Rigakos 2011: 20). This process has been referred to by critical or anti-security scholars as 'pacification'—a set of practices intended to establish a certain social order—'to produce undisruptive and unthreatening forms of collective action through a combination of coercion and consent' (Kienscherf 2014: 3; see also Neocleous et al. 2013).

We will return to the concept of 'pacification' at the end of this section. Before doing so, it is helpful to offer an example of how the distribution of food (or the lack thereof) can serve as an exercise of state power and subordination of populations. This was demonstrated extensively in earlier colonialist appropriations of land and expropriation of food. As Mutter and Barnard (2010: 276) remark, 'Historically, it is very possible that drought-induced food shortages combined with cruel, bias-based or malfeasant government actions have given rise to famines that have caused the greatest mortality of all disasters'. As Davis (2001) shows, however, in his detailed history of famines and droughts in late-nineteenth-century Brazil, China and India, there is more to 'drought-induced food shortages' than the forces or failures of nature. Davis argues that 'the division of humanity into haves and have-nots—was shaped by fatal interactions between world climate and world economy at the end of the nineteenth century' and that it is possible to identify 'Three waves of drought, famine and disease' that 'devastated agriculture throughout the tropics and northern China when the monsoons failed'. Davis estimates that the 'total human toll could not have been less than 30 million victims. Fifty million dead might not be unrealistic'. As has occurred in more recent years, the effects of the El Niño-Southern Oscillation on air mass and Pacific ocean temperature shaped 'these catastrophic climate disasters and crop failures', but as Davis remarks, 'nature alone is rarely so deadly'.

Millions of cultivators in India and China had been recently incorporated into webs of world trade as subsistence adversity, caused by various state and imperial policies, had encouraged them to turn to cash-crop cultivation. As a result, peasants and farmers became dramatically more vulnerable after 1850 to natural disasters such as extreme climate events and were at the same time whiplashed by long-distance economic perturbations whose origins were as mysterious as those of the weather (Davis 2002).

With the absence of the expected Monsoon in India in late 1876, a drought followed that resulted in a famine that led to inequalities in distribution of available food and water. Subsequently, the mechanisms of management of the starving and impoverished poor that were introduced by the colonial power involved what would today be regarded as indefensible abuses of human rights (Davis 2001: 38–40). As Davis records, in some parts of India at this time, rice and wheat production had been *above average* for the previous three years but with '[w]idespread unemployment and the high price of grain[,]... the spectre of hunger' began to intrude even into districts where rainfall had been adequate. So why did local populations starve? Where was the rice and grain? The answer, of course, is that 'much of the surplus had been exported to England. Londoners were in effect eating India's bread'. As we have noted earlier, this example illustrates how food has been a tool used at the discretion of those in charge of the administration of power. Their control of the supply of food has been one way of exercising power over life and death and proven a chillingly effective means of pursuing programmes of pacification of indigenous peoples (see Kienscherf 2014: 5; Neocleous 2011a: 198–201, 2013).

16.5 Conclusion

Neither food scarcity or famine need to happen. With regard to the former, White (2010: 10) observes that 'perceived national interests dictate that [food] scarcity is in part generated by efforts to control it for some population groups over and above use by others'. With regard to the latter, 'Modern famine scholarship regards famine mortality as entirely preventable by governments. If so, the question is why in the 20th century alone between 70 and 80 million people may have died in famines (Devereux 2000, p. 7)' (Plumper and Nuemayer 2009). We need, then, 'to understand security not as some kind of universal of transcendental value but rather as a mode of governing or a political technology of liberal order-building' (Neocleous 2011b: 26).

In sum, 'security' is, as Light (2014: 1792, 1797) argues, a *multifaceted* rather than *monolithic* concept. While it may mean many things in many different contexts, it can be—and often is—a means of pacification. Accordingly, when we speak of 'food security' [or *any* 'security', for that matter, we must ask (following Eman et al. 2013: 17 (citing Baldwin 1997) and Wall 2013: 41 (citing Rigakos 2011)]:

Security for whom?
Security for which values?
Security against which threats?
How much security?
Security at what costs?

References

Agnew, Robert. 2012a. Dire forecast: A theoretical model of the impact of climate change on crime. *Theoretical Criminology* 16 (1): 21–42.

Agnew, Robert. 2012b. It's the end of the world as we know it: The advance of climate change from a criminological perspective. In *Climate Change from a Criminological Perspective*, pp. 13–25. Springer New York.

Apodoca, Clair. 2014. The right to food. In *Handbook of human rights*, ed. Thomas Cushman, 349–358. London and New York: Routledge.

Baldwin, David. 1997. The concept of security. *Review of International Studies* 23 (1): 5–26.

Bennett, Nathan James, Hugh Govan, and Terre Satterfield. 2015. Ocean grabbing. *Marine Policy* 57: 61–68.

Berkowitz, Seth A., Hilary K. Seligman, and Niteesh K. Choudhry. 2014. Treat or eat: Food insecurity, cost-related medication underuse, and unmet needs. *The American Journal of Medicine 127*(4): 303–310.

Bickel, G., M. Nord, and Price, C., et al. 2000. *Guide to measuring household food security, revised 2000*. Alexandria, VA: U.S. Department of Agriculture Food and Nutrition Service.

Bohstedt, John. 2010. *The politics of provisions: Food riots, moral economy, and market transitions in England c. 1550–1850*. Burlington, VT: Ashgate.

Bradsher, Keith. 2008. From six-year drought in australia, a global crisis over rice. *The New York Times*. April 17: A1, A8.

Brisman, Avi. 2008. Fair fare: Food as contested terrain in U.S. prisons and jails. *Georgetown Journal on Poverty Law & Policy 15*(1): 49–93.

Brisman, Avi, and Nigel South. 2014. *Green cultural criminology: Constructions of environmental harm, consumerism, and resistance to ecocide*. London and New York: Routledge.

Brook, Daniel. 1998. Environmental genocide: Native Americans and toxic waste. *American Journal of Economics and Sociology* 57 (1): 105–13.

Brown, Lester R. 2012. *Full planet, empty plates: The new geopolitics of food scarcity*. New York and London: W.W. Norton.

Brown, Chip. 2014. Kayapo courage. *National Geographic* 225 (1): 30–55.

Butts, Kent Hughes, and Brent C. Bankus. 2013. Environmental change, insurgency and terrorism in Africa. In *Global environmental change: New drivers for resistance, crime and terrorism?*, eds. Achim Maas, Balázs Bodó, Clementine Burnley, Irina Comardicea and Roger Roffey, pp. 141–60. Baden-Baden: Nomos.

Carney, Megan A. 2014. The biopolitics of 'food insecurity': Towards a critical political ecology of the body in studies of women's transnational migration. *Journal of Political Ecology 21*: 1–18. Accessed at: http://jpe.library.arizona.edu/volume_21/Carney.pdf.

Cave, Damien. 2009. New license law in Florida divides shore anglers. *The New York Times*, 7 August, A9.

Coveney, John. 2014. *Food*. London and New York: Routledge.

Cullen, Francis T., and Robert Agnew. 2011. *Criminological theory: Past to present*. 4/e. New York and Oxford: Oxford University Press.

Dauvergne, Peter, and Kate J. Neville. 2010. Forests, food, and fuel in the tropics: the uneven social and ecological consequences of the emerging political economy of biofuels. *The Journal of Peasant Studies* 37 (4): 631–660.

Davenport, Coral. 2014. Climate change deemed growing security threat by military researchers. *The New York Times*. May 14: A18.

Davis, Mike. 2001. *Late Victorian holocausts: El Niño famines and the making of the third world*. London: Verso.

Davis, Mike. 2002. The origins of the third world: Markets, states and climate. *Corner House Briefing 27*. December 30. Available at http://www.thecornerhouse.org.uk/resource/origins-third-world#box-05-02-00-00.

Devereux, S. 2000. *Famine in the twentieth century*. Working Paper 105, Brighton: Institute of Development Studies.

Devereux, S. 2012. Famine. In *The companion to development studies*, ed. V. Desai, and R. Potter. London: Routledge.

Duffy, Rosaleen. 2010. *Nature crime: How we're getting conservation wrong*. New Haven, CT, and London: Yale University Press.

Eman, Katja, Gorazd Meško, and Charles B. Fields. 2013. Resistance to climate change policies: The conflict potential of non-fossil energy paths and climate engineering. In *Global environmental change: New drivers for resistance, crime and terrorism?*, ed. Achim Maas, Balázs Bodó, Clementine Burnley, Irina Comardicea, and Roger Roffey, 15–35. Baden-Baden: Nomos.

Ervine, Kate. 2011. Conservation and conflict: The intensification of property rights disputes under market-based conservation in Chiapas, México. *Journal of Political Ecology* 18: 66–80. Accessed at: http://jpe.library.arizona.edu/volume_18/Ervine.pdf.

FAO. 1996. *Declaration on world food security*. Rome: World Food Summit, FAO.

FAO. 2002. *The state of food insecurity in the world 2001*. Rome: FAO.

Fessler, Pam. 2014. Millions struggle to get enough to eat despite jobs returning. *National Public Ratio*. September 3. Accessed at: http://www.npr.org/blogs/thesalt/2014/09/03/345537318/ millions-struggle-to-get-enough-to-eat-despite-jobs-returning.

Friedman, Thomas. 2008. *Hot, flat and crowded*. London: Allen Lane.

Geary, Kate. 2012. *Our land, our lives: Time out on the global land rush*. Oxfam, Briefing Note, October. Available at: http://policy-practice.oxfam.org.uk/publications/our-land-our-lives-time-out-on-the-global-land-rush-246731.

Gillis, Justin. 2014. Restored forests breathe life into efforts against climate change. *The New York Times*. December 24: A1, A8–A9.

Green, Penny, Kristian Lasslett and Angela Sherwood. 2015. Enclosing the commons: Predatory capital and forced evictions in Papua New Guinea and Burma. In *The Routledge handbook on crime and international migration*, eds. Sharon Pickering and Julie Ham, 329–350. London and New York.

Hall, Matthew. 2013. Victims of environmental harms and their role in national and international justice. In *Exploring issues in green criminology*, eds. Reece Walters, Diane Solomon Westerhuis and Tanya Wyatt, 218–241. Basingstoke, Hampshire, UK: Palgrave Macmillan.

Hall, Matthew. 2015. *Exploring green crime: Introducing the legal, social and criminological contexts of environmental harm*. Basingstoke: Palgrave Macmillan.

Heinimann, Andreas, and Peter Messerli. 2014. Coping with a land-grab world: Lessons from laos. *Global Change* 80 (April):13–16. http://www.igbp.net/news/features/features/ copingwithalandgrabworldlessonsfromlaos.5.19895cff13e9f675e252ba.html.

Hauck, Maria. 2007. Non-compliance in small-scale fisheries: A threat to security?. In *Issues in green criminology: Confronting harms against environments, humanity and other animals*, ed. Piers Beirne and Nigel South, pp. 270–89. Cullompton: Willan

Hulme, Karen. 2008. Environmental security: Implications for international law. In *Yearbook of international environmental law*, ed. O.K. Fauchald, D. Hunter, and W. Xi. Oxford: Oxford University Press.

Mutter, John, and Key Mesa Barnard. 2010. Climate change, evolution of disasters and inequality. In *Human rights and climate change*, ed. Stephen Humphreys. Cambridge: Cambridge University Press.

Juniper, Tony. 2013. *What has nature ever done for us? How money really does grow on trees*. London: Profile Books.

Kareiva, Peter, Michelle Marvier, and Robert Lalasz. 2011. Conservation in the anthropocene: Beyond solitude and fragility. *Breakthough Journal 2 (Fall)*. Accessed at: http:// thebreakthrough.org/index.php/journal/past-issues/issue-2/conservation-in-the-anthropocene/.

Kaufman, Leslie. 2011. After years of conflict, a new dynamic in wolf country. *The New York Times*, 5 November, A9, A12.

Kienscherf, Markus. 2014. Beyond militarization and repression: Liberal social control as pacification. *Critical Sociology* 1–16. doi:10.1177/0896920514565485.

Klein, Naomi. 2014. *This changes everything: Capitalism vs. the climate*. New York: Simon & Schuster.

Kramer, Ronald C., and Raymond J. Michalowski. 2012. Is global warming a state-corporate crime? In Climate change from a criminological perspective, ed. Rob White, pp. 71–88. New York: Springer.

Kushkush, Isma'il. 2014. Food crisis worsens in South Sudan as civil war is displacing millions. *The New York Times*. May 20: A4.

Lee, James R. 2009. *Climate change and armed conflict: Hot and cold wars*. London and New York: Routledge.

Light, Sarah E. 2014. Valuing national security: Climate change, the military, and society. *UCLA L. Rev* 61: 1772–1812.

MacFarquhar, Neil. 2010. U.N. raises concerns on harvests. *The New York Times*. September 4: A4.

Mares, Dennis. 2010. Criminalizing ecological harm: Crimes against carrying capacity and the criminalization of eco-sinners. *Critical Criminology* 18 (4): 279–93.

Maas, Achim, Irina Comardicea, and Balázs Bodó. 2013. Environmental terrorism – a new security challenge? In *Global environmental change: New drivers for resistance, crime and terrorism?*, eds. Achim Maas, Balázs Bodó, Clementine Burnley, Irina Comardicea and Roger Roffey, pp. 203–20. Baden-Baden: Nomos.

Maupin, Jonathan N., and Alexandra Brewis. 2014. food insecurity and body norms among rural Guatemalan schoolchildren. *American Anthropologist* 116 (2): 332–337.

McCarthy, James, and Scott Prudham. 2004. Neoliberal nature and the nature of neoliberalism. *Geoforum* 35 (3): 275–83.

McNall, Scott G. 2011. *Rapid climate change: Causes, consequences, and solutions*. London and New York: Routledge.

Mooallem, Jon. 2013. Law & order: Endangered-species unit. *The New York Times Magazine*, 12 May, MM30–37, 46, 51, 57.

Neocleous, M. 2011a. 'A brighter and nicer new life': Security as Pacification. *Social and Legal Studies* 20 (2): 191–208.

Neocleous, Mark. 2011b. Security as pacification. In *Anti-security*, ed. Mark Neocleous, and George S. Rigakos, 23–56. Ottawa: Red Quill Books Ltd.

Neocleous, M. 2013. The dream of pacification: Accumulation, class war, and the hunt. *Socialist Studies* 9 (2): 7–31.

Neocleous, Mark, and George Rigakos. 2011. Anti-security: A declaration. In *Anti-security*, ed. Mark Neocleous, and George S. Rigakos, 15–21. Ottawa: Red Quill Books Ltd.

Neocleous, Mark, George Rigakos and Tyler Wall. 2013. On pacification: Introduction to the special issue. *Socialist Studies/Études socialistes* 9(2): 1–6.

Nord, M. 2009. *Food insecurity in households with children: Prevalence, severity, and household characteristics*. Washington, DC: Economic Research Service, U.S. Department of Agriculture.

O'Keeffe, Annamaree. 2010. Food security in the Arctic. *Griffith Review*. February 27. Available at: https://griffithreview.com/articles/food-security-in-the-arctic/.

Onishi, Norimitsu. 2011. Rich in land, aborigines split on how to use it. *The New York Times*. February 13. p. 6, 16.

Patel, R. 2010. Mozambique's food riots—the true face of global warming. *The Observer*. September 5. p. 31.

Plumper, Thomas, and Eric Neumayer. 2009. Famine mortality, rational political inactivity, and international food aid. *World Development* 37 (1): 50–61.

Revkin, Andrew. 2007. Poor nations to bear brunt as world warms. *The New York Times*. April 1. Accessed at: http://www.nytimes.com/2007/04/01/science/earth/01climate.html?_r=0.

Rigakos, George. 2011. 'To extend the scope of productive labour': Pacification as a police project. In *Anti-security*, ed. Mark Neocleous, and George S. Rigakos, 57–83. Ottawa: Red Quill Books Ltd.

Safo, Nova. 2014. Americans face food insecurity despite economic growth. *Marketplace*. September 4. Accessed at: http://www.marketplace.org/topics/life/americans-face-food-insecurity-despite-economic-growth.

Samson, Colin. 2003. *A way of life that does not exist: Canada and the extinguishment of the Innu*. London: Verso.

Sayer, Jeffrey, Jaboury Ghazoul, Paul Nelson, and Agni Klintuni Boedhihartono. 2012. Oil palm expansion transforms tropical landscapes and livelihoods. *Global Food Security* 1 (2): 114–119.

Scheffran, Jürgen, and Thomas Cannaday. 2013. Resistance to climate change policies: The conflict potential of non-fossil energy paths and climate engineering. In *Global environmental change: New drivers for resistance, crime and terrorism?*, ed. Achim Maas, Balázs Bodó, Clementine Burnley, Irina Comardicea, and Roger Roffey, 261–292. Baden-Baden: Nomos.

Slater, R., K. Sharp, and S. Wiggins. 2012. Food security. In *The companion to development studies*, ed. V. Desai, and R. Potter. London: Routledge.

South, Nigel. 2012. Climate change, environmental (in)security, conflict and crime. In *Climate change: Legal and criminological implications*, ed. S. Farrall, D. French, and T. Ahmed, 97–111. Oxford: Hart.

Stack, James P., Jacqueline Fletcher, and M. Lodovica Gullino. 2013. Climate change and plant biosecurity: A new world disorder? In *Global environmental change: New drivers for resistance, crime and terrorism?*, ed. Achim Maas, Balázs Bodó, Clementine Burnley, Irina Comardicea, and Roger Roffey, 161–181. Baden-Baden: Nomos.

The World Bank. 2006. *Repositioning nutrition as central to development: A strategy for large scale action*. Washington, DC: The World Bank. http://siteresources.worldbank.org/NUTRITION/Resources/281846-1131636806329/NutritionStrategy.pdf.

Trauzzi, Monica. 2014. CNA Corp's Military Advisory Board calls climate change a 'catalyst for conflict.' (Interview by Monica Trauzzi, OnPoint, with Sherri Goodman, senior vice president and general counsel at CNA Corporation and former deputy undersecretary of defense for environmental security.) *OnPoint/E&E Publishing, Inc.* May 14. Transcript available at: http://www.eenews.net/tv/videos/1827/transcript.

van Uffelen, Jan-Gerrit. 2013. The de-disasterisation of food crises: Structural reproduction or change in policy development and response options? A case study from Ethiopia. In *Disaster, conflict and society in crises: Everyday politics of crisis response*, ed. Dorothea Hilhorst, 58–75. London and New York: Routledge.

Wall, Tyler. 2013. Unmanning the police manhunt: Vertical security as pacification. *Socialist Studies/Études Socialistes* 9 (2): 32–56.

Walters, Reece. 2005. Crime, bio-agriculture and the exploitation of hunger. *British Journal of Criminology* 46 (1): 26–45.

Walters, Reece. 2006. Crime, bio-agriculture and the exploitation of hunger. *British Journal of Criminology* 46 (1): 26–45.

Walters, Reece. 2011. *Eco crime and genetically modified food*. Abingdon, UK: Routledge.

Warner, Jeroen. 2013. The politics of 'catastrophization'. In *Disaster, conflict and society in crises: Everyday politics of crisis response*, ed. Dorothea Hilhorst, 76–94. London and New York: Routledge.

Weeks, Linton. 2012. Championing life and liberty for animals. *National Public Radio (NPR)*, 25 October. Accessed at: http://www.npr.org/2012/10/25/158296711/championing-life-and-liberty-for-animals.

Wenar, Leif. 2016. *Blood oil: Tyrants, violence and the rules that run the world*. Oxford: Oxford University Press.

White, Rob. 2010. Globalisation and environmental harm. In *Global environmental harm: Criminological perspectives*, ed. Rob White, 3–19. Cullompton, Devon, UK: Willan.

White, Rob. 2011. *Transnational environmental crime: Toward an eco-global criminology.* London and New York: Routledge.

White, Rob. 2012. The criminology of climate change. In *Climate change from a criminological perspective*, ed. Rob White, pp. 1–11. New York: Springer.

Wight, Vanessa, Neeraj Kaushal, Jane Waldfogel, and Irv Garfinkel. 2014. Understanding the link between poverty and food insecurity among children: Does the definition of poverty matter? *Journal of Children and Poverty* 20 (1): 1–20.

Wilde, Parke. 2013. *Food policy in the United States: An introduction.* London and New York: Routledge.

Wutich, Amber, and Alexandra Brewis. 2014. Food, water, and scarcity: Toward a broader anthropology of resource insecurity. *Current Anthropology* 55(4): 444–468.

Chapter 17
Fence Posts to Blog Posts: An Exploration of the Classroom in Experiential Food Systems Education

Elissa Johnson

Abstract Food education for undergraduate and graduate students has grown far beyond the common land grant institution's agricultural goals and the culinary school's production methods. Access, justice, equity, and sovereignty have made their way into the lexicon of food. To this end, there is value in information sharing and experiential learning from experts in the field (quite literally: farmers, processors, distributors, food service workers, farmhands, community activists, and government officials) combined with the academy (researchers, academics, and scholars) that builds a framework to create a truly transdisciplinary education and critical reflection. I draw on my experiences as a student, and as teaching assistant and instructor in non-traditional food systems classrooms in this chapter. Ultimately, I urge the reader to consider an imperative direction in food systems education that may be the best platform for what follows: creatively structured, critically and community engaged, student-centered learning experiences.

Keywords Food systems · Experiential learning · Community partners · Food justice · Critical food systems education

Food education for undergraduate and graduate students has grown far beyond the common land grant institution's agricultural goals and the culinary school's production methods. Access, justice, equity, and sovereignty have made their way into the lexicon of food. Across the United States, universities title their innovative programs to etch their specific niche, such as food systems, food studies, or food policy. Advanced degrees allow for compelling, focused inquiry into one of the most abiding concerns of this generation, or any generation. As students seek out accredited education across scales and sectors specific to how food is grown, harvested, distributed, processed, accessed, consumed, and wasted, there exists opportunity for the colleges and universities to partner with local and regional

E. Johnson (✉)
Syracuse University, Syracuse, USA
e-mail: ejohns07@syr.edu

© Springer International Publishing AG 2017 201
I. Werkheiser and Z. Piso (eds.), *Food Justice in US and Global Contexts*,
The International Library of Environmental, Agricultural and Food Ethics 24,
DOI 10.1007/978-3-319-57174-4_17

entities engaged at every point in the system. This type of experiential, immersive, learning space facilitates a fusion of horticultural sciences and agricultural practice, with social and critical theory.

In this chapter, I discuss the value of information sharing and experiential learning from experts in the field (quite literally: farmers, processors, distributors, food service workers, farmhands, community activists, and government officials) combined with the academy (researchers, academics, and scholars) that builds a framework to create a truly transdisciplinary education and critical reflection. I illuminate some models at work throughout the country and why the idea of the classroom is being expanded to serve this engaged field of study as it challenges traditional ways of knowing and interacting in higher education. Further, the examples herein represent possible avenues for addressing complex issues of food access and food justice through models of education that include cultural narratives, personal histories, and challenges to traditional lecture style classroom power dynamics.

In addition to the diverse structures of these classrooms, there also exists an expansive network of digital platforms and options for community engagement that reach far beyond those in the typical seminar style classroom or science lab. Thus, these learning environments enable students to identify specific points of leverage within the often overwhelming grid of food "issues" through the rigorous, thought-provoking, and sometimes (literally) muddy, platform of experiential food systems education. I provide some examples of these multi-disciplinary, multi-media projects in this chapter.

I draw on my experiences as a student, and as teaching assistant and instructor in non-traditional food systems classrooms in this chapter. Ultimately, I urge the reader to consider an imperative direction in food systems education that may be the best platform for what follows: creatively structured, critically and community engaged, student-centered learning experiences.

17.1 What Is Food Education?

Beyond culinary training institutions, food is all over higher education. Historically, land grant universities cornered the market on agriculture and food education, often concerning specifically production and harvest. More recently foods courses have made their way into departments ranging from Biology and Natural Resources to Art and English.[1] Canonical disciplines like Geography, Philosophy, and

---

[1]For example, University of Pennsylvania's biology course "The Biology of Food", a variety of programs coupling food with Natural resources (University of Minnesota's College of Food Agriculture, and Natural Resources, Michigan State University's College of Agriculture & Natural Resources), Dartmouth's art history department's course "Food and Art: A Global Context", and University of California Berkeley's English department's special topics course on food writing.

Anthropology have academic associations directly concerning the study of food and its many components.[2] The connection between place, culture, reason, and food continues to entice scholars in these fields. Across the sciences food has permeated the practice as Agroecology, Plant and Soil Sciences, Entomology, and scores of other sub-disciplines emerge.[3] Important work is also being done in fields of Critical and Cultural theory, and Gender and Women's study related to food.[4]

If food education is applicable in every one of these seemingly disparate silos of education what, then, *is* food education? Perhaps this is what scholars saw as the conundrum when the first interdisciplinary food education began years ago. The current prevalence of programs encouraging multi-perspective pursuit of food education could indicate that food education is not a definable thing, but instead something that educators and students may apply across disciplines.

To this effect, the nomenclature of food education has shifted. Food studies, food justice, and food systems are common titles that aim to create space for specific aims within integrated inquiry. The study of food may seem vast, but inquiry into the elements independent of one another is moot. Without getting lost in the nuance of multi-, inter-, and trans-disciplinarily, but with great due respect to those authors who *have* to our benefit, I will say that the study of food in this form achieves an interdisciplinary mode of inquiry. Interdisciplinarity works by connecting the perspectives of disciplines that are typically housed in separate, but equally discerning, arenas of thought to comprehensively and thoroughly navigate complex issues (Stock and Burton 2011).

Although it would seem the popular way of speaking about the work coming out of these integrated programs is by speaking if the *trans*disciplinary nature of the work, there remains to be seen the extent to which these programs are truly achieving *trans*disciplinary work (Self et al. 2016). The pursuit of work of this nature is complex and intrepid enough on its own, "in this sense transdisciplinarity is the highest form of integrated project, involving not only multiple disciplines, but also multiple non-academic participants (e.g., land managers, user groups, the general public) in a manner that combines interdisciplinarily with participatory approaches" (Stock and Burton 2011).

By attempting a transdisciplinary approach to the study of food, we may begin to address the complexity therein (Burns 2012). This paper will build upon the idea

---

[2]Examples include the Geographies of Food and Agriculture Specialty Group within the Association of American Geographers, the University of Vermont's Food Ethics symposium, and Michigan State University's Food Justice Conference sponsored by the Department of Philosophy, and the Society for the Anthropology of Food and Nutrition.

[3]See University of California Santa Cruz's Center for Agroecology and Sustainable Food Systems, Cornell University's Crop and Soil Sciences courses "Sustainable Agriculture: Food, farming, and the Future", and University of Florida's Honey Bee Research Lab.

[4]See University of California Davis's "Critical Studies in Food and Culture" research cluster, Hamilton College's Women's Studies course "Kitchen Culture: Women, Gender, and the Politics of Food", Oregon State University's "Food and Ethnic Identity: Eating at the Border" course, and Syracuse University's "Food, Identity and Power" course.

that the study of food benefits from a range of academic thought and scholarly modes of inquiry.

Further, a "systems approach helps to define the object of inquiry as a larger set of relationships and factors than is typical among more disciplinary approaches, (Hilimire et al. 2014, 726–727). Here, I set out to show the importance of stepping away from a standard classroom environment and lesson structure, which are often linear or issue specific, and toward innovative classroom models for the sake of reflecting a *systems* approach to food in the actual structure of foods courses and programs. A systems approach to food, then, includes disciplines typically housed in nutrition as much as it does agriculture, and gender studies as much as it does economics, social sciences, natural resources, biological and physical sciences, humanities and medicine. With this in mind, the examples that follow are of experiential, immersion, and service-learning structured courses in *food systems education*. Through them I aim to illuminate some of the ways in which the classroom in food systems education is being explored and expanded.

17.2   Stalking the Wild in Vermont

In Summer, 2015, I was a co-instructor for the Vermont Food Systems Summer Study Tour (VFS3), the first of its kind in the state. Undergraduate and graduate students from Vermont, Pennsylvania, New York, New Hampshire, Virginia, and North Carolina embarked on a multi-week immersion learning course ready to engross themselves in a Vermont food systems experience. This course leapt from classroom lectures given by scholars of Economics, Agriculture, and Chemistry, to cutting edge culinary lessons from head chefs, world award-winning cheese makers, and farm-to-table restaurateurs, to a conference of food systems scholars and practitioners at the University of Vermont's Food Systems Summit. The class moved as a unit over the Vermont landscape to each consortium member school, absorbing pieces of that institution's food systems niche from scholars in their field. Additionally, this program partnered with food systems practitioners to ground the conversations in practice. This is how we came to eat "weeds" in a small southern Vermont town led by the skill and wisdom of our elders, and a precocious young boy.

With my "field eyes" on, I'm scanning the vegetation across this small opening in the trees for Oxeye Daisies and Milkweed buds. The daises have been difficult to see because I and the others in the class have been instructed to look for unopened buds which blend in easily with the lush mid-June greenery. Our host's young son, probably ten or eleven, possibly entertained by how long it has taken us to adjust our eyes to the landscape, shows the class where there is a large patch ready for collection. Once we all have had a chance to pick a few we'll focus on another wild edible plant, pick a few of that one too, and prepare them for dinner as part of our lesson on wild-crafting and forest edibles.

Our partners and educators in this venture were Nova Kim and Les Hook, internationally known wild-crafters, and long-time Vermonters. It is through this partnership that we examined the food growing right under our noses, often unnoticed and uneaten. Our host offered their land and homestead to us as a classroom for experiential education. Between answering the class's questions about every "LBM" (little brown mushroom) we saw and the cautions against eating plants and fungi you cannot identify with certainty, there were lessons on overharvesting, selective harvest, pollinators, species diversity, and a lot of plant identification.

The class (and instructors) were also challenged to consider intergenerational teaching and cultural ways of knowing that differed in familiarity from person to person. Our host's son led us on trails to fruitful areas for harvesting, challenging each of us to contemplate at what age one may be considered a teacher. Kim is of Osage heritage and teaches from the perspective of a land steward. Her seamless weaving of Osage traditional knowledge with horticulture and biological expertise instigated inquiry among the students into the ways science and tradition, western educational structure and Native narrative storytelling histories, complement and complicate one another. The intellectual critical complexity of this walk in the woods was raised by the connections the class had to the human cultural source(s) of information compared to the academic foregrounding they typically receive.

During this highlighted experience, the students learned formally through Kim and Hook's instruction, and informally from one another, and from our host's son. This type of experiential, immersion learning allows for intergenerational, multi-cultural, hyperlocal, and ecological perspectives to be shared and to develop, (Burns and Miller 2012). Particularly in this example, students benefitted from an exposure-first model of education that is highly impactful for certain lessons. When applied to food systems education, exposure-first teaching has been shown to incite curiosity about a subject, and to help the student develop the questions that need answering to go deeper, (Parr and Trexler 2011; Hilimire et al. 2014). This model includes field walks, but could also be as simple as a reading, documentary, or guest presentation on a topic that may be foreign to the students and not have an ostensibly clear connection to the larger picture. By exposing the student to new ideas or situations they learn how to develop their own interest in the subject, connect the content to that which they already know, and instigate a connection to the other themes from the course.

This method also allowed student engagement across their multiple "home departments" of study and life experiences. The students in this course ranged in age from less than 20, to over 70 years old and came from both urban and rural places, among other social locations discussed later in this chapter. They were met with the opportunity to apply their unique perspectives to the lesson at hand. Professional farmers, horticulture students and food media professionals alike were challenged to learn from one another as well as from their instructors. This peer-to-peer learning is indicative of cooperative learning methods which have been found to be important to food systems programs, (Parr and Trexler 2011; Hilimire et al. 2014). Each student, informed by their position and discipline, may pick up on different elements from the same lecture or field trip.[5] By drawing on a range of disciplines within the community partners and university lectures, also encouraging dialogue between students with varied interests and trainings, the outcome complements a systems approach to thinking. Students are encouraged to recognize and develop points of leverage in a broder system of issues and to make connections between their home disciplines to co-create knowledge around food systems inquiry.

17.3   Skills and Realities of Farming Minnesota

I'm walking next to a flatbed trailer pulled by a tractor across a field that is being prepped for planting. This class of nearly twenty is helping to rock pick—literally, picking up rocks from the tilled field to help ensure even rows of planting and lessen damage on farm equipment. For three summers, I was a student in, then a teaching assistant for, the Environment and Agriculture: Sustainable Food Systems (E&A) course offered through the Higher Education Consortium for Urban Affairs (HECUA). Though based in St. Paul, Minnesota, this was a similarly structured experiential and immersion learning course to the VFS3. HECUA has many foci, and has developed their programs to appeal to a range of students over the 40 plus years the organization has been running. The common threads are lessons through experiences and immersion learning programs with a community-centered approach to higher education and social justice. Creating this type of learning experience is an effective method of engaging students and crafting a reflective and flexible learning environment (Galt et al. 2012).

But what does rock picking have to do with this idea of complex, multifaceted, food systems? Or, put differently, is there educational value in this seemingly menial task? This is a legitimate consideration and one that instructors of these types of courses are always reflecting upon. Ultimately, the answer to whether there is a lesson to be derived from rock picking lies in the combination of core and practical skills development through the practice. Core skills include big picture maneuvers such as critical thinking, self-reflection, evidence-based decision making and argument building, interdisciplinary analysis and communication skills, (Hilimire et al. 2014). Practical skills are determined by their discipline more so, and might include specific instruction on a range of topics such as plant identification, effective community engagement, food hub operations, or crop planning (Hilimier et al. 2014).

---

[5]There are plenty of examples of this from VFS3. One student was particularly well-versed in horticultural sciences and often helped to explain these processes to the class on field walks, or related these processes to outcomes like how carbon storage might impact no-till vs. ridge-till practices. Another student with a business and finance background constantly made connections to economic viability. Students took on unassigned (or perhaps self-assigned) roles that empowered them to complicate and challenge their disciplines and ultimately build new perspectives on food systems issues.

Food systems education benefits from the combination of the two. Rock picking allows students who may not have experienced the practice to learn the skill, but also to understand the need and to place the information within the greater context of field prep, farm maintenance, food production and supply. By engaging in an active conversation and task oriented skill sharing, even repetitive tasks begin to take on meaningful educational value. Rock picking leads to discussions about ways that large-scale farms are attempting to become more sustainable through no-till and ridge-till practices to lessen soil erosion and disturbance. Carrot pulling leads to discussions about market value and the hardship of growing perfect-looking produce under less than ideal conditions. These experiences provide opportunities to open critical dialogue on farm labor and the economics of sustainable food systems. By connecting practical skills with core skills, food systems students engage deeply while connecting to a broad range of topics and issues.

Farm stays were a highlight of E&A. Students were placed in pairs at a range of host farms each with a particular focus—from dairy and beef production to permaculture vegetables, herbs and honey. This is a type of student-centered learning that allows students to identify what their interests are and the instructor places them with a host that will facilitate that type of learning endeavor (Self et al. 2016). Student interests may range from a specific production method or product (e.g.: certified organic CSA, pasture raised pork, free range chicken and egg production) to a set of ideals and practices of engagement (e.g.: agrotourism, festival planning, permaculture design, food access and justice). This course element draws on the importance of connecting core and practical skills, and further, reflects that food systems learning is an active process whereby students who participate in the farm stays are constantly assimilating new information and revising their paradigms.

This recalls the cognitive constructivist theory that suggests meaning and concepts are constantly revised and refined in relation to one's interactions with objects and situations in the world (Parr and Trexler 2011). Students who are engaged in a farm stay gain the opportunity to examine the day-in-day-out practice of farm operations, and the dynamics of family life on a farm in addition to the business. While the experiences are overwhelmingly positive, this doesn't always leave students with wide-eyed dreams of their own farm futures. Often students are critical of their hosts, sometimes due to the reality of some practices or of the questionable viability or volatility of the business. These are opportunities for dialogue on the commonly difficult economic realities of farming in the United States, and could instigate further study by students who wish to engage further. Students often come away from farm stays with a better understanding of some moral and ethical beliefs they hold regarding farms that raise animals for slaughter, or use animal labor. Questions arise regarding "truly" sustainable agricultural systems, equitable farm labor and issues of scale, and challenges specific to urban farms and food access. By engaging in these questions directly with practitioners, students approach new dimensions of food systems education.

17.4 Critical Dialogues on Race, Ethnicity, Development, and Privilege in Food Systems

The prior examples are of two experiential, immersion learning courses in food systems that set out to do what these types of programs were designed to do: engage students with community partners, academic instruction and participatory opportunities for food systems education. This model may not be feasible to many established programs due to the time and network intensive requirements (more on that later). Service learning allows for student engagement with a community partner through research, work, or consulting. These projects take on a variety of forms and are important endeavors in food systems education. Here I will focus on a consultant-based model for a service learning partnership between a graduate level course on food systems and the Vermont Land Trust (VLT). The class was tasked with addressing three case studies where VLT is engaged in a community land trust model. The aim was to illuminate the progress of the three sites and to situate that case study in the larger context of Vermont's environmental and social land history. In doing so, VLT would gain deeper insight into their projects and the community land trust model at work, and the students would develop their research, critical thinking, and presentation skills through the development of a day-long staff training where the information would be shared with the organization.

I was a student in the class, and with my two groupmates, chose to examine the Pine Island Community Farm (PI) for this project. The things we knew about the current state of the community land trust project came from our own experiences volunteering at PI, initial meetings with the VLT representative for this particular project, and press and publicity about the farm that preceded our involvement. We knew that, post auction sale to VLT, Pine Island housed a goat farm, focused on food security and making culturally appropriate food resources available to the substantial number of resettled refugees (New Americans) living in and around Burlington, Vermont. We knew that there were several shared garden plots being used by New Americans and while some of those people sell their surplus at area farmers markets, most growers used the plots for subsistence. At the outset, it seemed that our group would be researching the history of agriculture in the various regions from where New Americans were coming, and focusing on how VLT might assist in enhancing the already well-oiled machine that is Pine Island Community Farm.[6] After a bit of digging however, we became engaged in critical dialogues on race, ethnicity, development and privilege.

Moving forward with this project required our group to consider our social locations to recognize the power at play within the project itself. I am a white, female, with experience in both higher education and community based projects.

---

[6]Pine Island Community Farm sits in the middle of three Vermont towns, each with large New American populations. Since 1989, 6659 refugees have made Vermont their home. Numerous ethnicities come from Bhutan, Bosnia, Burma, Burundi, The Congo, Iraq, Sudan, Syria, Somalia, Rwanda, and Vietnam.

I speak English clearly, and consider myself to be an effective communicator and without any serious physical limitation. As part of this project I was encouraged to identify my whitened cultural ideas about what good food is and from where it comes (Slocum 2007; Meek and Tarlau 2015). I applied this lens to the project development and, with my groupmates, developed a set of ideas about how the project might be reaching its stated goals, or not.

These shifts in thinking reflect points made by Leslie Gray et al., regarding experiential learning and food justice: "Issues such as food access and inequitable distribution of resources are not issues at the forefront of most university gardens and farms.[7] These different types of "critical" experiential learning placements expose students to a different kind of food movement that goes beyond the promotion of sustainable agriculture and organic food to issues of racial, environmental, and economic justice, (Gray et al. 2012; Alkon and Agyeman 2011).

While Gray et al. refers specifically to the placement of students engaged in service-learning experiential education at the urban garden project at the center of the work cited here, the concept of critical dialogue and issues that may be decentralized from whitened food systems conversations are important and applicable. "Helping students to recognize the social reality of injustices in contemporary societies, including a realization of their privilege and the marginalization of other, is an important part of the learning experience (Gray et al. 2012).

As the aim of service learning projects develops, instructors and students must be cognizant of these issues of power and positionality and should make certain the needs of those on the host end of any service partnership are accurately represented in all aspects of the program. There has been a breadth of scholarship on the ways in which community-engaged learning, service-learning, and immersion learning experiences in food systems can in some cases lack resonance with the communities in which they are located (Guthman 2008; Gray et al. 2012; Bernard-Carreño 2015; Hilimire et al. 2014; Alkon and Agyeman 2011; Slocum 2007; Sommer et al. 2011; Meek and Tarlau 2015). Often this is due to the assumption of values and needs of the communities in which the projects take place. Ascribing "foodie" or "localvore" discourses to all food projects may not fit the desired end, however one must be careful to not assume that if, for example, access to food is a paramount issue in a community, that this means access to *any* food, without care being given to the cultural appropriateness and overall health of the food resource (Slocum 2007). While food systems instructors may advocate for learning through unfamiliar or uncomfortable situations and topics, there is no value to be found in the exploitation of a community partner for the benefit of a students' edification (Guthman 2008; Slocum 2007; Alkon and Agyeman 2011).

Instead, carefully constructed relationships may be built between community partners and institutions of higher education to ensure a mutually beneficial partnership and an opportunities for students to expand their perceptions of what food

---

[7]Here, Gray et al. is referring to the Bronco Urban Gardens (BUG) project, the subject of her article cited here.

systems work entails. Food systems education provides the option to become the catalyst for social justice education and may dissect complex issues of class, race and gender among others to form a truly systemic means of thinking about food.

17.5 Food Systems in the Cloud

Classrooms across disciplines are engaging students with multi-media experiences and responding to students' interests in a more digital means of connecting to scholarship. How does food systems education remain relevant to a technologically advanced student body? Are images of dirty hands and muddy boots too archaic to entice new food systems students? Do ideas of old farmhouses and rolling pastureland accurately represent the globalized food systems of which we are all a part? I argue that food systems education is uniquely positioned to incorporate the interests of a technologically inclined student population, while retaining the imperative perspectives of physical experience and critical thought.

It is not a question of whether technology plays a role in food systems education. We can look to advances in farm equipment, crop planning software, and organizational programs that streamline production, and make centuries-old agricultural practices viable in the twenty-first century. Here, I will focus on some of the web-based programs and multi-media endeavors that create a technologically adventurous food systems classroom, or supplement to food systems education, with applications to a unavoidably adaptable agricultural future. Horticultural and biological technology is advancing by leaps and bounds however, here I will focus on some "cloud-based" opportunities for engagement, information sharing, and knowledge creation with a food systems approach. Further, by embracing technology in the food systems classroom, educators are empowered to engage students in programs where textbook resources are low, garden education funding is nonexistent, and food access is a struggle inside and outside of school. Appreciating the reality that many students have greater access to a smartphone than to fresh foods may be the answer to outreach within a food justice framework.

17.5.1 Blogs for Reflection

Blogging comes to mind as an already widely-used option for web engagement. The VFS3 course set up its own course blog for family, friends and affiliated faculty to track the adventures of the course. The idea was that students would blog about the daily activities, integrating text, conversations, questions, experiences thus, in their own voice, would share their critiques of the process, the experience, and the materials and ideas that were brought about. An instructor may determine whether to make blogs public, open to comments from other students in the class, or kept private between instructor and student. There are benefits to each and a combination may be used to approach different topics.

The reflection experience is essential for experiential learning to be meaningful, and student's reflections may be shared with others through a course blog, or any number of formats. The importance of this reflective step in the process of experiential education cannot be understated. Without reflection, an experience could be reduced to contact with observations (Hilimire et al. 2014; Sommer et al. 2011). By enabling this interaction, food systems courses create space for meaningful contemplation and deepened critical thinking skills. Through reflection, a type of transformational learning is approached as students are prompted to reflect on their understanding, values, concerns, and questions regarding the complexities and realities of food systems that impact, and are impacted by them (Galt et al. 2012, 2013; Sommer et al. 2011; Hilimire et al. 2014; Self et al. 2016).

This type of critical reflection is being modeled by some of the many food-focused websites that take a food systems approach. Beyond recipe websites and cooking blogs, blog-style sites, such as the James Beard Award-winning *Civil Eats*, instigate a larger social critique anchored in food systems education (www. civileats.com). The various post authors breech cogent and political transdisciplinary issues which are made approachable by the author's personal voice that comes through in the informal blog/opinion format.[8] This allows for hybrid critical reflection, and integrated analysis of text, experience, and opinion.

17.5.2 Lexicon Mapping

The way scholars and practitioners talk about food systems adapts to the ways in which elements of food systems are practiced. To this end, the website *Lexicon of Food* tracks projects and areas of food systems work and scholarship through what it calls "channels" (lexiconoffood.com). Educators, students, and practitioners provide content for this relatively comprehensive crowdsourced collection based on many important principles, organized by terms, related to food.

Indeed, the language that academics use to talk, write, and work within their respective disciplines can be the very walls of the silos that house them. Food systems classrooms may refer to projects like the Lexicon of Food for an example of media that attempts to integrate disciplines, moving the conversation past an interdisciplinary approach toward a transdisciplinary visualization of what it means to do, discuss, and research food systems.

---

[8]Examples of recent posts titles: "What You Need to Know About the EPA's Assessment of Atrazine", "This Subsidized Bus Service is Helping Flint Residents Get to Grocery Stores", "How a Former Vegetarian Became a Butcher and Ethical Meat Advocate", "Can this Market Be a Model for Getting Good Food into Neighborhoods Shaped by Racism?".

17.5.3 Multi-Media Multi-Discipline

Multimedia projects incorporate blog posts, podcasts, Twitter, YouTube channels, photos, and film with travel, interviews, and ethnographic inquiry. These projects aim to highlight the transdisciplinary nature of food systems by examining the social locations of practitioners. The projects I highlight here start with a critique of systems of power and privilege, and combine it with the opportunities and challenges that are specific to food systems.

Visibility takes center stage in Marji Guyler-Alaniz's media and retail project, *FarmHER* (www.farmher.com). Through photographs, blogs, and a new television spotlight, self-identified women are made visible as land owners, operators, and workers in the agriculture industry. Issues of gender inequality are met with the empowerment of running a farm business.

Intersectional issues of culture, power, oppression, and food are the topic of the podcast *Racist Sandwich*. Hosts Soleil Ho and Zahir Janmohamed craft critical dialogue at the nexus of religion, body image, colonialism, social justice and food (www.racistsandwich.com).

Black Latino, Indigenous, and Asian farmers are the focus of Natasha Bowens' photographic storytelling project, *The Color of Food* (thecolorofood.com). With the same aim of empowerment, validation, and visibility, this project incorporates food activists and farmers, both urban and rural, to raise the visibility of farmers of color. Subjects of this project are empowered leaders of their communities who aim to challenge whitened cultural food histories (Slocum 2007).

Similarly, the Queer Farmer Film Project achieves visibility and empowerment through critical dialogue on place, space, and safety in an industry where queer and gender non-conforming individuals are seemingly absent (www.outheremovie.com). Jonah Mossberg's project is exhibited in the film Out Here, which brings to the forefront issues of sexual and gender identity, along with race, ethnicity, ability, urban and rural livelihoods, and the agriculture industry.

A North Minneapolis community organization found unexpected spotlight for a music video created by the kids, some not even twelve years old, addressing healthy food insecurity with a message as cogent as it is relatable. Kids active in a summer program through Appetite for Change, a North Minneapolis-based non-profit, "committed to using food as a vehicle for social change," (appetiteforchangemn. org) collaborated to write and perform the hip hop song titled "Grow Food" with limited time and tiny budget (www.youtube.com/watch?v=PqgU3co4vcI). This project shows the impact of a strong message as the video has spread across social media and had over 271,000 views on Youtube as of January 2017.

Food systems classrooms will benefit by engaging in these types of conversations, and may use these examples of multimedia formats to enhance the learning and reflection experience

17.6 Barriers and Complexity in Food Systems Education

The possibilities for engagement in food systems education are as limitless as the elements that food systems education may incorporate. That is to say, arguably, any issue or idea can be located somewhere within food systems. This can be an overwhelming, if empowering, realization. How does a food systems classroom engage in the most important aspects? What are those elements? How does one make time for all of that? Who will pay for that time? How does a food systems classroom cover the breadth of issues without sacrificing the depth necessary for critical engagement? These are legitimate concerns and ones to which I do not propose answers here. Instead, I will illustrate some of the barriers and complexities to food systems education in an effort to raise the dialogue among instructors, students, and community partners.

17.6.1 Time and Scope

The time commitment for students and for instructors is huge, especially in the developing phases of these kinds of experiential, immersion learning and service-learning course structures. Coordinating field trips and site locations, farm stay safety checks and insurance, drivers permits, scheduling and re-scheduling all take place before the course is even underway, and continue throughout (Hilimire et al. 2014). To make the process meaningful, reflections must be taken seriously, lessons on power and privileged must accompany lessons on horticulture and business. Community partners must be carefully cultivated after time has been spent fostering relationships between the university and the greater community.[9]

Curriculum development, and the daily work of the class itself can be a challenge as instructors and students identify issues and activities that are most important to them. Simply keeping on topic during any given class session may be difficult due to the application of intersectional impacts across food systems issues. This need for flexibility is extended from instructor to students and may result in more opportunities for student directed learning (Self et al. 2016). However, it is not enough to introduce critical issues to food systems education without making the time to fully address these connections (Gray et al. 2012). For example, if a lesson begins by introducing the benefit of gleaning projects to reduction of food waste on urban farms and to provide food access to urban food banks and shelter programs,

---

[9]A relationship that may be fraught with tension as the academy is often viewed as irrelevant to an applied community or disaffected by the community in which it is a part. Designers of such collaborative relationships must recognize the feelings of powerlessness that community partners can feel when partnering with a university due in part to these dynamics.

the time must be taken to examine how a project that effectively takes edible food from the fields of farms and gives it to others is (1) a viable organization, (2) attending to the needs of the urban hungry including who we mean by saying "urban hungry", or (3) fits into a larger plan for food access for marginalized communities.

17.6.2   Working with/in the Community

Many scholars and activists have written on the complexity of outreach, especially on the dynamic of well-intentioned white individuals and organizations doing work in communities of color. There is an idea that people just want to "help" those less fortunate, and thereby all agency is erased on the part of the receiving group (Guthman 2008; Alkon and Agyeman 2011; Bernard-Carreño 2015; Slocum 2007). The decision for an instructor to develop a course including community partners must be a conscious effort and so relationships must be thoroughly developed (Bernard-Carreño 2015). Community partners must be in on the lesson planning that is to incorporate their organization or work and must agree to the points and alignments that are made. Community partnerships as learning opportunities has the potential for exploitation of all participants. As discussed earlier, sustainable agriculture has a whitened cultural history, therefore the course content that wishes to engage farmers of color, or community organizations that work with marginalized populations must be aware of the potential to tokenize these groups and organizations. The course design must, again, take the time to have the important conversations about structures of power and privilege at play across all elements of food systems issues.

Participants in service learning and experiential education must ensure that both students and community partners are being served. This may be a matter of co-designing metrics of success for the partnership and course project in advance of the student's involvement (Gray et al. 2012). Successful partnerships have created pathways for communication and evaluation both during and after a course with the intention of developing the partnership outputs together (Burns and Miller 2012). Ultimately, the project and partnership must exhibit relevance for both the course, the students, and the partners. Students must be prepared to work with community partners whose ideas and practices around food may not match their own (Guthman 2008; Alkon and Agyeman 2011; Slocum 2007).

Instructors should carefully consider opportunities that the class may take to give back to the partner, or compensate them in some way for taking time from their workday (sometimes unpaid) to facilitate a lesson. Programs and departments should make the effort to financially compensate community partners foremost, but this is not always the case. The HECUA class often wove practical skills into the lessons, such as the rock picking example elaborated on earlier, which in many cases assisted the community partner with a task that would later be done on their own. Consideration must be given to what is actually useful to the host, what does not impede further on their time, or what is a general enough skill to be completed efficiently and effectively to scales of classes from five to twenty-five (or more) students.

17.6.3 Institutional Support

Given the time and resource commitment that must be made up front, instructors must rely heavily on the support of their departments. Transaction costs of curating an experiential learning or community integrated course in food systems are high and some programs and departments have little interest in supporting a style of course with such high financial and time need (Burns and Miller 2012; Galt et al. 2013). We've established here that food systems demolishes the silos of academic thought that once housed (protected?) different discourses from one another. Unfortunately, budgets and course development still, in many cases, adhere to those silos for the structure of funding within programs. Food systems courses, especially those that seek true transdisciplinarity, are part of a reformative movement in the classroom structure and educational partnership (Galt et al. 2013). After all the work to create equitable, meaningful, accountable projects and relationships to community partners, the future of food systems education may very well be determined by the support of a university through departmental buy-in (Self et al. 2016).

Due to the time and fiscally intensive nature of such programs, many experiential food systems education courses are billed to students as additional, unfunded, or domestic travel immersion courses. Frequently, these types of courses are offered over the summer months, or through consortiums or partner organizations that do not recognize the same institutional scholarships and funding that enable many students to participate in higher education. This limits the ability for low-income students to participate in such courses and so, perpetuates a kind of privilege that pervades sustainable food systems rhetoric, which I've covered in different ways herein. In addition to instructor and programmatic support for innovative food systems education and an expansion of the classroom, I call for a necessary inclusion of these programs within the standard school year or with recognition of funds made available to low-income students receiving scholarship, loans, and other funding. Only by ensuring equal access to students from all social locations may food systems classes be positioned to model the inclusivity and systems approach that it calls for in its curriculum.

17.7 Thoughts on Critical Food Systems Education and Expanding the Classroom

To achieve transdisciplinary knowledge creation, food systems education must fully embrace social justice education. By "helping students to recognize the social reality of injustices in contemporary societies, including a realization of their relative privilege and marginalization of others," (Gray et al. 2012) experiential, immersion, and service-learning platforms for food systems education will instigate dialogue necessary for students to become well-informed, critical thinkers who are poised to address the most pervasive problems of our time. Concepts of hunger and unequal access will complicate notions of overproduction, and waste. Visibility and viability of reproductive labor or "women's work" will be highlighted by experiences with in-home food businesses, and women farmers. The narrative of the invisible black farmer will itself be erased as students work alongside people of color across urban and rural landscapes. The food systems classroom is called upon to approach issues of social justice fully integrated with traditional agricultural, nutrition, and horti-cultural education. Through community partnerships, experiential, immersion, and service learning, students are encouraged to create these connections and build new knowledge in cooperation with practitioners. Food systems education then becomes a site to explore critical problems related to food, as well as an arena to develop transformative alternatives (Meek and Tarlau 2015, 2016).

David Meek and Rebecca Tarlau propose the concept of "critical food systems education (CFSE) as a theoretical framework, a set of pedagogies and pedagogical methods, and a vision for policy that can address the racialized narrative in the U.S. and international food systems education" (Meek and Tarlau 2015, 133). They pull from food justice frameworks, agroecology, and food sovereignty along with concepts of critical pedagogy, grounded in the educational ideas of Freire (2000). Indeed, the authors cited within this chapter have made progress in the direction of a new dialectic on food systems education.

With that, the definitions of educator, classroom, and reflection, are changing. Further still, concepts of food, place, justice, labor, and equity take on new meaning. Food systems classes are positioned with a unique opportunity to address systemic, wicked problems of inequality, and they may do that by pulling carrots, or by mapping systems dynamics, or by engaging in dialogue with community part-ners who are food systems practitioners. The examples I have provided here may be a platform for creativity. The future of the food systems classroom is as diverse as the students and practitioners of food systems education.

Note: The author does not speak as a representative of the Higher Education Consortium for Urban Affairs (HECUA) or the Vermont Food Systems Summer Study Tour (VFS3). This publication was conceived, researched, written, and submitted after her involvement in either program ended. The author does not speak as a representative of the Vermont Land Trust of any of its projects, nor does she write on behalf of the class consultant. The media examples are not to be considered an advertisement for any particular site or system, but instead, examples of various web-based and media food systems projects, and some that seem particularly innovative, in the author's opinion. The author writes here from her own perspective and from her own recollections of the programs and participants.

References

Alkon, A.H., and J. Agyeman. 2011. *Cultivating food justice: Race, class, and sustainability*. MIT Press.

Appetite for Change. Accessed January 10, 2017, http://appetiteforchangemn.org/.

Bernard-Carreño, R.A. 2015. Engaged advocacy and learning to represent the self: Positioning people of color in our contemporary food movement. *Journal of Agriculture, Food Systems, and Community Development* 5 (4): 189–193. doi:10.5304/jafscd.2015.054.028.

Burns, H., and W. Miller. 2012. The learning gardens laboratory: Teaching sustainability and developing sustainable food systems through unique partnerships. *Journal of Agriculture, Food Systems, and Community Development*.

Freire, P. 2000. *Pedagogy of the oppressed*. Bloomsbury Publishing.

Galt, R.E., S.F. Clark, and D. Parr. 2012. Engaging values in sustainable agriculture and food systems education: Toward an explicitly values-based pedagogical approach. *Journal of Agriculture, Food Systems, and Community Development* 2 (3): 43–54.

Galt, R.E., D. Parr, J.V.S. Kim, J. Beckett, M. Lickter, and H. Ballard. 2013. Transformative food systems education in a land-grant college of agriculture: The importance of learner-centered inquiries. *Agriculture and Human Values* 30 (1): 129–142.

Gray, L., J. Johnson, N. Latham, M. Tang and A. Thomas. 2012. Critical reflections on experiential learning for food justice. *Journal of Agriculture, Food Systems, and Community Development* 137–147.

Guthman, J. 2008. Bringing good food to others: Investigating the subjects of alternative food practice. *Cultural Geographies* 15 (4): 431–447.

Hilimire, K., S. Gillon, B.C. McLaughlin, B. Dowd-Uribe, and K.L. Monsen. 2014. Food for thought: Developing curricula for sustainable food systems education programs. *Agroecology and Sustainable Food Systems* 38 (6): 722–743.

Meek, D., and R. Tarlau. 2015. Critical food systems education and the question of race. *Journal of Agriculture, Food Systems, and Community Development* 5 (4): 131–135. doi:10.5304/jafscd.2015.054.021.

Meek, D., and R. Tarlau. 2016. Critical Food Systems Education (CFSE): Educating for food sovereignty. Agroecology and Sustainable Food Systems, (just-accepted).

Parr, D.M., and C.J. Trexler. 2011. Students' experiential learning and use of student farms in sustainable agriculture education. *Journal of Natural Resources & Life Sciences Education* 40 (1): 172–180.

Self, J.L., B. Handforth, J. Hartman, C. McAuliffe, E. Noznesky, R.J. Schwei, L. Whitaker, A. J. Wyatt, and A.W. Girard. 2016. Community-engaged learning in food systems and public health. *Journal of Agriculture, Food Systems, and Community Development* 3 (1): 113–127.

Slocum, R. 2007. Whiteness, space and alternative food practice. *Geoforum* 38 (3): 520–533.

Sommer, C.A., L.C. Rush, and D.H. Ingene. 2011. Food and culture: A pedagogical approach to contextualizing food-based activities in multicultural counseling courses. *Counselor Education and Supervision* 50 (4): 259–273.

Stock, P., and R.J. Burton. 2011. Defining terms for integrated (multi-inter-trans-disciplinary) sustainability research. *Sustainability* 3 (8): 1090–1113.

Chapter 18
Institutions and Solidarity: Wild Rice Research, Relationships, and the Commodification of Knowledge

Melanie Bowman

Abstract In this paper I express pessimism about the ability of universities and other knowledge-producing institutions to be in genuine solidarity with food justice and food-sovereignty movements, given the way these institutions treat knowledge as a commodity. Using a distinction between worldviews that treat knowledge as a commodity with worldviews that do not, I endeavor to understand a particular case concerning attitudes toward knowledge about wild rice and their role in the struggle to repair the relationship between University of Minnesota researchers and the Anishinaabe people. The tendency to treat knowledge as a commodity is hard to avoid within universities and other knowledge-producing institutions, given entrenched norms that support the colonizing role these institutions historically have played. Attention to the effects of commodifying knowledge ought to be a priority if we are interested in producing knowledge that supports rather than harms movements for liberation, sovereignty, and justice.

Keywords Knowledge · Commodification · Solidarity · Research practices · Wild rice

18.1 Introduction: Food Justice and Knowledge-Producing Institutions

Around the world, universities are paying attention to global demands for food security by offering new classes, majors, and graduate programs, and dedicating millions of dollars to food security and food justice research. At the University of Minnesota, "Feeding the World Sustainably" is one of five priority areas of research, and at least three of the other areas are closely related (they address water, health, and justice, respectively) (Regents 2016). This research crosses disciplines of agriculture, ecology, economics, and policy, and sometimes also includes input

M. Bowman (✉)
University of Minnesota, Minneapolis, USA
e-mail: bowma271@umn.edu

© Springer International Publishing AG 2017   219
I. Werkheiser and Z. Piso (eds.), *Food Justice in US and Global Contexts*,
The International Library of Environmental, Agricultural and Food Ethics 24,
DOI 10.1007/978-3-319-57174-4_18

from other social sciences and humanities. Yet the faith that funders place in technocratic and agroecological solutions to food insecurity avoids acknowledgment that the role occupied by these approaches within systems of power and knowledge production creates problems for food *justice*.

In this chapter I will express pessimism about the ability of universities and other knowledge-producing institutions to be in genuine solidarity with food justice and food-sovereignty movements, given the way these institutions treat knowledge as a commodity. I will examine a particular case in which, after causing harm, the University of Minnesota has tried to be in solidarity but at best struggles to really help. Then I'll suggest some theoretical tools that can help us understand this case as representative of some harmful norms for thinking about knowledge that exist within the university.[1]

18.2   Wild Rice Research and Relationships in Minnesota

Let us begin with a particular case concerning differing attitudes toward knowledge about wild rice. These attitudes have contributed to the history of harm, miscommunication, and the growth of the paddy-grown rice industry, but this section focuses on their role in the struggle to *repair* the relationship between University of Minnesota researchers and the Anishinaabe people, in light of this history. I want to be clear that the situation I describe here is not intended as an exemplar of the theoretical proposals discussed later; rather, my engagement with this situation as a member of the university, a non-Native resident of stolen land in Minnesota,[2] and a

---

[1]I will not provide a substantive definition of solidarity in this paper since I think such an account is a bigger project than can be addressed here. I mean the term to describe the relationship of trying to stand with or advocate with people who are fighting for their own rights, respect, or liberation when one is not a member of that group. This includes people oppressed in different ways by similar structures (e.g., Latinx and Black coalitions against racism) who wish to work in coalition, as well as a subset of people who benefit from the structures against which the group is struggling, but who nonetheless are interested (at least superficially, if not when it really comes down to it) in dismantling those structures. I am inclined to think that "solidarity" describes a relationship rather than a state, and that it includes reflexivity and growth. For one characterization of various kinds of solidarity, see Scholz (2008).

Food justice refers to "a transformation of the current food system, including but not limited to eliminating disparities and inequities" (Gottlieb and Joshi 2010, ix). *Food sovereignty* is the concept popularized by La Via Campesina, which defines it as "The right of peoples to healthy and culturally appropriate food produced through ecologically sound and sustainable methods, and their right to define their own food and agriculture systems. It puts those who produce, distribute and consume food at the heart of food systems and policies rather than the demands of markets and corporations" (2007, 1). For a discussion of the uses and meaning of "food justice" and "-sovereignty", see Cadieux and Slocum (2015).

[2]The University of Minnesota was funded by the 1862 Morrill Act which created land-grant universities by giving federal land to states, which were then to sell the land to settlers to create endowments for the universities. In Minnesota, this land was acquired by the federal government

person interested in fighting against domination and exploitation compel me to understand the epistemic dimensions of the misunderstandings that persist in this relationship.

Since its beginning, research on wild rice (manoomin, *Zizania palustris*) at the University of Minnesota has been intimately linked with the development of industry and the production of wild rice as a marketable crop (as, literally, a commodity). This research began in earnest in the late 1960s, when early paddy-growers persuaded state government funders that wild rice could be commercialized, after half a century of less-successful attempts to do the same (Doerfler 2003; Kahler 2015). Research focused on traditional breeding through phenotypic selection, hybridization, and later, genetic mapping in the interest of making wild rice more profitable to cultivate in paddies.

Since long before that time, the Indigenous peoples of the Great Lakes region have had a special relationship with sacred wild rice, which plays a central role in Anishinaabe culture. Manoomin has tremendous social, spiritual, and medicinal significance for Anishinaabeg. The tribe expressly reserved rights to wild rice in ceded territories in the treaties of 1837, 1854, and 1855, and has, since settler interest in wild rice began, resisted the commodification of the sacred manoomin. As Jill Doerfler (White Earth Nation) has articulated, "commercial exploitation of Manoomin by non-Indians is generally viewed by the Anishinaabe as a desecration." Though at the same time, "It is customarily considered acceptable for Anishinaabe and other Indigenous peoples to sell some extra manoomin after they have secured an adequate supply for their families" (2003). The Minnesota Cultivated Wild Rice Council now claims that 4–10 million pounds of cultivated wild rice is produced annually in Minnesota (with even more produced by California growers—the unintended beneficiaries of much of the breeding research in Minnesota). The successful commercialization of paddy-grown wild rice has flooded that market, and the excess wild rice that Anishinaabeg had used to supplement meager family incomes can rarely serve this purpose now.

In response to vocal resistance to wild rice breeding and gene-mapping research, as well as a leadership change at the university, in 2006 an ad hoc committee consisting of members of the university and Native communities was formed—the Nibi and Manoomin Bridging Worldviews Committee—with the goal of protecting natural stands of wild rice and improving the relationship between the University of Minnesota and the Anishinaabe of Minnesota and the Upper Midwest (Kokotovich 2014). As part of their work, the committee has organized four biennial symposia,

---

(Footnote 2 continued)

through a series of treaties with the Dakota, often procured under duress. The nominal payment to the Dakota was dramatically delayed, and the resulting privation led to the US—Dakota War in 1862, after which 38 Dakota were executed in the largest mass-execution in US history, and the Dakota were forced out of Minnesota.

with a fifth being planned. In spite of these efforts, a gulf of understanding between scientists and Native people[3] persists, and the manoomin continues to be at risk from environmental damage and biological contamination.[4]

At two of these symposia, I have observed a gulf between the ways many scientists and Anishinaabeg talk about the identity of wild rice. Anishinaabe participants in this conversation argue that the identity of wild rice is at risk of contamination by hybridized varieties of cultivated rice (Gunderson and Julin 2002). One premise for this claim is that cultivated wild rice is no longer genuinely wild and has lost its identity (Wild rice white paper 2011). Yet I have heard scientists dismiss this claim as scientifically naive, since, according to scientists, it fails to take account of the genetic variety within natural wild rice stands (protected by centuries of Anishinaabe stewardship), and ignores the similarities between cultivated and natural wild rice. It appears to me that scientists are thinking of identity in this context as genetic identity, and for this reason insist that cultivated and truly wild rice are the same, since they are the same species and have substantially similar genetic code.[5] This is in spite of the fact that there are quite significant phenotypic differences, including a desired trait in hybridized rice in which the amount of seed shed by wind and rain back into its growing water is minimized in order to maximize crop production (This trait—referred to as seed shattering—is how manoomin re-seeds for the next season in the wild, and thus represents not a loss of goods, but an essential part of the life cycle of the plant in natural stands). This insistence on genetic identity suggests a view of wild rice and of knowledge about wild rice that is starkly in contrast with Anishinaabe worldviews; it makes difficult effective communication with people who talk about identity in a more holistic and ecological sense that includes humans using traditional ricing practices and distinguishes between histories of phenotypic differences.

Because manoomin stands as the fulfillment of prophecy that foretold that the Anishinaabe should migrate west until they found the place where food grows on water, it has spiritual and medicinal value (Wild rice white paper 2011, 2). The centrality of wild rice is reflected in the Anishinaabe legal system, and

---

[3]Of course, these categories are not mutually exclusive. Indeed, the participation of people who straddle this boundary has been important and fruitful as they have knowledge from within both worlds. I will nonetheless continue talking about "scientists" and "Anishinaabeg" as synecdoche for a worldview characteristic of each.

[4]The risk of biological contamination is related in a different way to the gap in understanding at issue here: wild rice is at risk from sulfate runoff from mining operations, and many scientists are interested in doing research that can help "save" wild rice from this pollution. This has been one contested area of possible (depending who you ask) collaboration between scientists and Anishinaabeg.

[5]It is worth noting here, that the meaning and use of *species* is not uncontentious. There has been much discussion among philosophers and scientists about the concept (for one compendium of this discussion, see Wheeler and Meier 2000), so appeals to species membership are far from conclusive.

Anishinaabeg have always seen themselves as caretakers of the rice.[6] As one Native participant in the 2013 symposium commented, "Even though manoomin varies in size and color, it is always perfect" (NMBWC 2013, 2). These ideas refuse a conception of wild rice that reduces it to its genetic identity, nutritional, or commodity value. Manoomin is food, medicine, a gift from the Creator, variable, yet perfect, all at once. An Anishinaabe wildlife biologist shared the importance of imagining the role of wild rice as an organism, embedded in its ecology: "who are my allies, how do I serve of the earth for the purpose that I'm here for?" (8). Thus, the importance of connections and relationship are the starting point for learning about manoomin, rather than the interior patterns of its genetic identity.

From this perspective, the actions and intentions of the university can seem incomprehensible and downright sinister. Descriptions of attempts by the university to understand wild rice are frequently met by rancor and suspicion. At the same symposium described above, a Native participant remarked that "the university sees itself in the business of stealing life and patenting it to fund itself" (8). The university's appetite for information makes it inattentive to the particular state of the relationships we learn within. With manoomin, the state of our relationship changes what we can know; as another participant commented, "knowledge can be compared to picking berries; not all are ripe for the picking" (18). Here is a clear difference from the assumption that knowledge is contained within genetic identity, ready to be harvested by anyone; instead we have an expression that knowledge belongs to manoomin and has sense within our (healthy) relationship with manoomin—it cannot be claimed by just anyone and removed from its context.

18.3   Knowledge as a Commodity

An epistemologist's look at the dynamics of this relationship reveals that one persistent conflict centers on differing attitudes toward knowledge. In addition to the important non-epistemic reasons scientists and the university fail to understand and respond in ways that are satisfying to Anishinaabeg (e.g., funding ties to paddy-growers, legislative support for continued research, and a sense of entitlement supported by deep histories of colonialism and exploitation) underlying assumptions about what knowledge is play a role in this recalcitrant lack of understanding. I argue that in this case, scientists' tendency to view knowledge as a

---

[6]Here is a focus on governance that illustrates another important difference between commodified Western knowledge and Indigenous knowledge. As Kyle Powys Whyte explains in an effort to resist the instrumentalization of Indigenous knowledges by interested non-Native academics and policy-makers, one central feature of Indigenous knowledges is their role in Indigenous governance (Whyte 2015).

commodity contributes to their inability to understand or adequately respond to the concerns articulated to them by Native communities.

One way to begin thinking about the complex role commodities play in our lives is to consider what we mean when we talk about belonging. When I purchase a sweater, it is appropriate to say that the sweater belongs to me, and the same is true for a house, or a book, or a computer. In this way, belonging can refer to ownership of private property or commodities. Contrast this with a kind of belonging that designates membership in a group or a sense of place within community. Philosophy has at times offered me a sense of belonging that I had not experienced in other disciplines; I felt like I had found people who thought like I did, and I identified with the methods and interests of (some!) philosophers. Neither the commodity-sense or the identity-sense of belonging obviously applies to knowledge, but attention to the ways we actually treat knowledge in each of these senses can illuminate some of the difficulty plaguing researchers' attempts to be in solidarity with the communities they study, or with communities that have a stake in what they study.

We have some *prima facie* reasons to think that knowledge is not at all like a commodity. The most obvious one is that knowledge is not a material thing. I cannot purchase or exchange a bit of knowledge the way I can a sweater, in part because knowledge can be multiply instanced: I can acquire knowledge for myself without taking it away from anyone else. Likewise, I can share knowledge with someone else without losing it for myself. Another consequence of knowledge being multiply instanced is that it is not finite in the sense necessary for a commodity to be subject to supply and demand.

Nevertheless, we often do treat knowledge as though it were a commodity. For one thing, we appear to buy and sell it. Academics (especially those of us in the humanities, who rarely have anything material to show for our research besides articles and presentations) are, under a capitalist regime, paid and promoted on the basis of our production and attempted transmission of knowledge. The subscription fees to academic journals that open-access advocates challenge may also be enough to persuade us that what's being sold goes far beyond paper, ink, and binding. Patents, copyright, and other issues of intellectual property also point to an assumption that knowledge is the sort of thing that can be owned as a commodity.

The tensions between the *prima facie* reasons for thinking knowledge is not a commodity and the fact that we treat it as one create a difficulty: some work must be done to get knowledge into a state where it can be treated and traded as a commodity. In this way, knowledge functions as what the social theorist and economic historian Karl Polanyi described as a fictitious commodity. Polanyi counts land, labor, and money as fictitious commodities because they are not made for the purpose of being traded on the market and require regulation to prevent running out of natural resources, exhausting the labor force, and making money valueless (Polanyi 1944). Likewise, knowledge requires intervention in order to render it fungible. Data is excavated from its context and recontextualized using the theories of a particular discipline, within the background of the inherited histories of European imperialism and colonialism that found academia (Smith 2012), to make it legible within the academic context. Depending on the object of research, knowledge may need more or less intervention to render it a commodity. When research reaches across very different contexts, this processing can have dramatic effects, as when "the history and culture of the [global] South are discovered and translated in the journals of the North, only to come back, reconceptualized and repackaged, as development interventions" (Escobar 2012, 181; see also Smith 2012).

Obscuring the effects of this work on our relationships with each other and with knowledge itself creates serious barriers to working in solidarity with people who either do not see knowledge in general as a commodity, do not see the particular knowledge at play as a commodity, or who do not stand to benefit from the commodification of knowledge while others do (namely, researchers and various powerful institutions that rely on their work). It is unlikely that many academics explicitly think of knowledge as a commodity, but without an acknowledgement of this assumption, the possibility of solidarity is undercut by a refusal to see the ways others might think of knowledge differently. Treating knowledge as a commodity while maintaining otherwise creates a culture of denial that people with whom academics may want to be in solidarity can sense and that can create suspicion and distrust. This kind of ignorance can be a special barrier for researchers who take themselves to be doing work that benefits everyone, or is on behalf of a particular community, since faith in their own pure intentions creates no incentive to be critical of the harm they and their work may still cause (Smith 2012).

Consider the many projects which, though they engage the participation of community members in research within those communities, fail to address asymmetries of power and interest in the results produced. One such example of this process is articulated by Lorraine Code in her work on epistemic practices used by NGOs in Tanzania (Code 2008). Code praises the practice of incorporating local knowledge through participatory research on health initiatives for its ability to generate better (i.e., more reliable and accurate) knowledge about the needs of locals. Yet Kristie Dotson has noted that, in the absence of Tanzanian voices assessing the research programs, the possibility remains that the participatory research practices were epistemically effective, but not just (Dotson 2008). The mere fact of including some knowledge of community members within research is no guarantee that the research is genuinely in solidarity with the interests of those community members, and this is even more clear if we are alert to the ways that knowledge may be expropriated from the community and commodified in a way that benefits the researchers, even if their intentions are good. That as university researchers we are accumulating publications on the basis of such research, and that there are few, if any, norms for crediting the participation of community members, creates an imbalance that frequently goes unrecognized by researchers who take themselves to be doing the best work they can to involve communities and create knowledge on behalf of all of us.

18.4   Knowledge About Wild Rice

I now turn to a discussion of how the distinction between different senses of "belonging" may help explain the difficult relationship in the wild rice case: though scientists and Anishinaabeg are both talking about knowledge, scientists in general are working with a conception of knowledge (and wild rice itself) belonging as a commodity, and Anishinaabeg in general talk of knowledge (and wild rice itself) belonging—that is, having meaning—only within relationship.

There are several reasons I'm inclined to say that scientists are treating knowledge as a commodity. What follows is not so much an argument, but a list of these reasons, some of which appeal to the particulars of this case, and some of which are instances of the larger Western knowledge-producing paradigm described above. First, the structures of the university as a knowledge-producing institution strongly favor treating knowledge as a commodity and provide scant means to think of knowledge otherwise. Because academics are not typically aware of these assumptions, or of alternative ways of thinking about knowledge, they are not able to develop the skills to correct for them. Given the especially great value placed on objectivity within science and a persistent tendency to see knowledge as value-free (in spite of the substantial challenges to this idea within philosophy of science), there are even more reasons to think that scientists are especially susceptible to commodity-thinking, and especially unlikely to identify it as such.

Returning to the specific issue of genetic identity, I suggest that one reason this focus persists is that genetic identity is a tool to sort discrete items so that they can be treated as commodities. As knowledge in general must be extracted from its original context, and reprocessed to be traded on the market, those interested in commodifying knowledge of hybridized wild rice (i.e., the particular varietal) must create an object to be the commodity. Genetic essentialism is the vehicle for this process; collapsing the identity of the varietal itself with its genome allows scientists to point to a particular thing which they have created for the market and which is able to be their (or the university's or the state's) property.[7] It is thus ironic to refer to genetic similarity to dismiss Anishinaabeg concerns that hybridized wild rice has been changed, given that changing wild rice was the goal of hybridizing research.

It also must be noted that a turn from family lineage and history to genetic identity has played a different and historically relevant role in the relationship between Anishinaabeg and the European settlers of this land. The transition to blood quantum-based Tribal identification was used as a way for settlers to purchase land from "mixed-blood" Native people, which would have otherwise been

---

[7]In point of fact the University of Minnesota does not to my knowledge hold patents on the 6 varieties of wild rice developed as a result of its research. The harms to wild rice that concern Anishinaabeg are distinct from the sort of concerns characteristic of small farmers resisting the influence of patent-hoarding corporations like Monsanto. Nonetheless the habits of treating knowledge as property plays an important role in this harm.

illegal. As sociologist Eva Garroutte states, "The original, stated intention of blood quantum distinctions was to determine that point at which the various responsibilities of the dominant society to Indian people ended. The ultimate federal intention was to use the blood quantum standard as a means to liquidate tribal lands and to eliminate government trust responsibility to tribes, along with entitlement programs, treaty rights, and reservations" (Quoted in Doerfler 2015). This context is relevant to how contemporary emphases on genetic essentialism are perceived by Native people.

18.5 "Knowledge Belongs to Everyone"

Some scientists have a tendency to talk about data existing "out there" in the world, and as if it ought to be accessible to anyone. This perspective is reflected in governance institutions as well: In the unanimous U.S. Supreme Court decision declaring that naturally occurring DNA sequences cannot be patented (though closely related synthetic sequences can), Justice Clarence Thomas cites precedent that "manifestations ... of nature [are] free to all men" while arguing that the claim is about the *information* contained in the sequence, rather than the mere chemical structure (2013). These two claims together provide a compelling example of the belief that information exists in nature, for everyone, though they elide issues of who has the power, institutional support, money, resources, and authority to actually access that information and claim knowledge created from it.

This view of knowledge pervades even well-intentioned efforts to promote justice in knowledge production. Consider efforts to make knowledge and information more accessible, as we can see in the struggles for open-access publishing. For good reasons, many within and outside the academy are pushing against proprietary research and even publication in expensive, hard-to-access journals. At the University of Minnesota, and at other universities throughout the United States, research funded by private sources is strictly regulated to ensure that transparency and accessibility of the research meets standards for publicly-funded research, and are not distorted by the proprietary demands of funders (Regents 2010). Proprietary and otherwise inaccessible research has been criticized for restricting innovation and public deliberation, limiting who can create and consume research, and for keeping knowledge in the hands of the few, as it is often produced on the backs of people who are not able to supervise its production or access its results (e.g., Deibel 2014; Evans and Selgelid 2015; Parker 2013; Wellen 2004). As a general rule, it seems fair to say that knowledge should not be accessible only to a few privileged people; this creates serious concerns for democracy, and we have good reasons to challenge the practice.

Yet the response to problems of proprietary or inaccessible research is articulated within a paradigm that is itself problematic for many of the same reasons as proprietary research. A common solution to concerns about proprietary research is to offer open-source research and open-access publishing as necessary tonics that will make knowledge production and dissemination more democratic (e.g., Deibel 2014; Parker 2013). Many of these arguments are infused with rhetoric about knowledge belonging to everyone (e.g., Ernst 2012; Hansson et al. 2011). This response isn't merely theoretical: many universities are now encouraging and even requiring faculty to contribute their publications to open-access repositories (see Regents 2010 for one example). National initiatives with international cooperation, like the National Center for Biotechnology Information's GenBank maintain open access databases of millions of genetic sequences compiled from submissions from laboratories across the country, the U.S. Patent and Trademark Office, and genome-sequencing organizations (Benson et al. 2013). This information is public, purportedly accessible by anyone, thus apparently solving problems of unjustified possession of knowledge by only the privileged.

This response solves the problem of ownership of and access to knowledge by the rarefied few by claiming *collective* ownership. But the emphasis on collective ownership does nothing to challenge the assumption that knowledge is the sort of thing that can be owned. This is an important mistake, since the ways in which we treat knowledge as though it were a commodity, and the ways it in fact operates as one transform relationships and create barriers to solidarity. To put a more historical point on it, claiming that land in the first instance belonged to everyone has been an important step in European colonization of land inhabited by Indigenous peoples (Whitt 1998), and this practice continues as a tool for cultural appropriation (Wylie 2015; Hooks 1992). In a parallel fashion, the belief that knowledge belongs to everyone and exists out there in the world waiting to be harvested has served and continues to serve as a means to commodify that knowledge.

18.6   Knowledge as Relational

I've now painted a fairly bleak portrait of attitudes toward knowledge within the academy. Indeed, this view is somewhat totalizing; there are alternatives to thinking of knowledge as a commodity, and many of them do exist (though they often struggle for recognition) within the academy. Returning to 'belonging' allows us to see knowledge as something that exists within relationship, something that belongs within the worldview of a particular community, something that changes when it is appropriated into another worldview. Without endorsing any particular view, I suggest several variations of treating knowledge as other than a commodity. Since, as I claim, the tendency to treat knowledge as a commodity is deeply embedded in the customs of Western knowledge production, one place to look for alternative views is within indigenous methodologies and epistemologies.

While taking into account the enormous diversity of Indigenous peoples around the world, Cree scholar Shawn Wilson argues that a common feature of Indigenous worldviews is an emphasis on relationship and the centrality of interconnection (2008). Wilson argues that this perspective on metaphysics equally influences Indigenous epistemologies, since, by denying a clear distinction between observer and observed, an Indigenous researcher must engage in a reciprocal exchange with those from whom she hopes to learn. Wilson contrasts a Western worldview that treats knowledge as individual and able to be owned with "the Indigenous paradigm, where knowledge is seen as belonging to the cosmos of which we are a part and where researchers are only the interpreters of this knowledge" (38). Another way to see this is that in the reciprocal process of research and learning, the researcher belongs as much to the world she studies as it does to her.

Robin Wall Kimmerer, a botanist, poet, and enrolled member of Citizen Potawatomi Nation, articulates this reciprocal exchange in terms of gift-giving, though it is important to distinguish her claim from descriptions of gift-giving common to 20th century anthropology, in which gifts are exchanged as commodities to accrue debt to the recipient (2013). Kimmerer's conception of knowledge as a gift is meant to evoke the kind of relationship in which, though they inspire an obligation in the sense of strengthening the relationship, gifts require no particular exchange of goods as a response. Rather, this sense of gift-giving calls for attention and care as the prototypically human response, especially in light of the incommensurability of the wisdom we receive from our relations in the world, and what we are able to give in return.[8] For Kimmerer, gifts are inextricably identified with responsibility and relationship. She describes the clash of worldviews that stunned her family as they were forced to migrate away from their ancestral homeland: "In the settler mind, land was property, real estate, capital, or natural resources. But to our people, it was everything: identity, the connection to our ancestors, the home of our nonhuman kinfolk, our pharmacy, our library, the source of all that sustained us. Our lands were where our responsibility to the world was enacted, sacred ground. It belonged to itself; it was a gift, not a commodity, so it could never be bought or sold" (17). When we see the way knowledge is integrated into the material and spiritual relationships with land, nature, and human and nonhuman relations, commodifying any one of these appears incoherent.

Even from (somewhat) more central positions within the Western academy there is precedent for thinking of knowledge as other than a commodity. Feminist epistemologists and philosophers of science have argued that thinking about knowledge and research in a way that emphasizes the primacy of the relationships we learn within is a way not only to produce more just knowledge, but also to produce better knowledge. Alison Wylie has articulated this position in relation to archeological research in the midst of larger anxiety about pluralism and academic freedom. In response to negative reactions (articulated in terms of knowledge belonging to everyone) to the Native American Grave Protection and Repatriation Act (NAGPRA), Wylie claims that in many cases, deciding not to undertake a study in

---

[8]Tuck and Yang use incommensurability as a powerful tool to describe moves to settler-innocence that try to incorporate Indigenous worldviews into existing colonial epistemologies in their paper, "Decolonization is not a Metaphor" (Tuck and Wayne 2012). Incommensurability may be another productive way of thinking about the gulf between University of Minnesota researcher's and Anishinaabe attitudes toward knowledge.

the interest of preserving relationships with descendants of people studied can have positive effects on the knowledge we create (Nicholas and Wylie 2009; Wylie 2015).

One possible consequence if researchers genuinely engage with diverse communities' demands is that researchers may develop relationships that ensure more people from subordinated worldviews have access to and a hand in creating knowledge in the future. Naomi Scheman argues for this as a goal for academics and universities and thinks of it as a "sustainable epistemology" (Scheman 2012). This sustainable epistemology resists treating knowledge like a commodity because it acknowledges the importance of the context within which knowledge is produced: the people who produce it, their position within systems of power and domination, and their vulnerability to the consequences of that knowledge. Still, within the context of the Western university, this perspective is at risk of treating knowledge as a commodity in the long run, since, as Scheman notes, "Western science and the modern university are notoriously omnivorous, as adaptively capable of fattening on increasingly diverse epistemic diets as neoliberal regimes of multiculturalism are of commoditizing diverse cultures" (486). Given the weight of the historical political role knowledge-producing institutions play, even if put to use within the university these perspectives are no guarantee against commodifying knowledge, though for our purposes they do offer an alternative set of perspectives.

These alternative ways of thinking of knowledge within the university create some hope that collaborations between scientists and communities need not necessarily face the kinds of problems I've described in this paper. Many scholars are doing important and creative work within the space of these relationships, creating and revising methods of community-based participatory research, examining what it means to be an activist-scholar, and creating new ways to address issues of translation and collective authorship (De Leeuw et al. 2012; Derickson and Routledge 2015; Nagar 2014). These efforts face an uphill battle against the institutional inertia and incentives to treat knowledge as a commodity.

18.7   Conclusion

I have suggested that the difficulties plaguing attempts to improve the relationship between the University of Minnesota and the Anishinaabe people are symptomatic of larger structural issues within the academic system. Internal university structures entrench the commodification of knowledge. Disciplinary boundaries discourage collaboration and encourage ideas of purity within disciplines, and even interdisciplinary work in the sciences often focuses on the overlapping areas of sub-disciplines, as opposed to traversing larger differences between, for example, scientific disciplines and philosophy of science. Here, then, is one explanation for the ignorance by scientists in the wild rice case of prominent interventions in philosophy of science that challenge widely-accepted beliefs in the value-neutrality of science, as well as for the dearth of examples of scientists interested in working with Native communities seeking out the wealth of social science and humanities literature on these topics. This insularity protects disciplinary interests and fractures responsibility to the extent that no individual researcher feels empowered to take responsibility for research practices that appear increasingly outside his or her purview (Smith 2012, 71).

These structures are not limited to academia, but are a reflection of the great inheritance of Western universities: imperialism and universalism about the knowledge produced within such universities (as well as other privileged sites). As Linda Tuhiwai Smith articulates, this history and its continuing impact on the world reinforce and facilitate the enclosure and commodification of knowledge. She says, "The globalization of knowledge and Western culture constantly reaffirms the West's view of itself as the centre of legitimate knowledge, the arbiter of what counts as knowledge and the source of 'civilized' knowledge. This form of global knowledge is generally referred to as 'universal' knowledge, available to all and not really 'owned' by anyone, that is, until non-Western scholars make claims to it" (Smith 2012, 66). Within this context, we can see that the lasting impacts of imperialism and colonialism, and the governance structures supporting these forms of domination play an important role in the production of knowledge within universities and in other privileged contexts.

Finally, though universities are some of our most conspicuous knowledge-producing institutions, they are far from the only ones. Certainly, they are not our only institutions that treat knowledge as a commodity, as this is repeatedly affirmed within legislative and judicial bodies, and policy-setting agencies. These institutions, too, are in the business of producing knowledge, and in many cases are much more likely to translate that knowledge into practical results. The effects of these institutions treating knowledge as a commodity have ramifications beyond the (ir)responsible research relationships I have discussed as a primary target within academia. Attention to the extended effects of commodifying knowledge in these contexts ought to be a priority if we are interested in producing knowledge that supports, rather than harms movements for liberation, sovereignty, and justice.

These are some of the entrenched differences in attitude toward knowledge that contribute to my pessimism about the ability for even those scientists and university representatives (including myself) who desire to repair the relationship between the university and the Anishinaabe to be in genuine solidarity. For reasons outside the scope of this paper I do think that solidarity cannot require perfect understanding on the part of those wishing to be in solidarity (I'll call them allies), but an awareness of and sensitivity on the part of allies that they do not fully understand does seem like an important attribute of a relationship of solidarity. Thus, I suspect that unless academic attitudes toward knowledge are exposed, essential misunderstandings on the part of the scientists will persist and the effort to heal will be stymied. At the same time, addressing these assumptions will not be sufficient for relationship repair; there are a multitude of other historical and material harms that will need to be repaired before trust is restored. Yet even if we restrict our attention to the somewhat more surmountable epistemic issues addressed here, change from the university seems unlikely, given the structural and historical entrenchment of treating knowledge like a commodity. I hope that observing some of the social-epistemic barriers to doing responsible research can help us do more of it by bringing to light values to which we may have thought our research immune. Keeping these limitations in mind can help those of us interested in promoting genuinely collaborative and supportive work within the university to envision the space we must create to facilitate creative, critical, responsible research.

References

Association for Molecular Pathology et. al. v. Myriad Genetics Inc., et al. 569 U.S. 2013.

Benson, D. A., Cavanaugh, M., Clark, K., Karsch-Mizrachi, I., Lipman, D. J., Ostell, J., and Sayers, E. W. 2013. GenBank. *Nucleic Acids Research*, 41 (Database issue): D36–D42. doi:10.1093/nar/gks1195.

Cadieux, Valentine, and Slocum, Rachel. 2015. What does it mean to do food justice? *Journal of Political Ecology* 22: 1–26.

Code, Lorraine. 2008. Advocacy, negotiation, and the politics of unknowing. *The Southern Jounal of Philosophy* 46: 32–51. doi:10.1111/j.2041-6962.2008.tb00152.x.

De Leeuw, S., E. Cameron, and M. Greenwood. 2012. Participatory and community-based research, indigenous geographies, and the spaces of frienship: A critical engagement. *Canadian Geographer* 52: 180–194.

Deibel, E. 2014. Open genetic code: On open source in the life sciences. *Life Sciences, Society and Policy*, 10. doi:10.1186/2195-7819-10-2.

Derickson, Kate Driscoll, and Paul Routledge. 2015. Resourcing scholar-activism: Collaboration, transformation, and the production of knowledge. *The Professional Geographer* 67 (1): 1–7.

Doerfler, Jill. 2015. *Those Who Belong: Identity, Family, Blood, and Citizenship among the White Earth Anishinaabeg*. East Lansing: Michigan State University Press.

Doerfler, Jill. 2003. Where the food grows on water: The continuance of scientific racism and colonization. Unpublished manuscript.

Dotson, Kristie. 2008. In search of Tanzania: Are effective epistemic practices sufficient for just epistemic practices? *The Southern Journal of Philosophy* 46: 52–64. doi:10.1111/j.2041-6962. 2008.tb00153.x.

Ernst, David. 2012. Knowledge Belongs to Everyone. Presentation at TEDxKyoto. https://www.youtube.com/watch?v=eA9Tv-OvoZU. Accessed 5 June 2016.

Escobar, Arturo. 2012. *Encountering Development: The Making and Unmaking of the Third World*. Princeton: Princeton University Press.

Evans, N.G., and M.J. Selgelid. 2015. Biosecurity and open-source biology: The promise and peril of distributed synthetic biological technologies. *Science and Engineering Ethics* 21 (4): 1065–1083. doi:10.1007/s11948-014-9591-3.

Gottlieb, R., and A. Joshi. 2010. *Food Justice*. Cambridge: MIT Press.

Gunderson, Dan and Julin, Chris. 2002. Wild Rice at the Center of a Cultural Dispute. *Minnesota Public Radio*, September 24.

Hansson, B., H. Van Ditmarsch, P. Engel, S.O. Hansson, V. Hendricks, S. Holm, P. Jacobson, A. Meijers, S. Richardson, and H. Rott. 2011. A theoria round table on philosophy publishing. *Theoria* 77 (2): 104–116. doi:10.1111/j.1755-2567.2011.01097.x.

Hooks, Bell. 1992. Eating the other: Desire and resistance. In *Black looks: Race and representation*, 21–39. Boston: South End Press.

Kahler, Alex. 2015. University's Research History Involving Wild Rice. Presentation at Nibi miinawaa Manoomin: New Pathways to a Shared Future. Onamia, MN (September 28–29).

Kimmerer, Robin Wall. 2013. *Braiding Sweetgrass: Indigenous Wisdom, Scientific Knowledge and the Teaching of Plants*. Minneapolis: Milkweed Editions.

Kokotovich, Adam. 2014. Contesting Risk: Science, Governance and the Future of Plant Genetic Engineering. Doctoral dissertation, University of Minnesota.

La Via Campesina. 2007. Declaration of Nyéléni. https://nyeleni.org/IMG/pdf/DeclNyeleni-en.pdf . Accessed 16 Jan 2017.

Nagar, R. 2014. *Muddying the Waters: Coauthoring Feminisms Across Scholarship and Activism.* Chicago: University of Illinois Press.

Nicholas, G.P., and A. Wylie. 2009. Archaeological finds: Legacies of appropriation, modes of response. In *The Ethics of Cultural Appropriation*, ed. James O. Young, and Conrad G. Brunk, 11–54. West Sussex: Wiley-Blackwell.

NMBWC (Nibi and Manoomin Bridging Worldviews Committee). 2013. Working Group Notes. http://www.cfans.umn.edu/sites/cfans.umn.edu/files/NMS_2013_WorkingGroupNotes_0.pdf. Accessed 9 June 2016.

Parker, M. 2013. The ethics of open access publishing. *BMC Medical Ethics* 14 (1): 16. doi:10.1186/1472-6939-14-16.

Polanyi, Karl. 1944. *The Great Transformation: The Political and Economic Origins of Our Time*, 2001. Reprint, Boston: Beacon Press.

Regents of the University of Minnesota. 2016. Report of the Provost's Grand Challenges Research Strategies: Advancing the research goals of the Twin Cities Campus Strategic Plan. https://strategic-planning.umn.edu/sites/strategic-planning.umn.edu/files/gc_research_report_umn_strategicplan.pdf. Accessed 29 May 2016.

Regents of the University of Minnesota. 2010. Openness in Research. http://regents.umn.edu/sites/regents.umn.edu/files/policies/Openness_in_Research.pdf. Accessed 5 June 2016.

Scheman, N. 2012. Toward a sustainable epistemology. *Social Epistemology: A Journal of Knowledge, Culture, and Policy* 26 (3–4): 471–489. doi:10.1080/02691728.2012.727194.

Scholz, Sally J. 2008. *Political Solidarity*. State College: Penn State University Press.

Smith, Linda Tuhiwai. 2012. *Decolonizing Methodologies: Research and Indigenous Peoples*, 2nd ed. New York: Zed Books Ltd.

Tuck, Eve and Yang, K. Wayne. 2012. Decolonization is not a Metaphor. *Decolonization: Indigeneity, Education & Society*. 1 (1): 1–40.

Wellen, R. 2004. Taking on commercial scholarly journals: Reflections on the "Open Access" movement. *Journal of Academic Ethics* 2: 101–118. doi:10.1023/B:JAET.0000039010.14325. 3d.

Wheeler, Quentin D., and Rudolf Meier (eds.). 2000. *Species Concepts and Phylogenetic Theory: A Debate*. New York: Columbia University Press.

Whitt, Laurie Anne. 1998. Cultural imperialism and the marketing of native America. In *Natives and Academics: Research and Writing About American Indians*, ed. Devon A. Mihesuah, 139–171. Lincoln: University of Nebraska Press.

Whyte, Kyle Powys. 2015. What do Indigenous Knowledges do for Indigenous Peoples? In *Keepers of the green world: Traditional ecological knowledge and sustainability*, ed. Melissa K. Nelson and Dan Shilling. https://www.academia.edu/11293856/What_do_Indigenous_Knowledges_do_for_Indigenous_Peoples.

Wild rice white paper: Preserving the Integrity of Manoomin in Minnesota. 2011. http://www.cfans.umn.edu/diversity/web%20text/Wild-Rice/WhitePaper–Final%20Version2011.pdf. Accessed 9 June 2016.

Wilson, Shawn. 2008. *Research is Ceremony: Indigenous Research Methods*. Winnipeg: Fernwood.

Wylie, A. 2015. A Plurality of Pluralisms: Collaborative Practice in Archaeology. In *Objectivity in science, Boston Studies in the Philosophy and History of Science*, ed. F. Padoyani et. al., 198–210. doi:10.1007/978-3-319-14349-1_10.

Part IV
Food Justice and Animal Lives

Chapter 19
A Vignette from Michigan State University's Student Organic Farm

Laurie Thorp

Abstract Laurie Thorp explores the liminal space between vegetarian and non-vegetarian diets when she describes caring for pigs as ultimately "killed yes, but no, not killable," in an account that stimulates the kinds of reflection Heldke calls for.

Keywords Food justice · Animal agriculture · Eating animals · Organic animal husbandry · Animal welfare · Pig farming

Feminist authors Donna Haraway and Ursala Leguin implore us to begin telling our small stories from the field. These stories are important because they provide a counterbalance to the dominant narrative being authored by the power elite in agriculture. Indeed we had better start telling each other a new story, because the big boys are telling a monolithic story about food safety and GMOs feeding the world that just doesn't ring true for me. There are compelling food stories out in our communities but rarely are they part of our academic discourse, troubling, so allow me to tell you a story. This is my story of perseverance and love, caring, and joy catalyzed by pigs in communion with college students, the land and many care-filled colleagues. This is my fleshly practice, my pig/earth reunion story, told from the fertile margins of our university, from a place called the Student Organic Farm, a 15 acre farm—contested territory in many ways. This is where I go to sow my wild oats, and swing my handbag at the hoodlums of agribusiness, to dig in my heels and practice a different brand of science. Watching the spot fires on the back of my hands, I am filled with a sense of urgency for the plight of my companion critters. A gift/curse to feel the suffering of other living beings. My pulse quickens as I drive down the dirt road to be with our pigs, it is a visceral reaction, my body is sending me a message of kinship, I listen, I feel, I pine for these animals.

L. Thorp (✉)
Michigan State University, Lansing, MI, USA
e-mail: Thorpl@msu.edu

© Springer International Publishing AG 2017
I. Werkheiser and Z. Piso (eds.), *Food Justice in US and Global Contexts*,
The International Library of Environmental, Agricultural and Food Ethics 24,
DOI 10.1007/978-3-319-57174-4_19

Our companion species are Berkshire/Yorkshire crossbred sows, released from confinement to live out their life, brought to death, killed yes, but no, not killable, grazing on our pastures, all the while teaching us how to respond, to be responsible, to attend to, to care for and to be cared for.

I ask you, where do we learn this in school?

This began as a research project to study the grazing activities and crop preferences of these fine animals, to demonstrate how to bring animals back onto the land. Imagine that. Perhaps it was to repair what has been broken in two, to re-integrate animals in small scale farming systems. But there was so much more in this contact zone. You see, our students are starved for contact, students, hell, we all are starved for contact.

Contact with human and non-human critters.
Farm as contact zone. Farm as healing zone,
wet nose touch,
gentle head-butt hello
burst of hot breath, snort,
means special greeting of recognition, you are my kin.

It is farrowing time once again. The great cycle of life turns. As my farm colleague Dale once told me, this is the biorhythm of the farm. I am so grateful for these rhythms. Our students live in a time of great arrhythmia. The farm pulls them out of the so called virtual reality and draws them into a place of real consequence.

Terry Tempest Williams [A Voice in the Wilderness] says we are a people in a process of great transition, forgetting what we are connected to. We are losing our frame of reference. Birds pass by and we hardly know who they are, we don't know their stories. She observes that this is leading us to a place of inconsolable loneliness, of unspoken hunger and the only thing that can bring us into a place of fullness is being out on the land with the other. Our pigs brought us back to our senses, back to a place of fullness. Touch, taste, sight, sound, smell, remember we learned these in kindergarten. Did you know pigs bark? Have you seen pig play? Snouts dig through the soil, nibble on tender roots, they have a dunging pattern; feel the soft hairs around their nose. They have pads on their feet. Their skin smells sweet. No iPods, no cell phones, no ear buds, no texting here, no way. Pig life grabs your attention and doesn't let go. And if their life doesn't grab your attention their death certainly will. Joy, sorrow, joy.

This is where I grow my soiled wisdom, in the mud and the manure that feeds us, down among the hidden critters of the loam, where one can never forget that death and decay feed life. I've always been suspicious of anybody or anything that is too clean. Tired of all that goddamn Purell. The farm is a counterbalance to the cleanliness of the academy. Sweat and mud caked, tired muscles ache, we learn how to be fully human from our pigs. We come to know each other as more than purely rational beings. Pig work is hard work, difficult decisions, emotionally demanding, crossing disciplines, complex questions without any answers. These are the disruptive details of a story I don't know how to end. They are my excessive, transgressive, obsessive data. And so we grow in our love and our trust drawing a circle around our most cherished pigs. Ours is a shared and situated knowledge, soiled, blood soaked, tear stained, but resilient. This is the kind of knowledge production that I can believe in. Ours is a knowledge of becoming-with, staying with, accounting for, all the relationships in our contact zone. Indeed Donna, these species matter.

Chapter 20
Introduction to Food Justice and Animal Lives

Lisa Heldke

Abstract As Lisa Heldke points out in this introduction, much of the literature on food justice which touches on non-human animals focuses on divisions between those who see eating animals as an absolute injustice and those who do not—a division she describes as painful, counterproductive, and in need of philosophical reflection.

Keywords Food justice · Animal agriculture · Eating animals · Organic animal husbandry · Animal welfare · Pig farming

In an era in which alternative food and agriculture movements are making symbolically meaningful (though materially smallish) impacts on the industrial food system, it can be both difficult and painful to consider the issue of justice with respect to those non-human animals we call "farm animals".

Part of the painfulness of course arises from seeing the images of industrial-scale animal agriculture that fill our screens and haunt our dreams. It's difficult to contemplate industrial animal agriculture with anything like equanimity. It's hard to think about the fact that most of the animals that most Americans eat have lived their lives in vast confinement operations. When I contemplate scenes portraying the living conditions of factory-raised pigs, and the vast "lagoons" of their waste, I, for one, marvel that any parent who ever wanted their kid to eat another hot dog would allow that kid to see "Babe," a children's movie about a pig that begins in just such a factory setting, but manages to escape it to live out his life herding sheep. The movie begins in a stylized industrial barn, where piglets are being fed from mechanical teats. The voiceover explains that when piglets depart from the barn, they never return—leaving the remaining piglets to fantasize about the magical place they've gone; a place so wonderful the leavers never want to return

L. Heldke (✉)
Gustavus Adolphus College, St. Peter, USA
e-mail: heldke@gustavus.edu

© Springer International Publishing AG 2017
I. Werkheiser and Z. Piso (eds.), *Food Justice in US and Global Contexts*,
The International Library of Environmental, Agricultural and Food Ethics 24,
DOI 10.1007/978-3-319-57174-4_20

home. For a children's movie, it's a strikingly sardonic, shockingly blunt glimpse of an industry that brutalizes both animals and the people who work in it.

But no small part of the painfulness here comes from the fact that, even among advocates of alternative agricultural systems,[1] discussions about what justice demands of us vis a vis other animals often splits participants into two camps—those who eat animals and believe it can be justified,[2] and those who don't and believe it cannot be. Those two camps further splinter along questions of just which animals or animal products (eggs, milk, honey, wool) their members do or don't believe one can justifiably eat (free range, pastured, organic, wild caught, scavenged). Members of each group have often regarded the other group with suspicion and even hostility, each side accusing the other of acting in bad faith, or engaging in spurious reasoning to justify their position.

Such division is not only painful; it is counterproductive. Here, it seems, is a place where philosophical reflection can be of real use.

Presently, in the comparatively small body of philosophical literature that explores food and justice, a significant portion of that literature has been devoted to arguments about *what one ought not eat*—namely, animals. Philosophical arguments against eating animals can be grouped into several general categories. Environmental arguments hold that the consumption of animals is environmentally unjustifiable for any of several different reasons, including natural resource use, efficiency, and resource degradation. Arguments about the moral worth of animals hold that animals possess properties (sentience, capacity to feel pain, intelligence) that are, in and of themselves, deserving of moral consideration—consideration that includes, at a bare minimum, not being eaten by a human. (The most well-known of these arguments have been developed by philosophers Peter Singer and Tom Regan.) Another important strand of arguments, made most visible by Carol Adams, holds that the oppression visited on animals that are raised to be eaten parallels the oppression of women; rejecting meat is thus a feminist act that follows of necessity from understanding the logic of domination. Yet another suggests that vegetarianism is a situationally or contextually appropriate response to particular natural-cultural conditions. The work of Erin McKenna and Deane Curtin represent versions of this position.

Many ethical vegans and vegetarians not only argue *for* the moral benefits of refraining from eating meat, but argue *against* the very notion that there could *be* an argument from justice that advocates eating meat. From this vantage point, meat is murder and there can be no argument for that. Thus it is perhaps not surprising that the philosophical literature presently contains fewer works that develop arguments for an animal agriculture rooted in commitments to justice.

---

[1]Advocates, that is, who share the view that "factory farms" are unethical. In his essay in this section, Joseph Tuminello notes that, "opposition to industrial animal agriculture has become a point of ethical convergence for people upholding an array of moral principles…".

[2]Or, as I shall suggest below, who at least believe that arguments to show that meat eating cannot be just have fallen short of their mark.

Laurie Thorpe's paean to pigs suggests that we see this lack as a problem, a philosophical lacuna that ought to be filled, not an obvious consequence of the acknowledgment of animal killing. Thorpe writes that, for her and for her students, raising—and then slaughtering—pigs; living in close proximity with them, playing with them, caring for them, and then taking them to watch their throats be slit and their bodies carved into edible portions—teach us "how to respond, to be responsible, to attend to, to care for and to be cared for."

Thorpe does not avert her gaze from the fact that these pigs get killed and eaten; she reminds us of the fact in a phrase like "blood-soaked." All the same, she does not, in this essay, attempt to justify, or to argue for, the fact that these pigs become pork. She invites us to contemplate the kinds of learning that can only happen in such close proximity to these animals. She does not go on to tell us how to think about their deaths.

Thorpe gives us a hint about what that argument would be, when she writes that the pigs are "killed, yes, but no, not killable." How can we understand the gap between *killed* and *killable*? Can we make sense of the distinction between "killing" an animal, and that animal being "killable?" And does this distinction create a space big enough to accommodate a theory of just animal eating?

Among meat eaters, arguments have not so much attempted to justify the eating of animals, as to acknowledge that (many) humans do it, and to ask "how can we do it with a minimum of harm?" A mainstream, theoretically-rich work entitled *The Compassionate Carnivore* does a good job of this task. The author, Catherine Friend, sets out from her matter-of-fact admission that she *is* going to eat meat to ascertain how she might do so in ways that manifest the greatest respect for the animals she eats. She is not apologetic, but neither does she attempt to do anything like argue *for* the legitimacy of meat eating; instead, she simply states that she does it.

From the vantage point of many versions of ethical vegetarianism, such an argument can look like a cop-out; that, unless one grasps the nettle of arguing for the in-principle right to eat other animals, one is engaging in moral sleight-of-hand.

The essays in this section suggest otherwise. They explore the notion that it is possible to speak meaningfully and substantively about just relationships between humans and the animals we eat. Anne Portman, in her piece "Agriculture, Equality, and the Problem of Incorporation," puts the issue this way: "One might…object that the *problem of incorporation* [whereby a subordinate group is defined entirely in relation to the capacities of a dominant group] makes agriculture fundamentally incompatible with interspecies justice." In response, she argues that "defining ourselves in terms of our relationship to the nonhumans that we raise and consume goes a long way toward addressing the problem of incorporation (a problem that, in the case of animals raised for meat also refers to the fact that we take them into ourselves, literally). Thorpe and her students come to understand—to "ken"—the head butts and snorts and nose prints and barks with which pigs interact with their humans. They come to "define themselves" in relation to the pigs with which they spend their days.

Portman draws our attention to the necessity of eating and the continuity of planetary life, to argue that what are called for are "practices of ethical consumption." "We can," she writes, "identify the ethical commitments of reciprocity that underlie such practices: that a good human life acknowledges *kinship* with nonhuman life while at the same time *respecting* the other as independent, different, and always 'more than food.'" Or, in Thorpe's words, killed, yes, but no, not killable. In the end, Portman suggests, not only is animal agriculture not incompatible with interspecies justice, it might in fact be one of the most important sites from which to practice it. When we eat as an "agricultural act" (in Wendell Berry's powerful phrase), we "have the opportunity to address incorporation on an immediate and daily level."

Jonathan McConnell, in his "An Ecophenomenological Approach to Hunting, Animal Studies, and Food Justice," picks up some of the threads from Portman's and Thorpe's hands. Taking an ecophenomenological approach, he agrees with Portman that the great ethical harm to be avoided is to conceive of another animal as merely existing instrumentally for our benefit, something he describes as "ontological violence." Yet rather than look at agricultural animals as Portman and Thorpe have done, he examines the possibility of ethical, respectful hunting.

For animal rights activists, it is not always clear whether hunting is worse than animal agriculture. On the one hand, it seems better for an animal to have a good life cut short rather than a painful one with the same end, yet many vegetarians seem more bothered by hunting than eating a burger, perhaps because of what it signifies about the person who could do such a thing. McConnell complicates this picture. He does not simply assert that hunting requires or entails respecting one's prey; indeed he discusses several examples (like so-called "canned hunting," where deer are bred in pens and shot in those pens by trophy hunters) which are objectionable precisely because of their disrespectful ontological violence. Instead, McConnell offers very personal stories to suggest that hunting can be a part of an orientation to other animals based on respect and understanding. He also suggests that being prey, or at least potential prey, is another part of that orientation. McConnell suggests that perhaps most people's unfamiliarity with the thought of ever being food for another animal, combined with our unfamiliarity with seeing a particular living being and deciding that it will be our food, leads us to countenance the kinds of ethical violations we find in high-capacity animal production.

Joseph Tuminello's essay directly challenges the absolutism and theoretical "abstractionism" from which many arguments for vegetarianism begin. His question is not the "can we ever justify the eating of meat?" question I posed at the outset, but rather the pragmatist's "how can we creatively address the current conditions in which we raise animals, so as to improve their welfare?" On Tuminello's view, Thorpe's silence on the matter of how to justify eating-animals-in-general is not a skulking avoidance; it is a rejection of the legitimacy of starting from a question so abstract and hypothetical as "can eating animals ever be justified?"

Tuminello credits pragmatism's commitments to situatedness, pluralism, and amelioration as the source of the building blocks with which to formulate this theoretico-practical approach to animal welfare. He identifies the Farm Forward movement as embodying this pragmatist animal ethic, and articulates the ways they develop a "balance between setting long-term goals and pursuing amelioration through embracing multiple viewpoints...." Farm Forward's commitment to multiple viewpoints dramatically and concretely challenges vegetarian/meat eater polarization; while it advocates veganism, it "embraces a plurality of ways that people can positively respond to the concerns raised by industrial animal agriculture"—ways that include meat eating.

R.W. Mittendorf continues this effort to ground ethical debate about eating non-human animals in concrete practices. He expands the discussion to include the ritual slaughtering of animals for religious purposes. Mittendorf examines the debates that have arisen in the European context, where animal welfare regulations require stunning animals before slaughtering them, a requirement that challenges both Halal and Kosher laws and practices concerning slaughter—laws that make stunning problematic, at minimum.

Mittendorf documents the ways in which the clash between these two bodies of policy have resulted in animal welfare advocates and those he terms "religious traditionalists" seeing each other as the enemy. It's another example of the ways a sharp fault line can develop between those who argue for an end to all animal killing, and those who argue that killing and consuming animal flesh is justifiable in at least some cases for at least some purposes. Mittendorf argues that, rather than regarding each other as adversaries, members of these two groups should regard each other as sharing the commitment to improving animal welfare. The two groups will clearly *not* share many goals. Nonetheless, they have pragmatic reasons to work together, because doing so has the chance of advancing the interests of animals that they do share.

These philosophers show that it is both possible and desirable to destabilize the sharp divisions separating those who oppose all animal consumption, and those who condone it, or even advocate it for specific reasons. Their arguments help us see the power and effectiveness of combining forces to work *in* the chasm that too often separates food justice advocates who stand on different sides of the animal agriculture issue and thwarts their efforts.

Chapter 21
Farm Forward: A Pragmatist Approach to Advocacy in Agriculture

Joseph A. Tuminello III

Abstract In this chapter, I examine the philosophical and ethical views under-girding Farm Forward, a nonprofit animal advocacy group focused on reforming agricultural practices and ending factory farming. Specifically, I frame Farm Forward as an organization that embodies key dimensions of the pragmatist philosophical tradition. I begin by providing a brief overview of existing pragmatist work within animal ethics and describing the pragmatist concepts of pluralism, particularism, and amelioration. After placing Farm Forward's general vision for agricultural reform within this philosophical context, I review a number of the organization's recent projects to illustrate their commitment to the above pragmatist principles, as well as the progress that they have attained through particular advocacy strategies including education initiatives, public engagement, negotiations with multinational corporations, and building coalitions with other advocacy groups. Farm Forward's pragmatist approach to advocacy in agriculture, I argue, serves as an exemplary model for achieving positive change in concrete and inclusive ways.

Keywords Animal ethics · Agricultural ethics · American pragmatism · Animal advocacy

21.1 Agriculture, Advocacy, and Pragmatism

In this chapter, I examine the philosophical and ethical views undergirding Farm Forward, a nonprofit animal advocacy group focused on agricultural issues. Specifically, I frame Farm Forward as an organization that embodies key dimensions of the pragmatist philosophical tradition. I begin by providing a brief overview of existing pragmatist work within animal ethics and describing the pragmatist concepts of pluralism, particularism, and amelioration.

J.A. Tuminello III (✉)
Department of Philosophy and Religion, University of North Texas, Denton, TX, USA
e-mail: joseph.tuminello@unt.edu

© Springer International Publishing AG 2017
I. Werkheiser and Z. Piso (eds.), *Food Justice in US and Global Contexts*,
The International Library of Environmental, Agricultural and Food Ethics 24,
DOI 10.1007/978-3-319-57174-4_21

247

After placing Farm Forward's general vision for agricultural reform within this philosophical context, I review a number of the organization's recent projects to illustrate their commitment to the above pragmatist principles. These projects include: (1) consultation efforts with the American Society for the Prevention of Cruelty to Animals (ASPCA) on specific welfare issues in industrial animal agriculture, (2) the success of the "Buying Mayo" campaign in calling attention to common practices in egg production, (3) the launch of the website BuyingPoultry.com as an educational tool that circumvents ambiguous labels and marketing claims on retail products, (4) Farm Forward's interactive webinar series on food production ethics with author Jonathan Safran Foer, and (5) the organization's efforts to raise awareness of the importance of farmed animals' genetic welfare. Farm Forward's pragmatist approach to advocacy in agriculture, I argue, serves as an exemplary model for achieving positive change in concrete and inclusive ways.

21.2 Key Dimensions of Pragmatism

Before framing Farm Forward as a pragmatist animal advocacy group, I provide a brief overview of existing pragmatist work within animal ethics, as well as a description of the particular dimensions of pragmatism that I will be utilizing in my subsequent discussion. Pragmatism developed in the late nineteenth and early twentieth centuries, and its emergence as a philosophical school is closely connected with the work of American philosophers Charles Sanders Peirce, William James, and John Dewey. Many elements of pragmatism emerged as critiques of absolutist Enlightenment philosophy and the quest for certainty, as well as considerations regarding the nature of inquiry, democracy, and education. One of the cornerstones of pragmatist philosophy is its acceptance of "fallibilism," the epistemological view that it is always possible for a given belief to be false—that "our knowledge is never absolute but always swims [...] in a continuum of uncertainty and indeterminacy" (Peirce 1940, 356). Adopting the view that absolute certainty is not a worthwhile goal to pursue, pragmatists generally favor the examination of particular concrete problems, employing democratic deliberative techniques to allow for a multitude of possible perspectives with the aim of working together to improve a given situation (as well as to define what should *count* as a problem in the first place), acknowledging that any single perspective will not be able to provide the means to reach a desired collective end on its own. Just as there are many more facets of pragmatist thought than are within the scope of this chapter, there are many more pragmatist philosophers and iterations of pragmatism as well. Taking these stipulations into account, I will focus on the concepts of pluralism, particularism, and amelioration which are pervasive in pragmatism, and will demonstrate the ways that Farm Forward embodies and exhibits these qualities through its advocacy projects and philosophical underpinnings. While these dimensions of pragmatism overlap significantly with one another, I will both discuss them separately and reflect on the ways that they operate in tandem.

Philosophers have made significant contributions to animal ethics since the mid-1970s, establishing it as a legitimate area of inquiry under the larger umbrella of "applied ethics." The two dominant ethical theories that scholars advanced in the early stages of animal ethics were utilitarianism and deontology. Utilitarianism defends sentience and the capacity to suffer as the central properties that are required for a being to have moral standing. Thus, utilitarian theorists working within animal ethics favor actions that lead to the reduction or elimination of unnecessary suffering, avoiding charges of "speciesism" by considering all sentient beings (human and nonhuman) to be worthy of moral concern (Singer 1975). Above and beyond calls for the reduction of suffering and the utilitarian emphasis on the consequences of actions, animal ethicists operating within a deontological ethical framework argue that any exploitation of human and (at least some) nonhuman animals is unethical because of their nature as "subjects-of-a-life" (Regan 1983).[1] While there are palpable differences between these two theories, both are undergirded by ethical monism, the view that morality can be reduced to a single action-guiding principle which can adequately be applied in specific instances to yield prescriptions for morally acceptable behavior. This is important to note because of ethical monism's contrast with pragmatism's emphasis on ethical pluralism, as I discuss below. Besides utilitarian and deontological approaches, animal ethics discourse has expanded in recent years to consider virtue ethics, contractarian, ecofeminist, semiotic, and pragmatist frameworks (Aaltola and Hadley 2015).

Animal pragmatism emerged within animal ethics and philosophy in the early 2000s, as scholars began explicitly examining the relationship between pragmatist views and various human-nonhuman animal relationships (Tuminello 2014). Key animal ethics issues that pragmatists have looked at thus far include the use of animals in food production, strategies for animal advocacy, the tension between animal and environmental activism, biomedical research on animals, and the relationship between humans and companion animals. While not disparaging previous efforts to reflect on ethical decision-making in regards to human-animal relations, some contemporary pragmatist philosophers have noted that the sustained relevance of philosophy in the debate over topics such as animal welfare and animal rights is "yet to be determined" (Light and McKenna 2004, 4). In their edited volume *Animal Pragmatism*, Andrew Light and Erin McKenna look to pragmatism as a philosophical tradition that has great potential to yield important practical contributions: "[E]ven with the proliferation of the literature in animal welfare and rights, the fact is that 10 billion birds and mammals were raised and killed for food last year in the United States without much by way of an ethical quibble or second

---

[1]Here, I refer specifically to Tom Regan's rendering of deontology. Regan modifies Immanuel Kant's formulation of deontology such that it is nonanthropocentric, taking nonhuman animals into account as moral patients.

thought by most of those benefiting from their consumption" (Light and McKenna 2004, 4). While there are many facets of pragmatism that can serve as useful pathways for ethical deliberation, in the rest of this section I will focus on describing pluralism, particularism, and amelioration as key elements of pragmatism. I will also address and respond to some challenges to pragmatist ethics, and defend pragmatism as an appealing model for the theory and practice of food justice.

21.2.1 Pluralism

Whereas many philosophers begin their work by developing theoretical principles in the abstract and then applying them to examine, critique, and provide suggestions regarding what ought to be done in a given case, pragmatism promotes acknowledgment of the unity of theory and practice, emphasizing attentiveness to context in an attempt to develop consensus from a variety of viewpoints. In this vein, Light reflects on the potential for pragmatism to be brought to bear on issues of conflict among diverse communities. For example, while a number of environmental ethicists and activists have de-emphasized concerns for the treatment of individual animals in favor of a more ecologically oriented ethical perspective, Light argues: "The environmental consequences of factory farming are too well documented to be ignored. [...] I think there is something intellectually suspicious, if not completely dishonest, about the wholesale rejection of animal welfare positions by many environmental ethicists under the guise of holism" (Light 2004, 128). Opposition to industrial animal agriculture has recently become a point of ethical convergence for people upholding an array of moral principles, due to its practices that involve both harm to and exploitation of human and nonhuman animals, as well as other living beings and the environment.

Ethical pluralism, this aversion to reducing all moral judgments to a single principle (in contrast to the aforementioned ethical monist stance), is a fundamental characteristic of pragmatist ethics. In his essay "Three Independent Factors in Morals," John Dewey reflects on what he takes to be an oversimplification in monist approaches to ethical deliberation: "Whatever may be the differences which separate moral theories, all postulate one single principle as an explanation of moral life. Under such conditions, it is not possible to have either uncertainty or conflict: *morally* speaking, the conflict is only specious and apparent. [...] Intellectually and morally, distinctions are given in advance [...]" (Dewey 1998, 315). Such an approach to ethical decision-making, according to Dewey, betrays the actual experiences of conscientious moral agents who are able to note the complexities regarding questions of what is good, what end(s) should be pursued, what duties obligate us, and what habits should be cultivated or avoided. Ethical pluralism calls for theorists to abandon attempts to discover a unifying ethical principle (Fesmire 2004). Pragmatist ethics, then, is anti-foundationalist in the sense that it does not advocate relying on a single absolute principle which can be applied in all possible situations to yield the most morally acceptable course of action. While monist moral theories can provide guidelines that are worth considering, questions of what should be done are best settled in the moment, through attending mindfully and seriously to the complexities and particularities of a given scenario. Thus, pragmatism does not yield a ready-made "final answer," but a framework for deliberating with others by emphasizing context-sensitivity.

21.2.2   Particularism

It is this attention to context and particularity in pragmatist thought that I refer to as its focus on "particularism." Because ethical decision-making is not reducible to a single absolute moral principle, it is important to pay close attention to the uniqueness of each scenario. This includes the unique socio-historical and environmental contexts in which we find ourselves with others, as well as the unique needs and natures of living beings. In his essay "Getting Pragmatic about Farm Animal Welfare," Paul B. Thompson addresses animal welfare scientist David Fraser's criticism of animal ethics as being too exclusively concerned with abstract theoretical issues regarding the treatment of animals (Thompson 2004). While many of the issues tackled by animal ethicists do have real-world implications, there is a shortage of thinkers who are well-versed in particular technical practices which are staples of industrial agriculture, instead largely concerning themselves with individual dietary choice.

Fewer philosophers have looked at specific ways to improve the lives of farmed animals, as well as how to delineate the criteria that could be used to signify such improvement. These tasks, as well as the recognition of the unique needs, desires and natures of individual animals and species of animals are important tasks for the pragmatist ethicist, and Thompson looks to the philosopher Bernard Rollin as someone who has stepped up to this challenge. Bracketing the absolutist question of whether or not animals should ever be used in food production, and accepting that they are currently being used in this way and are often mistreated, Rollin looks at specific ways that their lives can be improved. In his book *Farm Animal Welfare*, Rollin catalogues a number of standard procedures regarding animal treatment in industrial agriculture, such as debeaking chickens and castrating cattle. He suggests ways that these methods can be modified or replaced to increase animal welfare, and also argues that certain practices should be discontinued entirely due to their being harmful and unnecessary (Rollin 1995).

Exhibiting the pragmatist emphasis on attending to the particularities and contexts of animals in their environment, another hallmark of Rollin's work in animal ethics is his adaptation of the Aristotelian concept of *telos* to refer to the nature of an animal—the set of activities that make it what it *is*. As Rollin illustrates using the example of a spider, "[t]he *telos* of the spider is its own, imposed upon it by nature, encoded in its genetic blueprint, and protected by a thousand activities that evidence a struggle to actualize that *telos* and preserve its life" (Rollin 2006, 101). Each animal, on this view, has a *telos* that is in part genetically dictated. Rollin argues that treating an animal in such a way that it can fulfill its *telos* will generally result in higher welfare. Understanding how a given animal should be treated, then, depends on acquiring an understanding of an animal's *telos*, which subsequently depends on both direct experience with and education regarding the animal as a biologically, environmentally, and socially situated being. Pragmatism takes attendance to particularities and the uniqueness of organisms and situations to be fundamental in determining appropriate courses of action, and Rollin's approach is emblematic of this perspective. The incorporation of particularism in pragmatist approaches to animal ethics entails that certain animals may be more or less suited to certain environments and activities, sometimes in ways that are extremely detrimental or beneficial.

Erin McKenna defends the importance of these species- and breed-specific nuances in her book *Pets, People, and Pragmatism*, where she argues against the outright rejection of human-animal relationships in favor of a view that sees many human-animal relationships (particularly, regarding relationships between humans and companion animals) as having the potential to be mutually beneficial, rather than necessarily involving the exploitation of animals on the part of humans. Regarding the involvement of horses in specific activities, McKenna writes:

> Not all horses are equally suited to pulling a cart—physically or mentally. Not all horses want to look a steer in the eyes. Not all horses have the suspended gaits for dressage. Not all horses have the temperament to enjoy parades, or the desire to jump, or the drive to run. There are also differences in personality by breed as well as by individual. Some breeds are more clownish, some are more serious, some are more absent-minded, some are intent on learning. (2013, 48)

On a more negative note, the recognition of the moral importance of context and nuance also allows for the possibility that, while factors such as activities and environmental conditions do play an important role in welfare, there are also cases (as I illustrate below) where the alteration of an animal's *telos* through practices such as selective breeding will lead to compromised welfare, even in the presence of optimal environmental and social factors.

Before moving on, it is worth noting that pragmatist ethics do not necessarily lead to the moral acceptability or unacceptability of consuming animal flesh or other products, or to absolute moral pronouncements of any sort. Rather, by adopting a pluralist perspective and taking into account the idiosyncrasies of living beings, environments, and situations, pragmatism provides a framework for identifying and working toward the eventual resolution of problems. At the same time, it may turn out that a person employing dimensions of pragmatism in their decision-making decides to cease all consumption of animal products, or decides only to purchase animal products from higher welfare farms. The point is that she would be attending to the particularities of the situations in which she finds herself, and involving multiple perspectives in making her determination without the simple application of a pre-formed absolutist principle.

21.2.3 Amelioration

The aforementioned pragmatist technique of collectively identifying and resolving problems warrants a discussion of the related concept of "amelioration," the process of working to make a situation better. Akin to and overlapping with ethical pluralism, amelioration entails the recognition that it is impossible to discover a final, perfect solution to any problem, and that people should instead work to seek improvement while being guided by provisional "ends-in-view" (McKenna 2001, 97, 2013). In this way, amelioration involves compromising and establishing areas of agreement with those whom you may strongly disagree with in many ways; it calls for openness toward the views of others even when acting from your own deeply-held convictions. While some courses of action may be more beneficial than others, operating under the assumption that you have the final answer to a given problem can lead to closing off potentially fruitful possibilities and opportunities for collaborative inquiry and collective achievement (McKenna and Pratt 2015).

In the context of agricultural and animal ethics, amelioration calls not only for advocacy groups with disparate views to work together for common interests, but also for advocacy groups to work and dialogue with scientists, business professionals, policymakers, and the public to engage collaboratively in making the world a better place. Members of these groups often have extremely disparate ideas regarding ideal conditions for human-animal relations, as well as the steps that should be taken in achieving those conditions. Some hold the view that any use of nonhuman animals is justified due to their alleged inferiority to humans, but they are concerned about human and environmental impacts of industrial agriculture; some believe that industrial agriculture is unethical but that certain forms of "humane" animal farming can be justified; others are opposed to any use of animals whatsoever, and view these practices as inherently exploitative and unethical. Taking these differences into account, a pragmatist approach calls for advocates to find common ground and to respectfully include diverse perspectives in working to improve and reconsider current methods of food production.

21.2.4 Challenges to Pragmatist Ethics

Having reviewed three of the key elements of pragmatism above and their roles in ethical decision-making, it is also important to take some potential challenges to pragmatist ethics into account. One could argue, for instance, that a pragmatist approach to ethics risks engendering a certain degree of complacency in those adopting this approach. This challenge is especially worth considering in the contexts of social justice and agricultural ethics, where people often interpret incremental reforms as indicators of progress. To elaborate on this challenge a bit, I will invoke the aforementioned concept of amelioration in pragmatist philosophy. If a key dimension of pragmatist ethics involves doing what one can to improve a given situation, rather than possibly avoiding action in a given case because such action does not meet one's ideal aims, then it is also possible that one may begin to "accept" less-than-ideal practices which are nevertheless better than the worst existing practices.

Responding to this challenge calls for emphasizing the pragmatist concept of an "end-in-view" which is also compatible with ameliorative ethical practices. Often, pragmatist ethics will call for *some* ameliorative action to be taken, rather than calling for the agent to refrain from acting for the sake of personal purity (or for other reasons). However, prescriptions of action will also be dependent on the degree to which those actions appear to align with the agent or group's end-in-view, a decisive goal which is also malleable and open to revision. Taking the concepts of amelioration and ends-in-view together, a pragmatist may work towards incremental progress, giving due praise to those involved while simultaneously declaring that these ameliorative actions are "not enough." Thus, a fuller picture of pragmatist ethics shows that it is oriented against agential complacency while also encouraging the agent to do what she can to improve a given situation. I illustrate this point below in discussing Farm Forward's ameliorative approach to assessing victories and working toward further change.

An additional challenge to pragmatist ethics lies in assessing the limits of pragmatism's inclusive, democratic approach to ethical decision-making in terms of defining and achieving social progress. One could look to the phrase "too many cooks spoil the broth" to better grasp the force of this challenge. Pragmatism's pluralist approach to ethics, in tandem with its aversion to ethical monism, also encourages collaboration in ethics. This collaboration is not limited to diverse groups working together to solve problems, but (as mentioned above) also working together to deliberate over and identify what should *count* as a problem, and what should *count* as progress in the first place. Thus, a potential disadvantage of this approach is that pluralism and democracy could be emphasized to the degree that real progress is ultimately inhibited or drastically slowed due to the need to include multiple viewpoints in defining ends-in-view and practical courses of action.

While this potential challenge to pragmatism should be taken seriously and weighed against other interests in specific cases, it is also important to point out that it assumes that there is some mind-independent, a priori standard of ethical progress that agents either meet or do not meet, and that such a standard could be inhibited by a pluralist outlook which emphasizes democratic deliberation. However, one of the very reasons that pragmatists have emphasized inclusivity in the determination and solution of social issues is due to a denial of such a mind-independent standard. While it is true that specific cases may call for the judgment of participants in determining the time at which goals are sufficiently specified to merit action, this need not entail an outright rejection of pluralism and inclusivity based on the above concern.

21.2.5 Pragmatism as a Model for Food Justice

Challenges to pragmatism, such as those addressed above, should be taken into account by any person adopting a pragmatist approach to ethics. In fact, taking into account potential problems with one's framework can be considered to be part and parcel of pragmatism, given its emphasis on fallibilism and its other aforementioned facets. In my view, this openness combined with a dedication to positive action lends strength to the consideration of pragmatism as a model for food justice, including justice for humans and nonhumans that are part of or otherwise impacted by agricultural practices.

Pragmatism's ethical pluralism not only entails the rejection of monism, but also the importance of taking monist ethical theories into account in larger discussions of ethics and justice. While not accepting the idea that ethics can be boiled down to a single universally applicable principle, this also entails that various monistic perspectives also should not be ignored or cast aside because of their adherence to monism. In this way, scholars and practitioners of food justice can employ a pragmatist framework in determining what justice *is*, as well as *who* should be included when seeking justice over injustice, rather than relying on predefined but exclusionary notions on these topics.

Pragmatism's emphasis on particularism also entails that what counts as "justice" in one context may not count similarly in others. Thus, those working together on issues of food justice should both decide on guiding ends-in-view while also being open to challenging these goals depending on specific situational parameters. Finally, pragmatism's emphasis on amelioration calls for those seeking food justice to make judgments about the degree to which certain actions align with their ends-in-view, while also being open to critiquing certain forms of progress as "not enough" or "falling short" of larger goals. For these reasons, pragmatism presents a vital and useful set of tools for the determinations and actions which are embodied in food justice discourse and practice.

To conclude this section, I have described and examined three key dimensions of pragmatism and pragmatist ethics, addressed and responded to two challenges to pragmatist ethics, and have also defended pragmatism as a model for food justice. In what follows, I will describe the projects and aims of the nonprofit animal advocacy group Farm Forward through a pragmatist lens, demonstrating their embodiment of the above dimensions of pragmatism. Ultimately, my aim in doing this is to consider Farm Forward as a model for the importance of pragmatism in animal and agricultural advocacy.

21.3   Farm Forward: History, Projects, and Philosophy

While Farm Forward was incorporated in 2007, many of its members have been working in animal advocacy since the early 1990s (About Us 2016). The organization's board, a testament to ethical pluralism in itself, is comprised of two animal welfare experts (one an animal scientist, the other a philosopher who is cross-appointed in animal and biomedical sciences), the CEO of Whole Foods Market, a rabbi, a bestselling author, and a preindustrial (or "heritage") breed poultry farmer (The Board 2016). Steven Jay Gross, the Chairman of the Board, is a pioneer of animal advocacy, with decades of experience negotiating with food retailers on behalf of animals. Aaron Gross, the CEO and founder of Farm Forward, is an experienced activist as well as a professor of religious studies, with expertise on the relationship between animal ethics and Jewish thought (Gross 2014).

According to their website, Farm Forward's mission is to implement "innovative strategies to promote conscientious food choices, reduce farmed animal suffering, and advance sustainable agriculture" (Mission 2016). In this way, Farm Forward can be seen as an agricultural advocacy group rather than working strictly in animal advocacy. Their attention to the myriad possible ethical perspectives on industrial agriculture is evinced in their social media outreach and on their own website, where they provide information on worker exploitation and public health, as well as issues regarding the lives of animals, the environmental impacts of industrial agriculture, and discussions on the ways that these areas of concern intersect with one another (Ending Factory Farming 2016).

While Farm Forward is involved in coalition-building efforts and negotiations to improve the lives of animals in factory farms, they are also explicitly opposed to this system and work toward ending factory farming. Their fundamental message is encapsulated in board member and author Jonathan Safran Foer's nonfiction work *Eating Animals*. The book chronicles Foer's own oscillation between vegetarianism and meat-eating throughout his life, as well as the evolution of his ways of seeing and relating to other animals. In the midst of his impending fatherhood, Foer set off on a journalistic and exploratory journey to learn about where his food comes from, and the nature of the system in which he, his family, and future child were complicit. Sneaking onto factory farms, visiting slaughterhouses and heritage poultry farms, consulting with various experts and reflecting on related philosophical concepts such as tradition, shame, and the nature of morality, Foer ultimately becomes a committed vegetarian and advocate against industrial agriculture. While developing and refining his own positions, he very clearly gives due attention to the perspectives of others by devoting space to their voices within *Eating Animals*. In their own words, Foer includes significant amounts of direct prose from an animal rights activist, a factory farmer, a heritage poultry farmer, a vegetarian rancher, and a vegan who builds slaughterhouses. In a manner exemplary of pragmatism's resistance to dichotomous thinking, as well as the feminist tradition of the ethics of care, Foer's own perspective in the end becomes a critique of the idea that reason and emotion are separate, or that we should rely solely on one or the other when engaging in ethical deliberation: "Responding to the factory farm calls for a capacity to care that dwells beyond information, and beyond the oppositions of desire and reason, fact and myth, and even human and animal" (Foer 2009, 263).

In what follows, I provide an overview of five of Farm Forward's recent and ongoing projects in order to elucidate the ways that the organization embodies the pragmatist dimensions of pluralism, particularism, and amelioration. To summarize briefly: (1) Farm Forward's consultation services with the ASPCA illustrate its commitment to ameliorative practices through the organization's emphasis on ending factory farming while also providing suggestions for incremental change within the current system; (2) The "Buying Mayo" campaign is pluralist in its coalition-building with other advocacy groups as well as the public, and ameliorative in its calling for unification to end some of the cruelest practices within industrial agriculture; (3) the launch of BuyingPoultry.com is pluralist and ameliorative in its advocacy for higher welfare poultry as well as plant-based poultry alternatives; (4) in the annual "Jonathan Safran Foer Virtual Classroom Visit" webinar series, Foer takes a pluralist approach by avoiding absolutist ethical prescriptions and encouraging participants to think critically for themselves about the nature and ethics of food production; finally, (5) Farm Forward's attentiveness to the importance of genetic welfare illustrates the organization's commitment to particularism by taking into account the unique ways that farmed animal welfare has been sacrificed within industrial agriculture through selective breeding for economic gain.

21.3.1 Consultation with the ASPCA on Industrial Agriculture

Farm Forward's balance between setting long-term goals and pursuing amelioration through embracing multiple viewpoints is illustrated in their consultation with other advocacy groups. In 2011, the ASPCA approached Farm Forward with interest in developing and presenting information on and nuanced positions regarding a variety of general and species-specific practices within industrial animal agriculture. These practices included confinement, genetics, transport, slaughter, medication, and routine surgical mutilations such as castrating pigs and cattle and debeaking poultry.[2] Besides surgical mutilations, other practices of interest to the ASPCA included the use of electric prods on cattle and pigs and the provision of furnished cages as opposed to open cage systems for laying hens.

---

[2]Routine mutilations are performed on animals within industrial agriculture to reduce profit-loss incurred through destructive behaviors (e.g. feather-pecking, cannibalism, and tail-biting). These destructive behaviors are exacerbated through overcrowded conditions in barren environments. Rather than altering the conditions that lead to these behaviors, it is considered more economically feasible to perform the aforementioned surgical mutilations.

Through consultation, experience, and research within the relevant animal science literature, Farm Forward drafted position papers on each of these issues, providing executive summaries of ethical concerns, reasons for general opposition to these practices, as well as different degrees of recommendations, given that many of the practices were likely to continue. To illustrate this process, below are highlights from Farm Forward's position paper to the ASPCA on debeaking chickens and turkeys:

> Welfare experts and organizations agree that debeaking [...] causes acute pain and stress and that there are other successful means of reducing feather pecking, cannibalism, and other undesirable behaviors that debeaking aims to limit. [Compassion in World Farming] and [World Animal Protection] both support a ban on debeaking, while [the Humane Society of the United States] advocates that it should be used only as a last resort when other methods to control feather pecking and cannibalism have failed. Alternatives including increasing space, providing foraging opportunities, enrichment of cages, and altering the chickens' diets to promote pecking at food rather than other birds. Most significantly, experts agree that debeaking can be made unnecessary through selective breeding.

> Given that the practice of debeaking is likely to continue, it is important to note that welfare groups and experts agree that hot-blade amputation is unacceptable. Debeaking with an infrared laser [...] has some advantages over hot-blade amputation but still causes acute pain during the procedure. [...]

> We suggest that the ASPCA oppose debeaking for all species of fowl in favor of the alternatives identified above (Farm Forward 2012, 1).

In this and other position papers, Farm Forward expresses their own recommendation (that the ASPCA oppose all poultry debeaking practices) in light of their values and empirical research while also designating between higher- and lower-welfare methods, as well as presenting perspectives from other leading advocacy groups. This strategy exemplifies their pursuit of ameliorative practices, their incorporation of a plurality of viewpoints, as well as their own aim of ending unnecessary, exploitative, and cruel surgical mutilations in animal agriculture. The ASPCA has since incorporated Farm Forward's agricultural recommendations within their own policy and position statements, which include perspectives on companion animals, farmed animals, animals in research and education, animals in sport and recreation, and wild animals (ASPCA Policy and Position Statements 2016). For instance, in the ASPCA's positions on physical alterations of farmed animals, they write:

> The ASPCA opposes the practice of debeaking [...] and recommends the use of husbandry techniques and breed selection that eliminate the feather pecking and cannibalism that debeaking aims to prevent. Further, the ASPCA supports research into less invasive alternatives to debeaking such as beak blunting. Until debeaking is eliminated, debeaking must include appropriate analgesia, and be performed only by those who are properly trained, properly monitored and using proper equipment (Physical Alterations 2016).

To some degree, the ASPCA's own position reflects Farm Forward's multi-tiered strategy for providing recommendations. While Farm Forward has a clear end-in-view (in this case, opposition to surgical mutilations in keeping with the larger goal of ending factory farming), they (and the ASPCA) also take ameliorative positions which reflect an acknowledgment that these sorts of practices will likely persist in the near future.

21.3.2 The 'Buying Mayo' Campaign

The success of Farm Forward's "Buying Mayo" campaign serves as further evidence of the possibility of effecting positive change through education, collaboration, compromise, and negotiation. While many people have a vague awareness of the exploitation and suffering that factory farmed animals endure, the public at large lacks (and is often not interested in seeking) information regarding the set of specific practices that comprise the intensive production of animal products. One example of this is the practice of "chick culling" in the egg industry. Due to their lack of economic value to the egg industry (since only female chickens lay eggs, and male layer chickens are not selectively bred for optimal growth for flesh production), hundreds of millions of live male chicks are ground up alive (in a process known as "maceration") or suffocated each year (Halteman 2011). Given that, on most views, the production of mayonnaise requires the use of eggs, and the United States Food and Drug Administration's definition of "mayonnaise" legally mandates that any product labeled as such *must* contain eggs, major mayonnaise producers such as Hellmann's and Best Foods, both of which are owned by the multinational corporation Unilever, are directly complicit in these practices.[3]

In an effort to raise awareness and work toward the resolution of this ethical issue, Farm Forward launched the website BuyingMayo.com in June 2014. The website consisted of a video educating consumers about chick culling and the complicity of the egg and mayonnaise industries, as well as a petition specifically targeting the aforementioned mayonnaise brands, requesting that they end support of this practice. Nearly half a million people signed the petition within the first two months of the campaign's launch. Uniting with the Humane Society of the United States (HSUS) and Compassion in World Farming (CIWF), Farm Forward began negotiations with Unilever. As described on Farm Forward's website,

> Farm Forward offered to place a moratorium on our campaign if Unilever: (1) agreed that the routine practice of macerating day-old chicks did not conform to their animal welfare policy, (2) committed to investing in identifying and implementing an alternative to the practice, and (3) pledged to explore egg-free alternatives for their products, particularly in the event that an alternative to maceration was not found (Victories 2016).

---

[3]Mainstream perspectives on mayonnaise currently and historically entail that eggs are a necessary ingredient for a product to appropriately count as "mayonnaise." However, the increasing popularity of vegan products such as Hampton Creek's Just Mayo, as well as the FDA and Unilever's recent challenges to Hampton Creek regarding the legal definition of mayonnaise, raise important ontological questions regarding the nature of this and other food products.

By August 2014, Unilever agreed to these terms, updated their welfare policy, began researching alternatives to maceration, and has also agreed to update Farm Forward on their progress on a regular basis.

While not bringing chick culling to an immediate halt, the success of this campaign lies in Farm Forward's ability to exercise a pluralist approach, joining forces with groups (HSUS and CIWF) who have praised Unilever's animal welfare policies in the past in order to open negotiations and have the company make an ameliorative commitment to phasing out practices such as maceration. The campaign also saw success as a venture in education and public engagement, providing information to raise awareness regarding practices of which many people were previously unaware. Besides putting pressure on industry to adopt stricter animal welfare policies, the Buying Mayo campaign has prompted interest in legislative and policy changes. In mid-2015, Germany announced advancements in research which would facilitate their elimination of chick culling by 2017, and possibly within 2016 (Nagesh 2015). Also, in June 2016, the United States trade group United Egg Producers committed to completely phasing out chick culling by 2020 (McKenna 2016).

21.3.3 BuyingPoultry.Com

After years of research and consultation with poultry welfare experts, Farm Forward launched the website BuyingPoultry.com in 2015. Currently in beta-testing (as of Summer 2016), the website serves as an educational resource for consumers interested in both higher-welfare poultry as well as plant-based alternatives. Users of BuyingPoultry.com can search thousands of brands and retailers, and can also search by location to gain insightful information that cuts through the ambiguous, often empty ethical claims that brands make for purposes of marketing their poultry products. When searching for a particular brand of poultry, users will see the expert grade (ranging from A to F) that Farm Forward has assigned to the particular product, as well as information on how they grade products, and their suggested alternatives. Alternatives typically include preindustrial poultry breeds raised on higher welfare farms and vegan poultry substitutes from brands such as Beyond Meat and Gardein. The vast majority of products on the website receive a low grade of D or F, evincing the extreme lack of welfare experienced by factory farmed poultry.

Farm Forward has recently begun organizing an educational program centered around the implementation of BuyingPoultry.com as a basis for classroom service learning projects. In one iteration of these projects, once students educate themselves about the various welfare certification standards and make their own informed judgments regarding ethical issues in poultry production, they are encouraged to learn about the sources of food in their university dining halls, and to meet with dining service employees within their own school to think about ways that the school could transition to higher welfare and vegan products. For example, in 2015 students at the University of San Diego worked successfully to have multiple campus dining halls replace their standard mayonnaise products with Hampton Creek's Just Mayo, a vegan mayonnaise substitute (Student Sustainability Projects on Campus 2016).

Thus, one of the key merits of BuyingPoultry.com is in its pluralist approach, making information on both higher welfare and vegan products readily available for consumers without also presenting veganism as the only possible consumer choice aside from factory farmed poultry. However, it is also important to keep in mind that an extreme minority of poultry farmers currently raise preindustrial breeds of poultry, so this is not necessarily a viable alternative for many people who are unwilling to place large poultry orders online.

21.3.4   The Jonathan Safran Foer Virtual Classroom Visit

In 2012, Farm Forward began sponsoring, organizing, and hosting a webinar series known as the "Jonathan Safran Foer Virtual Classroom Visit." This series takes place one day per year, and is typically comprised of 8–10 sessions within which high school and college classes participate from around the world. The event features the aforementioned author Jonathan Safran Foer, and the main purpose is to provide a space for students to collaboratively engage with Foer (rather than merely listening to him lecture) in a dialogue about the nature and ethics of food production, as well as to inquire about the key themes that he discusses within *Eating Animals*. Each session is also recorded and made available to participants for further educational use.

From 2012 to 2015, an estimated 11,100 students have participated in the Virtual Classroom Visit within about 330 total classes. Participation in this event is free of charge, as Farm Forward covers all costs incurred in its production. While participating classes are not required to have read *Eating Animals* prior to participation, many instructors use the event as an opportunity for their students to interact in meaningful ways with an author whose work they have read, and to facilitate their critical thinking about issues at the nexus of food, animal, and environmental ethics, as well as related social justice issues. In addition, Farm Forward has donated over 350 copies of *Eating Animals* to high schools which are interested in participating and reading the book, but lack room in their budget to purchase their own class sets. The Virtual Visits are heavily interdisciplinary events, with students and teachers participating from fields as varied as Philosophy, English, Anthropology, Environmental Studies, Animal Studies, Cultural Studies, Sociology, Sustainability, Theology, and Nutrition.

Embodying Farm Forward's pluralist approach to advocacy, Foer engages in critical dialogue with the Virtual Visit participants while avoiding an absolutist, unwavering ethical stance. Acknowledging the various perspectives from which factory farming can be seen as ethically problematic, Foer does not present one single option as the only possible alternative, instead encouraging students to continue to do their own research and to make informed choices that are plausible and meaningful within their own lives.

21.3.5   Emphasis on Genetic Welfare

Farm Forward is one of very few animal advocacy groups that places heavy emphasis on the importance of genetic welfare, setting themselves apart from organizations that are mainly concerned with external factors affecting welfare. As discussed above, a particularist approach entails taking into account the unique variables of each situation, rendering moral judgments that are context-sensitive and based on empirical knowledge of particulars that are often neglected by sweeping, absolutist, and universal moral codes. Within the context of animal agriculture, judgments rendered in this way will sometimes be made in regards to the amount of space which an animal has, and the way that the environment and activities that she endures are conducive or inhibitive in the fulfillment of her *telos*. However, because particularism also entails familiarity with the genetic predisposition of animals, this will sometimes result in judgments that condemn the existence of certain types of animals based on the way that people have altered their *telos*, often for the sake of profit.

Within contemporary industrial animal agriculture, there are only two types of chicken: layers, bred for their egg production, and broilers, bred for their flesh. The proliferation of monocultures and the subsequent narrowing of gene pools of industrial livestock have severely impacted biodiversity in the name of increasing productivity and efficiency. Intense selective breeding of animals has progressed throughout the 20th century into the present, and farmers have bred livestock with the narrow goal of increasing the production of animal protein as quickly as possible, which has also resulted in increased animal suffering. Broiler chickens, for instance, regularly endure muscular and skeletal problems, their bodies struggling under the weight of breasts that have been selected to grow larger than ever, and in as small a time period as possible. In 1956, broilers required 84 days to reach 1.82 kg. In 2000, broilers took only 34 days to reach this same weight (Hafez and Hauck 2005).

Attending to the genetic uniqueness of modern-day broiler chickens entails recognizing that there are real limits to the degree to which environmental conditions can enhance welfare. That is, even if a broiler chicken is raised in an ideal environment and treated well during its lifetime, she will still suffer immensely simply due to her breeding history. This is particularly problematic because many "humane" poultry farmers are still raising poultry with industrial genetics. As Foer points out, "[b]ecause the average farmer can't run his own hatchery, concentrated industry control of genetics locks farmers and their animals into the factory system" (Foer 2009, 235). However, while the number of poultry breeds in industrial agriculture bottle-necked immensely in the 20th century, a small amount of hobbyists retained poultry with preindustrial genetics. For consumers who are interested in alternatives to factory farmed poultry but who are unwilling to end their consumption of poultry, Farm Forward has consulted with higher welfare heritage poultry farmers such as Frank Reese at Good Shepherd Poultry Ranch, who raises preindustrial turkey and chicken breeds within his network of farms. Further, data collected by Farm Forward indicate that sales of Good Shepherd's poultry are directly replacing purchases of poultry raised in intensive industrial facilities:

> When a family orders a turkey from Good Shepherd for Thanksgiving, it means they don't buy a factory farmed bird. When Chipotle ordered thousands of pounds of Good Shepherd chicken, they also ordered that much less meat from factory farms. Our analysis also suggests that as individuals and companies move to higher welfare birds, the increased cost mitigates towards an overall reduction in meat consumption (About Us 2016).

Farm Forward's mindfulness of the genetic health of farmed animals and the way that this has been compromised within the industrial agricultural system is emblematic of its particularist and pragmatist approach to animal and agricultural advocacy. This concern also illustrates their commitment to a pluralist approach, promoting veganism and the end of factory farming while also cultivating alternatives that avoid alienating farmers as well as consumers who continue to consume animal products. While it is important to note that the current and growing worldwide demand for animal products cannot feasibly be met through a full transition to high welfare nonindustrial farms, this approach acknowledges that there are multiple paths forward, and that it is important for a diverse array of stakeholders with a diverse array of values to come together for the sake of non-human animals, the environment, social justice, and public health.

21.4 Farm Forward as a Pragmatist Model for Advocacy

Farm Forward claims that it is "a new kind of nonprofit advocacy and consulting group at the forefront of pragmatic efforts to transform the way our nation eats and farms" (About Us 2016).

Farm Forward works through social media, education initiatives, and other projects to promote veganism. At the same time, the organization works to end factory farming while providing consultation with and encouraging high welfare farmers and promoting plant-based alternatives to animal products. Rather than merely paying lip service to veganism, Farm Forward promotes this as a powerful stance in the face of industrial agriculture and animal exploitation. However, as illustrated in many of the examples above, the organization does not treat this as the only possible option. Instead of alienating people who are interested in making gradual changes regarding their consumption of animal products, or people who are interested in reducing their meat consumption and/or sourcing the animal products that they purchase from places that they judge to be more morally acceptable, Farm Forward embraces a plurality of ways that people can positively respond to the concerns raised by industrial animal agriculture. This multi-faceted approach allows Farm Forward to identify as "the first centrist organization where disparate interests opposed to the abuse of animals on factory farms can unite in coordinated and effective ways" (About Us 2016).

In conclusion, I have provided an overview of pragmatist thought in relation to animal ethics in order to frame Farm Forward's approach to animal and agricultural advocacy as embodying key dimensions of pragmatist philosophy, specifically regarding its emphases on pluralism, particularism, and amelioration. I have discussed several main projects from Farm Forward in order to illustrate this point, including their consultation with the ASPCA, the Buying Mayo campaign, the launch of BuyingPoultry.com, the Jonathan Safran Foer Virtual Classroom Visit, as well as their emphasis on genetic welfare and consultation with higher welfare heritage poultry farmers. It is my hope that the success of Farm Forward's campaigns in spreading awareness and encouraging systemic and individual change in a non-alienating manner serves as a model for the viability and importance of approaches to animal advocacy that embody elements of the pragmatist philosophical tradition.

References

Aaltola, Elisa, and John Hadley. 2015. Introduction: Questioning the orthodoxy. In *Animal ethics and philosophy: Questioning the orthodoxy*, ed. Elisa Aaltola, and John Hadley, 1–11. London: Rowman and Littlefield.
About Us. 2016. Farm Forward. http://farmforward.com/about-us/. Accessed 3 June 2016.
ASPCA Policy and Position Statements. 2016. ASPCA. http://www.aspca.org/about-us/aspca-policy-and-position-statements. Accessed 12 June 2016.
Dewey, John. 1998. Three independent factors in morals. In *The essential Dewey: Ethics, logic, psychology*, ed. Larry A. Hickman, and Thomas M. Alexander, 315–320. Bloomington: Indiana University Press.
Ending Factory Farming. 2016. Farm Forward. http://farmforward.com/ending-factory-farming/. Accessed 12 June 2016.
Farm Forward. 2012. Specific position: Debeaking chickens and turkeys. *ASPCA Position Papers*.
Fesmire, Steven. 2004. Dewey and animal ethics. In *Animal pragmatism: Rethinking human-nonhuman relationships*, ed. Erin McKenna, and Andrew Light, 43–62. Bloomington: Indiana University Press.
Foer, Jonathan Safran. 2009. *Eating animals*. New York: Little, Brown and Company.
Gross, Aaron S. 2014. *The question of the animal and religion: Theoretical stakes, practical implications*. New York: Columbia University Press.
Hafez, H.M., and R. Hauck. 2005. Genetic selection in turkeys and broilers and their impact on health conditions. In *World Poultry Science Association, 4th European Poultry Genetics Symposium*, Dubrownik, Croatia.
Halteman, Matthew C. 2011. Varieties of harm to animals in industrial farming. *Journal of Animal Ethics* 1: 122–131.
Light, Andrew. 2004. Methodological pragmatism, animal welfare, and hunting. In *Animal pragmatism: Rethinking human-nonhuman relationships*, ed. Erin McKenna, and Andrew Light, 119–139. Bloomington: Indiana University Press.

Light, Andrew, and Erin McKenna. 2004. Introduction: Pragmatism and the future of human-nonhuman relationships. In *Animal pragmatism: Rethinking human-nonhuman relationships*, ed. Erin McKenna, and Andrew Light, 1–16. Bloomington: Indiana University Press.

McKenna, Erin. 2001. *The task of utopia: A pragmatist and feminist perspective*. Lanham, MD: Rowman & Littlefield.

McKenna, Erin. 2013. *Pets, people, and pragmatism*. New York: Fordham University Press.

McKenna, Maryn. 2016. By 2020, male chicks may avoid death by grinder. *National Geographic*, June 13. http://www.nationalgeographic.com/people-and-culture/food/the-plate/2016/06/by-2020–male-chicks-could-avoid-death-by-grinder. Accessed 13 June 2016.

McKenna, Erin, and Scott L. Pratt. 2015. *American philosophy: From Wounded Knee to the present*. New York: Bloomsbury.

Mission. 2016. Farm Forward. http://farmforward.com/mission/. Accessed 10 May 2016.

Nagesh, Ashitha. 2015. Germany becomes the first country to ban disturbing 'chick shredding' practice from egg industry. *Metro*, October 24. http://metro.co.uk/2015/10/24/germany-becomes-the-first-country-to-ban-disturbing-chick-shredding-practice-from-egg-industry-5459759/. Accessed 12 June 2016.

Peirce, Charles S. 1940. *The philosophy of Peirce: Selected writings*, ed. J. Buchler. London: Routledge and Kegan Paul.

Physical Alterations. 2016. ASPCA. http://www.aspca.org/about-us/aspca-policy-and-position-statements/physical-alterations. Accessed 12 June 2016.

Regan, Tom. 1983. *The case for animal rights*. Berkeley: University of California Press.

Rollin, Bernard. 1995. *Farm animal welfare*. Ames: Iowa State Press.

Rollin, Bernard. 2006. *Animal rights and human morality*. New York: Prometheus Books.

Singer, Peter. 1975. *Animal liberation*. New York: Avon Books.

Student Sustainability Projects on Campus. 2016. Office of Sustainability—University of San Diego. http://sites.sandiego.edu/sustainability/student-sustainability-projects-campus/. Accessed 12 June 2016.

The Board. 2016. Farm Forward. http://farmforward.com/meet-the-board/. Accessed 10 May 2016.

Thompson, Paul B. 2004. Getting pragmatic about farm animal welfare. In *Animal pragmatism: Rethinking human-nonhuman relationships*, ed. Erin McKenna, and Andrew Light, 140–159. Bloomington: Indiana University Press.

Tuminello III, Joseph A. 2014. Environmental and animal pragmatism. In *The encyclopedia of food and agricultural ethics*, ed. Paul B. Thompson, and David M. Kaplan, 566–573. New York: Springer.

Victories. 2016. Farm Forward. http://farmforward.com/victories/. Accessed 10 May 2016.

Chapter 22
Agriculture, Equality, and the Problem of Incorporation

Anne Portman

Abstract This paper argues for the possibility of ethical agriculture by employing ecological feminist philosophical frameworks in articulating and responding to two important objections. First, the *problem of equality* expresses the objection that a just, egalitarian relationship is incompatible with the agricultural project of cultivating nonhumans for human consumption. I argue against this seeming incompatibility by reformulating equality as incommensurability; we might think of two entities as equal not because they occupy the same position along a scalar measure of value, but because they ought not be ranked against each other along such a scale. To me it is plausible that a high degree of avoidance of ranking is possible in agricultural practices that are authentically dialogical. I suggest that when practical ranking is unavoidable, then dialogical relationships allow for responsible ranking. Second, the *problem of incorporation*, expresses the objection that in agricultural contexts, nonhuman beings are defined by reference to the human ends they will serve, and the meaning of such beings is incorporated into human meaning, reflecting a lack of respect for the beings themselves. In response I suggest that defining *ourselves* in terms of our relationships to the nonhumans that we raise and consume goes a long way toward addressing this problem. I argue that the ecological feminist commitment to an understanding of self as fundamentally ecological and ethically open to communication across species can be meaningfully embedded in our understanding and undertaking of agricultural projects.

Keywords Ecological feminism · Val Plumwood · Interspecies justice · Ethical agriculture · Agrarianism

A. Portman (✉)
University of Georgia, Athens, GA, USA
e-mail: anne.portman@gmail.com

© Springer International Publishing AG 2017
I. Werkheiser and Z. Piso (eds.), *Food Justice in US and Global Contexts*,
The International Library of Environmental, Agricultural and Food Ethics 24,
DOI 10.1007/978-3-319-57174-4_22

22.1   Introduction: Justice in Relationships of Production and Consumption?

At the Workshop on Food Justice and Peace in 2014, the first that I attended, conversation between sessions frequently circled back to the idea that the moral concepts underlying our common understanding of justice—like respect and equality—cannot be applied in relationships where one party intends to eat the other party. In other words, if the other (especially the sentient other) is regarded as edible, then one does not regard the other respectfully or as equal. I pushed back against such a view in much the same way that the philosopher Val Plumwood engaged similar objections in feminist animal ethics in the 1990s–2000s, by suggesting that the objection relies on a denial of ecological selfhood. Ecological selfhood is a concept that captures the co-constituting relationship between self and earth others, recognizing necessary dependencies across species. As articulated by ecological feminists, like Val Plumwood, this alternative conception of selfhood opens up possibilities for imagining radical alternatives to existing perspectives on ethical relationship. My partners in conversation rejected the idea that one can simultaneously respect and consume other beings. I found myself responding that their position either denies the possibility of ethical living, or it denies our ecological embodiment—our reliance on other beings to sustain our lives, and our own human position as, ultimately, consumable. I wanted to clarify my position and so expanded my response into this paper, presenting an early version at the Workshop in 2015. I embraced the opportunity to foreground a marginal(ized) theoretical approach in a way that emphasized the radical alternative conceptual understandings that the approach illuminates. I also aimed to demonstrate how the approach provides a vocabulary for situating some existing marginal(ized) agricultural projects in theoretical contexts. Thus this theoretical work is intended as a small piece of a larger public philosophy project of envisioning significant transformation in our understandings of human-nonhuman ecological interrelationship.

The paper begins with an overview of the ecological feminist philosophical frameworks from which I draw an understanding of interspecies domination and the nature of ecological selfhood. Inspired by Plumwood's (2000) arguments regarding the possibility of respectful hunting, I use this framework to think about whether *agriculture* is necessarily a practice of domination. In doing so, I divide the general concern regarding the possibility of manifesting justice in agricultural contexts into two more specific objections, what I term the problem of equality and the problem of incorporation.

The *problem of equality* expresses the objection that an egalitarian relationship with nonhumans is incompatible with the agricultural project. After all, one might say, agriculture is about cultivating and directing nonhuman life under human terms of use and ownership. It might seem that regarding nonhuman life with equal respect and care in such contexts would be nearly impossible. In response I will argue that, although for most of us the idea of equality connotes something like equal rank, a counterhegemonic reinterpretation opens up new possibilities for the concept. As Plumwood (1998, 2002) suggests, we might think of two entities as equal not because they occupy the same position along a scalar measure of value, but because they are incommensurable and ought not be ranked against each other along such a scale. I will argue for the plausibility of a high degree of avoidance of ranking in agricultural practices that are authentically dialogical.

Even assuming this possibility, there is still the apparent problem that the very existence of the nonhuman beings in question is defined by reference to the human ends they will serve, and the meaning of such beings is *incorporated* into human meaning. Plumwood uses the term "incorporation" to describe a hegemonic form of relational definition wherein the subordinated group is defined only by reference to the dominant group's traits and projects, and most often merely as lacking what the master has (1993, 52). One might again object that the *problem of incorporation* makes agriculture fundamentally incompatible with interspecies justice. In response I suggest that defining ourselves in terms of our relationship to the nonhumans that we raise and consume goes a long way toward addressing the problem. The responses that I provide to these important objections are intended to alleviate their severity, if not fully refute them. What remains of the objections will be the product of some remaining fundamental disagreement regarding the role of human beings in the ecosphere, and role of agriculture in human and nonhuman life.

22.2 Ecological Feminism on Species Domination and Counterhegemonic Ethics

In thinking about these problems, I begin with the claim that industrial agriculture (at its extreme) relegates nonhumans to purely economic valuation, reflecting a denial of noninstrumental moral standing for nonhumans. Ecological feminist philosophers, like myself, would explain this as a devastating byproduct of value dualistic conceptual frameworks. By that I mean either/or frameworks that construct dichotomous pairs with fixed identities that are defined in opposition to one another and hierarchically ordered. The male/female binary under patriarchy is a paradigmatic example of value dualism, wherein male and female are presumed to have fixed identities that are defined in opposition to one another and hierarchically ordered. Ecological feminists argue that human/nature is a parallel dualism, one that reinforces and is reinforced by patriarchal frameworks. Human and nonhuman, or culture and nature, are understood to be radically separate categories, with the human being defined by our rationality and the qualities that emerge from it, and the nonhuman defined by its supposed lack of these qualities (Plumwood 1993). The nonhuman is understood as subordinate to the human, often merely as a resource for the fulfillment of rational human ends. For ecological feminists, these dualisms reinforce one another insofar as nature is feminized and women are naturalized (Warren 2000). I hold that historically and conceptually this can be demonstrated, and other thinkers have done the hard work of doing so (for preeminent examples see Merchant 1983; Plumwood 1993, 2002; Warren 2000). For the purposes of this chapter, I will just assert that on Plumwood's view and on my view, these mutually reinforcing dichotomies function to create a hegemonic framework with human reason at the center and nonhuman beings and their ends always on the periphery.

The norms and practices of industrial agriculture can only be regarded as just if the supposed superiority of the dominant group is assumed to entail the legitimacy of exploiting the supposed inferior group. In generic form, the idea is that for any oppositional categories, if one is superior to the other then it is morally justified in dominating the other. Warren (1990, 2000) has famously called this the "logic of domination." Specifically with regard to human-nonhuman relations the logic runs such that: because humans are superior to nonhumans, humans are morally justified in dominating nonhumans. In the context of industrial agriculture, this logic has been taken to its utmost extreme in denying nonhumans any noninstrumental significance. It is the shared position of my partners in conversation and myself that, in fact, the denial of moral standing for nonhumans is itself fundamentally an injustice; that the logic of domination and the dualistic framework that are central to the denial are oppressive and unjust.

It is my view that an ecological feminism, principally inspired by the life work of Val Plumwood, offers a robust, counterhegemonic understanding of interspecies justice. The virtue of justice is tethered to the simultaneous recognition of continuity and difference. Responding to the hegemonic dualism by embracing the extremes of utter sameness or radical difference would ultimately maintain the "all or nothing" structure of dualism. A counterhegemonic response requires embracing ambiguity, fluidity, and overlap—and hence both continuity and difference between and within categories. With regard to human-nonhuman relationship, this dual recognition requires adopting an understanding of oneself as an ecological being, predicated on the openness of what Plumwood (1993, 2002) calls the "intentional recognition stance."

Like other feminist theorists I want to reject the idea of the human self as fundamentally autonomous and disembedded in favor of a view of self as "self-in-relationship." The "ecological self" is in co-constituting, non-instrumental relationship with human and with earth others, or nature (Plumwood 1993, 154). The identity of ecological selfhood grounds the commitment to non-instrumental relationship with others insofar as the ecological self recognizes others in mutual, or reciprocal, interrelationship. This recognition is possible through the adoption the "intentional recognition stance;" in order to include the flourishing of others among one's own primary ends, and for their own sakes, one must conceive of others as intentional (in the broadest sense), and as potentially communicative ethical subjects (Plumwood 2002, 177). As an alternative to the colonizing view of the supposedly non-rational as an undifferentiated resource, in adopting the intentional recognition stance one recognizes the nonhuman other as agentic, as a source of striving, and as potentially communicative.

This understanding of human identity and its appropriate posture in the world grounds certain ethical attitudes and commitments toward the others with whom one is in relationship. These responses are fundamentally dialogical rather than monological, meaning that they are genuinely informed by the other and not prescribed "top-down." As commitments of ecological selfhood, these virtues must be understood as applicable across not just social boundaries but species boundaries as well. This is not a difficult proposition to embrace when these virtues are understood as the byproducts of the posture of recognition that the moral agent adopts in inhabiting the world. Limiting or enclosing the predisposition to be open to communication from others on the basis of species could only be arbitrary. To draw such an arbitrary boundary is fundamentally inconsistent with the recognition of ecological selfhood, the recognition that one is in co-constituting relationship with earth others. Here we can see why Plumwood (2000, 2002) argues that ecological feminism prompts an interspecies egalitarianism. If one adopts another's flourishing among one's own primary ends, that flourishing would be deemed valuable alongside the other ends one pursues. In any prioritization scheme, its not the case that the flourishing of others would always be secondary to one's own immediate goals.

Plumwood (2002, 171) provides the example of thinking through the respectful way to handle a potentially dangerous other that has settled on one's front porch; in one case the other is a human vagrant, and in the second the other is a venomous snake. She suggests that in either case an excessively punitive reaction would be unjust. The species of the other seems to be one of the least relevant factors in making a decision about how to handle the situation. More relevant on Plumwood's view are questions about the context. Is the other on the porch because of an emergency or temporary situation like a flood? Did you have any part in the other's homelessness? Might you be able to find alternative accommodation? And so on. It seems that we don't need to resort to dualistic human/nature hierarchies to illuminate general ethical approaches to such cases (Plumwood 2002, 171). Dismantling the dualistic framework allows for the rebuilding of conceptual structures that no longer rigidly divide the world into human and other, rational and non-rational, allowing for a robust, contextual interspecies politics and ethics. Of central importance for my purposes here is the implication that a new egalitarian ethical framework will prompt us to rethink the way nonhumans are cultivated for consumption with an eye toward interspecies justice.

Plumwood herself engages this question in the paper "Integrating Ethical Frameworks for Animals, Humans, and Nature" (2000). In it she criticizes a position held by some ecofeminists, notably Carol Adams (1994), that she calls "ontological vegetarianism." It asserts that nothing morally considerable should ever be ontologized as edible or as available for use, and so feminists committed to the moral recognition of nonhuman animals ought to be vegan (Adams 1994; Plumwood 2000, 287). Plumwood articulates and rejects the position's grounding claim, the "exclusion assumption": "that because food is inevitably a site of domination, degradation, and exclusion, ethical food practice consists in ensuring that nothing that is morally considerable can ever become our food or be ontologized as edible" (2000, 295). In other words, the claim asserts that if a being is regarded as food, that being is ontologized as edible and nothing more; appropriate for consumption and use but as a result excluded from moral consideration.

Plumwood argues that animal welfare oriented approaches, feminist or not, that seek to extend the sanctity of human life to nonhuman animal life by employing this assumption, are well meaning but misguided. Plumwood argues that the exclusion claim *assumes* that if a being is ontologized as edible it cannot, at the same time, be respectfully ontologized as more-than-food. Thus the exclusion assumption begs the question regarding whether or not we can produce and consume anything and remain ethical by assuming at the outset that food is a site of domination rather than reciprocity. It does so by relying on a reductive understanding of eater as subject and food as object that maintains dualistic conceptual frameworks.

Ultimately, the exclusion assumption denies human beings' own position in the food chain as both eaters and edible. Some of Plumwood's (1995) most powerful writing, in her description of her own experience as crocodile prey, serves as a reminder that we fail to ontologize ourselves as edible at our own risk! It has the effect of reorienting our perspective, from that of the master to that of the "meat," and this reorientation reveals the fact that complex, morally valuable beings (humans included) are edible food for others. It is worth reminding ourselves of less extreme examples of human edibility as well: the mosquitos, fleas, ticks, and leaches of the world, not to mention the variety of parasitic worms that make human beings their home.[1] This denial represents a dangerous rejection of our ecological embeddedness and prevents a full recognition of one's ecological selfhood. Moreover, the exclusion assumption begs the question regarding whether or not we can produce and consume anything and remain ethical by assuming at the outset that food is a site of domination rather than reciprocity.

As an alternative to understanding food relationships as a dichotomy between unedible/edible, Plumwood suggests a "reciprocity model" as the appropriate model for ethical consumption relationships. She turns to indigenous sources of knowledge on relational hunting and gathering as a resource for thinking about reciprocal food relationships. Non-dominant cultural traditions serve as alternative models for food relationship, "which see the food chain in terms of reciprocity rather than domination or alienation…in which all ultimately participate as food for others, and the 'moreness' of all beings is recognized" (Plumwood 2000, 299). On Plumwood's view (2000, 300), such frameworks and practices are respectful of animals' individuality and species life, as well as compatible with an egalitarian ethic of honesty, gratitude and reciprocal ecological benefit.

Plumwood reminds us that arguments for veganism that rest on the exclusion assumption will always require that a line be drawn between what is morally considerable and what is not. For at the end of the day, we all have to eat something! Either we maintain the exclusion assumption, draw a line of inedibility between animals and plants, and ignore "the continuity of planetary life" (Plumwood 2000, 301), *or* we take account of those morally considerable beings

---

[1] I thank Lisa Heldke for reminding me of these smaller examples. As she puts it, we are always chomping and being chomped on. Remembering parasites is a way to rethink the false dichotomy between the chompers and the chomped.

who are our food and attempt to build ethical structures around our consumption. The "sacred eating" framework of relational hunting and gathering may provide a model for how to "take account." We can identify the ethical commitments of reciprocity that underlie such practices: that a good human life acknowledges *kinship* with nonhuman life while at the same time *respecting* the other as independent, different, and always "more than food."

Plumwood is not alone in appealing to the practice of hunting in order to illustrate an alternative human-animal relationship.[2] But as Kheel (1995) points out, there are distinct categories of hunting practice that rely on distinct forms of ethical justification. The relational, or what Kheel calls the "holy hunt," remains undefined in most invocations, but is presumed to involve respect, a sense of sacred identification, and to be primarily based on spiritual and physical need. Support for the possibility of such a hunting practice and for a reciprocal moral orientation is frequently garnered from illustrations of Native American or other indigenous hunting cultures. There are two problems here. First, I share the concern that such illustrations tend toward the appropriation and homogenization of non-dominant cultures (Adams 1994, 105; Kheel 1995, 101). Adams worries that environmentalists further only their own self-interest when appealing to indigenous hunting cultures, for the focus tends to be on hunting to the exclusion of other aspects of indigenous food culture. Variation in population density, settlement, and reliance on agricultural production between indigenous cultures tend to be overlooked, and the narratives of Native relational hunting appropriated to suit the argument's needs. I think this worry can be alleviated by more careful description of the contexts in which relational hunting was/is practiced and the additional contexts in which it may be appropriate (Warren 2000, 134–137; Plumwood 2000). It is also necessary to open a space in the conversation for advocacy of indigenous rights (Adams 1994, 105). But second, even if that worry is alleviated, one can still ask, as feminist vegetarian Carol Adams does (1994, 104), "what does the animal who dies receive in this exchange?" Does the hunter's victim care what the attitude of the hunter is at the moment of slaughter? For Kheel the "holy hunt" concept is nothing but a romanticized "legitimization of violence and biocide" (1995, 88).

In response to such objections Plumwood's view (2000) prompts us to think about reciprocity in the context of existing within the exchange cycle of the food chain. What the animal received is life and the condition of life is embodiment and all of its strengths and vulnerabilities. I suggest that the objection presumes that we are justified in radically reducing the context of the action in order to judge it. Adams' questions ask us to focus telescopically on the moment of the animal's death at the expense of a wider picture of the conditions of animal life. Maybe it is true that "it doesn't matter to the animals" whether they are killed by a relational hunter, or sport hunter. But it likely does matter for the conditions of life that the animal leads up until that moment, and that its progeny will lead in the future.

---

[2]See Leopold (1989), Rolston (1988), and Snyder (1969) to take just a few early prominent, and disparate, examples.

The kinds of activities that we do shape the context in which we do them, and vice versa. It is not just in death and consumption that the relational orientation is manifest. The orientation itself prompts a consideration of the wider conditions of ecological stability and resilience, while maintaining an ability to oscillate readily between care for ecological system health and respect for individual animal beings. I agree that the relational hunt can be invoked inappropriately. But the practice captures a relationship, not just between hunter and hunted, but between the hunter and the broader ecological reality of mutual exchange.[3] Thus I dismiss Adam's point, but don't want to go so far as to assume that how and why you die are unimportant to one's conscious experience of death. Plumwood reminds us that at least in the human case this is plainly wrong: "As political activists have long recognized, it can make a big difference whether you die in struggle or in submission, for yourself or in solidarity with others, in changing the world or uselessly" (Plumwood 2000, 319).

Insofar as we are able to discern the virtues of care and respect in particular attitudes and practices, like the "relational hunt," we can at least see the possibility of ethical *eating*. But, even while highly critical of current factory farming methods, Plumwood does not fully address the question of the possibility of ethical *agriculture*. I turn to that question now, using the above arguments as a starting point for my own articulation and rejection of two potential objections to the assertion that agriculture is compatible with respect across species.

22.3   The Problem of Equality

In shifting from the question of ethical eating to the question of ethical agriculture, we encounter the *problem of equality*, that is, the seeming incompatibility between interspecies egalitarianism and the agricultural project. While the objector might concede that it is wrong to assume that *food* is a site of domination, it still seems the case that *agriculture* is an obvious site of domination. In the end, the objector might say, agriculture is about cultivating and directing nonhuman life under human terms of use and ownership. It depends on the rational control of human beings and, in so doing, assumes human superiority. This assumption is either unjustified, in which case agriculture will always be a site of exploitation, or it is justified, making the appeal to reciprocity unnecessary. To the skeptical this would suggest that attempting to use egalitarian reciprocity to guide the structure of our agricultural food systems is futile.

My response to the *problem of equality* is to suggest that the seeming incompatibility between interspecies equality and agriculture is due to a misreading of

---

[3]Truly embracing a reciprocal framework would require major adjustments in our practices connected to human death. The law and our own squeamishness prevents dead human bodies being accessible to other animals as food. In many ways the question of our response to our own death is a necessary corollary to the question of how we can responsibly feed ourselves.

what equality means, or rather what it ought to mean if we are seriously engaged with the project of deconstructing value dualism. Plumwood (2002, 173) concedes that "true" interspecies egalitarianism, under which all natural entities, including humans, carry the same moral weight, is both implausible and unworkable. I agree with her on that point for the reason that, even in the human case, having to give exactly the same weight to each person's needs and interests requires us to ignore particularistic claims and responsibilities. This runs contrary to the fundamental notion that appropriate ethical responses must be context-sensitive. But for my purposes here, the more important problem with an equality based on sameness is that many interests and qualities are incommensurable, so to attempt to rank them along a scalar axis in order to determine their value is irresponsible.

Because resisting the structure of value dualism requires recognition of both continuity and difference, we cannot premise egalitarianism on a simple under-standing of sameness. So we are not looking for the kind of framework that many environmental ethicists seem to seek, where we identify some criterion of value (maybe rationality, or sentience) and then rank beings on a scale according to their possession of the criterion trait, the result being the equality of the humans and nonhumans that occupy the same position along that scale. For example Singer's (1974, 2002) utilitarian ethical framework prompts the ranking of all individual beings on a scale according to their capacity to experience pleasure and pain. The interests of those nonhumans who experience pleasure and pain comparable to that of humans ought to rank equally in humans' moral considerations. Singer char-acterizes the denial of equality of interests on the basis of species as a kind of bias akin to sexism and racism. But this utilitarian version of equality is not counter-hegemonic; it is a reproduction of center/periphery thinking. Rather than under-mining the conceptual framework of exclusion, the utilitarian account extends moral consideration to include those species whose conscious experience is most like our own, while simultaneously excluding the majority of nonhuman life. With the performance of this "double gesture" (Plumwood 2002, 151) of inclusion and exclusion, the structure of value dualism is maintained. A counterhegemonic response requires the recognition of both continuity and difference, but scalar equality requires only the recognition of sameness. Truly recognizing difference requires recognizing that the attempt to rank incommensurable qualities is inappropriate.

The virtue of non-ranking is an important part of the content of equality and respect in the human case. Familiar democratic struggles remind us that ranking is morally problematic when it is unnecessary, invariant, and context-insensitive; the more generalized the ranking is, and the closer its connection to moral value, the more reason we have to be wary of hidden hegemonic agendas (Plumwood 2002, 173). And at the personal level, we don't regard two humans as morally equal because they are factually equal with regard to some criterion; we regard them as morally equal because we refuse to rank them in such a manner. Despite its clarity in the human case, this conception of equality runs contrary to the way we tend to think about interspecies equality. The most prominent attempts to include nonhuman interests in our moral deliberations, such as the utilitarian efforts of Singer (2002), do rely on a conception of scalar equality. But counterintuitiveness or unfamiliarity from the point of view of tradition cannot be good reasons to reject the reconceptualization of equality. As Plumwood insists,

> Between categories of very different beings, many of whose capacities the ranker may not be in a position to know, insistence on ranking (on a scale of superior/inferior which includes the case of equality) is both poor methodology and symptomatic of an arrogant stance of closure which is impoverishing and limiting for both human self and non-human other (2002, 173).

Once we reject the stance of closure toward nonhumans by adopting the intentional recognition stance, the avoidance of ranking becomes as important a virtue for an ecological ethic as it is for human-human ethics.

Perhaps it is the case that the process of decision-making necessarily involves the ranking of possible benefits, harms, consequences, commitments, and so on. The attempt to make context-sensitive decisions may even require us to engage in ranking more frequently than we would if a scale of value was set and consistently recognized. It might even seem like context-sensitivity requires the ranking of contexts themselves, from morally optimal to morally stifling. If so, it would seem that non-ranking is an impossible virtue to uphold both practically and meta-ethically. Plumwood does use the term non-ranking to identify the commitment to recognize incommensurability across species, and hence moral equality. But on my view these concerns prompt a change in emphasis from Plumwood's virtue of non-ranking to the need to develop the ethical sensitivity to apply ranking wisely/responsibly. To me this remains compatible with continuing to avoid a fixed scalar ranking at the level of fundamental moral value.

The ecological feminist ethical framework that I am working with should be regarded as virtue-based insofar as ethical commitments are understood as inherent to a particular kind of identity, in this case one's identity as an ecological self. Yet these commitments remain generic, or "determinable," until applied in particular contexts. Thus knowing when and how to manifest these commitments appropriately requires wisdom. The virtue of non-ranking is a generic virtue that may not be appropriate in all contexts. For example, in a triage situation, the needs of individuals must be ranked in order to provide good care, and this may result in great harm befalling some individuals. When faced with the fact that I will destroy something in order to feed myself, I must at the very least rank my options for meeting my needs. But even in contexts of conflict or scarcity where ranking seems inevitable and/or necessary, one can proceed in context-sensitive ways that avoid generalizations of rank (Plumwood 2002, 174).

Contrary to the objection under consideration, I argue that agriculture is a context in which it is possible to reflect a responsible ranking. The agrarian wisdom of Wendell Berry and his contemporary disciples offers plenty of examples of agricultural practices that reflect this virtue. Agrarianism has been described as a comprehensive worldview that is grounded in a "synoptic vision of the health of land and culture" (Wirzba 2003, 5). It provides a normative vision of the good life as rooted in land and tradition, while simultaneously providing a discursive context for sharing the practical knowledge of healthy farming techniques. Such techniques include, for example, diversification of crops and rotations, small-scale production, radically reduced chemical use, waste-minimization, and in general farm practices that are carefully tailored to suit the particular characteristics of the land in a given place (Berry 1978; Jackson 1980, 1994; Kline 2001; Logsdon 2000). The "good farmer" as Berry calls him [sic] does not rank himself and his own short-term needs or desires as more valuable than the long-term health of the land, crops, and animals that he works with. And "with" is the proper word for the good farmer whose "kindly use depends upon intimate knowledge, the most sensitive responsiveness and responsibility" (Berry 1978, 31).

The Ohio farmer and writer Gene Logsdon has taken up the role of giving voice to the "good farmer" in the American agrarian tradition. Logsdon doesn't actually label himself the "good" farmer, but rather the "contrary" farmer, a figure who offers place-specific advice on organic gardening, pasture farming, and other homesteading skills from a perspective that is largely critical of dominant culture and industrial agriculture. Logsdon's many essays reflect the difficulties, pleasures, limitations and possibilities of living and farming "at nature's pace" (2000).[4] Logsdon's approach to pasture farming provides an example of interspecies "dialogue" producing a responsive practice (2001, 2004). In pasture farming, animals are let out on pasture to forage and graze rather than confined and fed a diet of harvested grains. The success of the pasture farming enterprise depends on careful attention—to the growth cycles, nutritional requirements, and response to grazing of the ground cover crops in a given place—and on adjusting one's practices in response to the information this careful attention provides.[5] This kind of pasture farming reflects the intentional recognition stance at work. The pasture's plant and insect species, the grazing animals, and even the watershed must all be recognized as capable of flourishing and as potentially communicative in order to develop authentically responsive, place-based pasturing practices. Logsdon's understanding of "all flesh [as] grass" provides a particularly interesting example of non-ranking. Not only does Logsdon avoid ranking his own aims over or against the long-term health of his land or the health of the animals in his care, but he also avoids making a rigid animal-plant moral distinction of the sort that remains common in arguments for the moral equality of animals.

Such agricultural practices are recognized as, largely, human-driven, and there is no shortage of argument for the "practical" economic benefits of pasture farming (Logsdon 2001, 155). When the economic benefits are outlined, the nonhumans raised for consumption are often included in strictly economic terms (Logsdon 2001, 162).

---

[4]Logsdon contributes weekly musings on farming and pasture farming, gardening, homesteading, and rural life to a weekly blog, also entitled "The Contrary Farmer." Find it at: https://thecontraryfarmer.wordpress.com.

[5]For example, Logsdon writes, "I then learned that I could make a crop of hay from a stand of oats before it went to head and, when the oats regrew and headed out, pasture it as a grain supplement. And then, to my utter amazement, a third crop emerged from the oat grains that the sheep missed. This third crop provided green pasture even into early January" (2001, 167–168).

By no means am I suggesting that Logsdon's perspective reflects a perfect avoidance of ranking, for in many of his essays, the human-focused reasons for upholding agrarian values dominate. Neither am I suggesting that one can always attain a perfect avoidance of ranking, as my shift to the concept of *responsible* ranking above indicates. Yet, in his work Logsdon's language often does reflect a sense of partnership and an ever-present recognition of the human as one amongst many species that contribute fundamentally to the character of a given place. Take, for example, the following description:

> Walk with me over our little farm where biological diversity is our first order of business... On this farm lives a human family along with several families of corn, oats, wheat, orchard trees, grasses, legumes, berries, and garden vegetables, the whole domestic tribe living in a sort of hostile harmony with the wild food chain: animals, insects, and plants in such diversity that I have not been able to name them all. On our little farm, I have identified 130 species of birds, 40 species of wild animals (not counting coonhunters), over 50 species of wildflowers, at least 45 tree species, a myriad of gorgeous butterflies, moths, spiders, beetles, etc., and about 593,455,780 weeds (Logsdon 1995, 7).

And more directly:

> Pasture farming... requires a humble dependence on forces beyond tillage machines, and a recognition that humans are not really in control, no matter what. Pasture farming recognizes that our survival depends upon our ability to stand by patiently and *work in partnership with nature, not in domination over nature* (Logsdon 2004, 5–6, *my emphasis*).

Here Logsdon's writing reflects his posture of openness toward his nonhuman partners and their independent projects. The humility inherent in this posture is a reflection of the refusal to rank the human farmer as always above and in control of the others.

Additionally, the work of the agroecologist[6] and crop scientist Wes Jackson, and his pioneering colleagues at The Land Institute in Salina, Kansas, provides another illustration of the intentional recognition stance at work in an agricultural context. The Land Institute's central project is to develop agricultural systems that mimic the perennial polyculture of the native prairie as a viable alternative to the soil-depleting, annual monocultures of the dominant farming techniques in the United States (Jackson 1980).[7] This work takes "nature as a standard" and looks to prairie wisdom over "human cleverness" in developing cultivation techniques

---

[6]Agroecology is the application of knowledge of ecological processes to the design and management of agroecosystems (Altieri 1987). The aim is a whole-systems approach to agriculture and food systems development. That agroecology exists as a legitimate field of study and practice reflects the idea that many agriculturalists adopt a posture of openness to knowledge from "agentic" nonhuman sources.

[7]An annual is a plant that performs its entire lifecycle from seed to flower to seed within a single growing season. All of the roots, stems, and leaves of the plant die annually and the dormant seed bridges the gap between growing seasons. Perennial plants maintain a viable root system even if the plant dies back after the growing season. Monoculture farming produces a single crop in a given space while polyculture imitates other ecosystems by integrating multiple crops in a given space.

(Sanders 2001, 11). It regards appropriate agriculture as "community ecology in which human-directed arrangements are featured" but those arrangements mimic as much as possible a so-called "natural ecosystem" (Jackson 1994, 49). This work requires that nonhuman species (plants, insects, birds, microbes) and systems (the native prairie ecosystem) be understood as sources of knowledge, knowledge that is only available to us if we adopt the openness to interspecies communication of the intentional recognition stance. Like Logsdon's pasture farming, the efforts of The Land Institute provide an interesting example of non-ranking insofar as the general importance of soil retention and fertility is emphasized as strongly as the productivity of any individual plant species.

What I want to suggest is that the practices of small-scale sustainable farming and pasturing provide opportunities to identify the kinds of activity that would be consistent with reciprocity and responsible ranking. Yet, some objectors will remind us that the animals raised for grazing provide not just wool or dairy products, but are killed for food. Is agriculture rightly regarded as a partnership when one partner gets eaten? Certainly. I have already gestured to my answer to this important question above and I will return to it below. Recognizing that one will have to eat other morally valuable beings, including (perhaps *preferably*) those that one is close to, is central to acknowledging one's embodiment and ecological embeddedness. In an agricultural context, the good farmer would develop the ethical sensitivity to apply ranking wisely in determining how to farm and what to eat. Presumably this ranking would not occur universally or forever after, would not be a fixed scalar ranking of moral value across time and place.

22.4 The Problem of Incorporation

Assuming that it is possible to structure agricultural practices to reflect a high degree of interspecies equality, there is still the apparent problem that the very *purpose* of agriculture must be understood in terms of human ends. The *problem of incorporation* has an appropriate double meaning in the context of agricultural food relationship. First, the nonhuman other is being cultivated and raised with the explicit purpose of being (literally) incorporated into a human body. Second, on Plumwood's theory, "incorporation" is a technical term used to describe a hegemonic form of relational definition wherein the subordinated group is defined only by reference to the dominant group's traits and projects (1993, 52). A structure or practice that relies on such a dualistic relational definition cannot possibly be counterhegemonic. Thus, one might object that because the ultimate end of agriculture is to kill and consume nonhuman beings (animal or plant), the existence of the nonhuman beings in question is defined by reference to the human ends they will serve, and their meaning is incorporated into human meaning. The *problem of incorporation* thus suggests that agriculture is fundamentally incompatible with a respect.

Any problem that is tied to literal incorporation has already been addressed with the rejection of the exclusion assumption as discussed above. The physical incorporation of others' bodies through consumption is only necessarily unjust if we are forbidden to conceive of morally valuable entities as ultimately consumable. As I've already noted, this assumption must be rejected if the truth of our ecological selfhood is to be fully embraced. Thus the literal incorporation of morally significant others is only a problem when ecological selfhood is denied.

The seriousness of the second dimension of the problem can be mitigated when one realizes that recognition of ecological selfhood significantly reduces the dangers of incorporation. For the problem is not merely the fact that the other is defined in relation to the self, but that the other is defined *only* and *always* in relation to the self conceived as center. Ecological selfhood is a relational selfhood that attempts to correct this imbalance such that human self and earth others are recognized as co-constituting, and as irreducible to an assimilated definition. Likewise, the danger of incorporation is not in the use of the other, but in defining the other only as something to be used, or always by their role in the master subject's projects. Incorporation results in a lack of space to encounter the other fully, richly, dialogically. It undermines one's ability to be attentive and responsive to the other's needs and ends. In so doing it precludes the possibility of responding to others with genuine respect or care. Thus countering the problem of incorporation begins with openness to encountering the other in their fullness of being, and letting such encounters be transformative of our outlooks and behaviors.

Plumwood gives examples of specific ways to counter incorporation through biological education, local wildlife preservation, and so on (2002, 112–113). These practices successfully address and/or resist assimilated definition when they contest the traditional view of nonhuman difference as a lack, replacing it with an affirmation of the richness of nonhuman life and "a view of nonhumans as presences to be encountered on their our terms as well as on ours" (Plumwood 2002, 112). Surely there are ways to minimize incorporation in agricultural contexts. The examples I refer to above, of practices that manifest responsible ranking, might also serve as a list of practices meant to address incorporated definition. For example, pasture farming counters incorporation by recognizing the health benefits, of grazing and foraging rather than consuming a diet of harvested grain, *for the animals themselves*. The work of The Land Institute counters incorporation by creating an understanding of, and attempting to develop cultivation practices that sustain, nature's own complex ecological order.

It is important to note, as Thompson (1995) does, that many farmers who adopt sustainable, appropriately responsive practices will be the first to describe themselves as "self-interested." They farm the way that they do primarily for their own or their successors' own long term benefit, in service to the self-interest created by land ownership (Thompson 1995, 74). The question of ownership is critical to some animal rights advocates, in particular, who worry that ownership is a necessary site of domination; animal agriculture (at least) is premised on owning beings that ought not be owned. This objection most forcefully motivates Stanescu's (2013) critiques of "happy meat" marketed as the product of love and care. Stanescu writes, "what concerns me most is that these expressions of feelings of care for animals serves to mask the simple reality that for the entirety of their lives, these animals live only as buyable and sellable commodities, who exist wholly at the whim of their 'owners'" (2013, 108). Thus we might worry that, in the end, even the practices that appear to mitigate incorporation ultimately fail to do so. But it seems that, for many farmers, it is possible to regard the animals in one's care not as flesh-producing machines but as partners in the project of sustaining a robust agroecosystems; recall Logsdon's explicit appeal to partnership above. If so, then despite the fact that the farmer legally owns the land, animals, and crops, it is simply not the case the animals exist "only" or "wholly" as commodities. Moreover, the self-interest of good farming is tempered by recognizing that the self in question is defined as *ecological* at the most fundamental level, thus requiring a different understanding of interests. Rather than leading to a maximizing principle, ecological self-interest of the sort at the heart of agrarian thinking leads to a keen understanding of ecological limits and a principle of cyclical reciprocity.

On my view, cultivation and consumption are the primary sites for recognizing, engaging, and acting on our ecological embeddedness on a daily basis.[8] Rather than proving devastating for the possibility of ethical agriculture, thinking through the problem of incorporation reveals that agriculture may in fact be the most significant area of human life in which incorporation can be obviously addressed. If we heed Wendell Berry's urging (2002) and recognize eating as an "agricultural act," then we have the opportunity to address incorporation on an immediate and daily level. This leads me to conclude that the best way to counter the objection under consideration is to propose (optimistically) that it mischaracterizes the agricultural project. In one sense, of course, the end of agriculture is food provisioning which necessarily involves cultivating nonhuman beings and then consuming them. In a more expansive sense though, the end of agriculture might be understood as the sustained health of human and nonhuman communities in interrelationship. Or the creation of human cultures that are responsive and responsible to their places. The problem of incorporation is alleviated insofar as human beings come to understand and define themselves in terms of the nonhumans that co-constitute their lives and culture. It is alleviated insofar as nonhumans and place are recognized as creative, transformative, and having meaning that cannot be reduced to their usefulness toward human beings' short-term aims.

---

[8]Obviously, fishing and hunting traditions are sites for encountering and organizing life around our ecological embeddedness as well, as the authentic practice of relational hunting described above indicates. The prominence of farming is contingent on historical relationships to land and the historical settlement and colonization of the American continent (Thompson 2010).

References

Adams, Carol J. 1994. *Neither man nor beast: Feminism and the defense of animals*. New York, NY: Continuum.

Altieri, Miguel A. 1987. *Agroecology: The scientific basis of alternative agriculture*. Boulder, CO: Westview Press.

Berry, Wendell. 2002. *The art of the common place: The agrarian essays*. Berkeley, CA: Counterpoint Press.

Berry, Wendell. 1978. *The unsettling of America*. New York, NY: Avon Books.

Jackson, Wes. 1994. *Becoming native to this place*. Lexington, KY: The University Press of Kentucky.

Jackson, Wes. 1980. *New roots for agriculture*. Lincoln, NE: University of Nebraska Press.

Kheel, Marti. 1995. License to kill: An ecofeminist critique of hunter's discourse. In *Animals and women: Feminist theoretical explorations*, ed. Carol J. Adams, 85–125. Durham, NC: Duke University Press.

Kline, David. 2001. Great possessions. In *The new agrarianism: Land, culture and the community of life*, ed. Eric T. Freyfogle, 181–195. Washington, DC: Island Press.

Leopold, Aldo. 1989. *A Sand County almanac and sketches here and there*. New York, NY: Oxford Paperbacks.

Logsdon, Gene. 2004. *All flesh is grass: The pleasures and promises of pasture farming*. Athens, OH: Swallow Press.

Logsdon, Gene. 2001. All flesh is grass: A hopeful look at the future of agrarianism. In *The essential agrarian reader: The future of culture, community and the land*, ed. Norman Wirzba, 154–170. Lexington, KY: The University Press of Kentucky.

Logsdon, Gene. 2000. *Living at nature's pace*. White River Junction, VT: Chelsea Green Publishing.

Logsdon, Gene. 1995. *The contrary farmer*. White River Junction, VT: Chelsea Green Publishing.

Merchant, Carolyn. 1983. *The death of nature: Women, ecology and the scientific revolution*. San Francisco, CA: Harper San Francisco.

Plumwood, Val. 2002. *Environmental culture: The ecological crisis of reason*. London: Routledge.

Plumwood, Val. 2000. Integrating ethical frameworks for animals, humans and nature: A critical feminist eco-socialist analysis. *Ethics & the Environment* 5 (2): 285–322.

Plumwood, Val. 1998. Intentional recognition and reductive rationality: A response to John Andrews. *Environmental Values* 7 (4): 397–421.

Plumwood, Val. 1995. Human vulnerability and the experience of being prey. *Quadrant* 39 (3): 29–34.

Plumwood, Val. 1993. *Feminism and the mastery of nature*. London: Routledge.

Rolston III, Holmes. 1988. *Environmental ethics: Duties to and values in the natural world*. Philadelphia, PA: Temple University Press.

Sanders, Scott Russell. 2001. Learning from the prairie. In *The new agrarianism: Land, culture and the community of life*, ed. Eric T. Freyfogle, 3–15. Washington, DC: Island Press.

Singer, Peter. 2002. *Animal liberation*. New York, NY: Ecco Press.

Snyder, Gary. 1969. *Earth house hold: Technical notes and queries to fellow Dharma revolutionaries*. New York, NY: New Directions.

Stanescu, Vasile. 2013. Why 'loving' animals is not enough: A response to Kathy Rudy, locavorism, and the marketing of 'humane' meat. *The Journal of American Culture* 36 (2): 100–110.

Thompson, Paul. 2010. *The agrarian vision: Sustainability and environmental ethics*. Lexington, KY: The University Press of Kentucky.

Thompson, Paul. 1995. *The spirit of the soil: Agriculture and environmental ethics*. New York, NY: Routledge.

Warren, Karen J. 2000. *Ecofeminist philosophy: A western perspective on what it is and why it matters*. Lanham, MD: Rowman & Littlefield.

Warren, Karen J. 1990. The power and promise of ecological feminism. *Environmental Ethics* 12 (2): 125–146.

Wirzba, Norman. 2003. *The essential agrarian reader: The future of culture, community and the land*. Lexington, KY: University of Kentucky Press.

Chapter 23
Religious Slaughter in Europe: Balancing Animal Welfare and Religious Freedom in a Liberal Democracy

R.W. Mittendorf

Abstract In February 2014, Denmark joined a growing list of European nations to ban *kosher* and *halal* butchering methods without pre-stunning. While this policy is a victory for the animal welfare and rights movements, it is also part of a troubling trend of European states limiting religious freedom for religious minorities. This violation of minority rights is not just a problem for liberal democracy, but I suggest it is also a challenge to animal welfare and rights globally. In this chapter, I will look at the ethical, political, and theological issues surrounding the banning of non-stun religious slaughter to underlie a practical policy recommendation that members of a liberal democracy, including proponents of animal welfare and animal rights, ought to oppose bans on *kosher* and *halal* butchering without pre-stunning for the sake of religious freedom and animal welfare. First, I will describe the requirements of religious slaughter in Judaism and Islam along with the controversy in Europe over the banning of non-stun religious slaughter. Next, I will challenge the claim that stun slaughter is more humane than non-stun slaughter. In the third section, I will discuss religious freedom and the moral status of animals within the liberal requirements of moral pluralism and secularism. Finally, I will argue that religious traditionalists and animal welfarists ought to work together to promote overall animal well-being. My purpose in this essay is not just to discuss liberal democratic principles but to engage in the deliberative process of exchanging reasons and finding common ground. Far too often groups with opposing interests work to improve their arguments against one another, but genuine deliberation involves understanding the underlying motivations for policy positions. The goal here is not to downplay the differences between animal movement proponents and religious traditionalists but to highlight the areas where their interests overlap for the benefit of animal welfare, religious freedom, and liberal democracy.

R.W. Mittendorf (✉)
Chaffey College, Rancho Cucamonga, CA, USA
e-mail: will.mittendorf@gmail.com

© Springer International Publishing AG 2017
I. Werkheiser and Z. Piso (eds.), *Food Justice in US and Global Contexts*,
The International Library of Environmental, Agricultural and Food Ethics 24,
DOI 10.1007/978-3-319-57174-4_23

285

Keywords *Kosher* · *Halal* · Animal welfare · Liberal democracy · Religious freedom

In February 2014, Denmark joined a growing list of European nations to ban *kosher* and *halal* butchering methods without pre-stunning. While this policy is a victory for the animal welfare and rights movements, it is also part of a troubling trend of European states limiting religious freedom for religious minorities. This violation of minority rights is not just a problem for liberal democracy, but I suggest it is also a challenge to animal welfare and rights globally.

In this chapter, I will look at the ethical, political, and theological issues surrounding the banning of non-stun religious slaughter to underlie a practical policy recommendation that members of a liberal democracy, including proponents of animal welfare and animal rights, ought to oppose bans on *kosher* and *halal* butchering without pre-stunning for the sake of religious freedom and animal welfare. First, I will describe the requirements of religious slaughter in Judaism and Islam along with the controversy in Europe over the banning of non-stun religious slaughter. Next, I will challenge the claim that stun slaughter is more humane than non-stun slaughter. In the third section, I will discuss religious freedom and the moral status of animals within the liberal requirements of moral pluralism and secularism. Finally, I will argue that religious traditionalists and animal welfarists ought to work together to promote overall animal well-being.

My purpose in this essay is not just to discuss liberal democratic principles but to engage in the deliberative process of exchanging reasons and finding common ground. Far too often groups with opposing interests work to improve their arguments against one another, but genuine deliberation involves understanding the underlying motivations for policy positions. The goal here is not to downplay the differences between animal movement proponents and religious traditionalists but to highlight the areas where their interests overlap for the benefit of animal welfare, religious freedom, and liberal democracy.

23.1 *Kosher* and *Halal* Slaughter in Europe

In Judaism, the religious requirements of religious slaughter, known as *shechita*, are governed by the Jewish religious law, *Halacha*, which can be divided into two subsets: the written law and the oral law. The written law comes from the *Tanahk*, or Hebrew Bible, with the first five books of Moses referred to as the *Torah*. According to Jewish tradition, the Oral Law was passed down from Moses, teacher to student, until leaders began transcribing and commenting on it between the 2nd and 5th centuries C.E. This collection of writings is referred to as the *Talmud*. It is through this Talmudic tradition where the requirements for *kosher* food preparation are found. These *Kashrut* laws are not governed by a central religious body but are instead a historical development that has widespread agreement across Jewish movements. As is the case across all religions, one does not need to follow all traditions to be considered a member of the faith, and many secular and Reformed Jews do not follow *Kashrut* law; however, for those Orthodox Jews who do follow the practice, *shechita* is an important aspect of their religious observance.

Shechita is a method of animal slaughter where the throat is cut using a *chalef*, which is a long, sharp, non-serrated knife. The cut severs the jugular veins and carotid arteries, and the animal is quickly rendered unconscious through exsanguination and is meant to die painlessly. After death the animal can then be slaughtered and consumed. The *shechita* must be performed by a trained professional called a *shochet*. The *shochet* must be trained in scripture and in the slaughtering technique at a Jewish seminary. The *chalef* must be examined by the *shochet* before and after each cut and to ensure there are no nicks, and must be presented to a local rabbi regularly for additional inspection (Zivotofsky 2010, 13). After the animal's death, the *shochet* must inspect the body for *treifa*, which is any physical defect of the animal, including signs of disease or injury.

The presence of *treifa* is integral to the problem of pre-stunning slaughter. For meat to be *kosher* it must be free of defects before the *shechita*. Bolt or electrical stunning causes an injury to the animal pre-*shechita*, or makes the inspection for *treifa* difficult because of the effect the stun has on the muscles and organs, which renders the animal non-*kosher*.

In Islam, the religious requirements of religious slaughter, known as *dhabiha*, are governed by the Islamic religious law, known as *Sharia*. *Sharia* is not governed by a central religious body but exists in several iterations depending on various schools of Islamic thought. Dietary restrictions are derived from three sources: the *Qur'an*, which is the main sacred text, the *Sunnah* and *Hadith*, which are the example and the sayings, respectively, of the prophet Mohammed, and the views of religious scholars, many times in the form of a *Fatwa*, a religious decree.

For meat to be *halal*, meaning permissible, *dhabiha* must be performed on animals that are *halal*, such as cattle, fish, and not on animals that are *haram*, impermissible, such as swine and donkey. Furthermore, the animals must be healthy and alive before slaughter, blood must flow sufficiently out of the animal, and the slaughterer must recite the name of *Allah*. There seems to be variance in the practice regarding the requirement of a Muslim slaughterer, with some allowing for any slaughterer by 'people of the book' (meaning a Christian or Jew), whether or not the animal needs to be *kible* (facing Mecca at the time of *dhabiha*), and whether or not the animal can be stunned before *dhabiha*.

Dhabiha is a cut to the neck which severs the jugular vein, carotid artery and windpipe. There is debate over the method of the cut and the necessity of a human handler, so while some practitioners require a human to perform the cut with a sharp knife, recently the use of a machine with an electric blade has become permissible (Anil et al. 2010, 6). There is a plethora of opinions on whether or not pre-stunning is permissible. For example, in 1978 a *fatwa* was issued by Al-Azhar University in Cairo, Egypt, which allowed for pre-stunning, but in 1995 Al-Azhar issued another fatwa which disapproved of mechanized bolt pre-stunning

(Anil et al. 2010, 7). According to the Halal Food Authority (HFA), stunning is permissible if it does not kill the animal, and electric pre-stunning has become a widespread global practice. Nonetheless, there are many Muslims whose faith requires them to abstain from meat that has been pre-stunned before slaughter. Throughout this chapter, I will refer to these Jewish and Muslim practitioners of non-stun slaughter as religious traditionalists.

Shechita and *dhabiha* have been practiced in Europe for centuries; however, beginning in the mid-19th century the rise in animal welfare concerns and anti-Semitic attitudes led to a series of bans on religious slaughter throughout the 19th and 20th centuries. In contemporary Europe, ritual slaughter is considered a religious act covered by Article 9 of the *European Convention for the Protection of Human Rights and Fundamental Freedoms*, so each European Union member state is required to permit religious slaughter with pre-stunning. However, in the European Union, Council Regulation 1099/2009, *On the Protection of Animals at Time of Killing*, and the *Welfare of Animals (Slaughter or Killing) Regulations of 1995* forbid slaughter without pre-stunning, but individual member states can set their own exemptions to the law to permit slaughter without pre-stunning to accommodate religious ritual practices. So there is currently a patchwork of regulations across Europe on a country-by-country basis, some allowing for non-stun slaughter, e.g. France, Spain, Italy, Poland, and the U.K., while others forbid the practice but allow for non-stunned meat to be imported, e.g. Denmark, Switzerland, Norway, Iceland, Liechtenstein and Sweden.

In Europe, religious slaughter without pre-stunning is a contentious debate between those concerned with animal rights and welfare and religious traditionalists. For example, Poland banned the non-stun slaughter in 2013 citing animal welfare concerns but then overturned the ban in 2014 citing the importance of religious freedom. Bans on religious slaughter without pre-stunning are supported and advocated for by the Royal Society for the Prevention of Cruelty to Animals (RSPCA), the British Veterinary Association (BVA), the Farm Animal Welfare Committee (FAWC), the Federation of Veterinarians of Europe (FVE) and the Humane Slaughter Association (HAS), the People for the Ethical Treatment of Animals (PETA), the National Secular Society (NSS), The British Humanist Association (BHA), and Compassion in World Farming (CWF).

Meanwhile, many Muslim and Jewish religious and political leaders from around the world have denounced these bans on religious slaughter as anti-Semitic and Islamophobic. Israel's Deputy Minister of Religious Services Rabbi Eli Ben Dahan responded to the ban in Denmark by saying, "European anti-Semitism is showing its true colors across Europe, and is even intensifying in the government institutions" (Nelson 2014). Danish Halal said the ban is a "clear interference in religious freedom" (Withnall 2014). Ashkenazi Chief Rabbi of Israel, David Lau, said the ban was "a serious and severe blow to the Jewish faith and to the Jews of Denmark" (Jewish Telegraphic Agency 2014).

Despite these religious objections, animal rights and welfare advocates support bans on non-stun slaughter and use two different types of arguments to underlie their positions. First, that slaughter with pre-stunning is more humane than non-stun *halal* or *shechita* slaughter. Second, that because the moral status of animals ought to be equivalent to the moral status of humans (either as equal rights or equal consideration), there should be no exemption given to religious traditionalists when it comes to non-stun slaughter. In the next two sections, I will challenge the efficacy of these two types of arguments.

23.2 Is Slaughter with Stunning More Humane Than Non-stunning?

There are two major intellectual threads to the animal movement: those who are interested in improving animal welfare and those who advocate for animal rights. Of the various animal welfarists, the most well-known is Peter Singer, whose book, *Animal Liberation*, is seen as one of the most important philosophical foundations for improving the lives of animals. Singer argues for a Utilitarian approach to animal welfare and claims that humans ought to give equal consideration to the interests of animals and humans. The utilitarian position is that people ought morally to minimize pain and maximize the greatest amount of happiness and interests for the greatest number (including animals). So it is not surprising that Singer has argued that non-stun religious slaughter is not a morally acceptable practice. He claims that "cut[ting] the throat of a fully conscious animal is not the most humane way of killing an animal" thus, religious traditionalists ought to refrain from eating meat altogether (Singer 2012).

I agree with Singer that it seems like non-stun religious slaughter is a less humane slaughtering option—although there are a number of scientific studies which challenge this assumption (Zivotofsky 2010, 2012; Grandin 2010)—nonetheless, I suggest Singer's position will not have a positive practical effect on animal welfare. While it may be the case that, in a moral vacuum, Singer's argument is correct that it is better for animal welfare for religious traditionalists to simply become vegetarians, from a practical standpoint religious traditionalists disagree with Singer and believe that *shechita* and *dhabiha* are humane options for slaughter and will continue the practice. The issue I am addressing in this chapter is whether or not animal welfare is improved by banning non-stun religious slaughter in Europe, and I argue the Utilitarian ought not to support these bans—despite the theoretical position that fewer animals slaughtered is better for global animal welfare—because allowing non-stun religious slaughter has the practical effect of improved global animal welfare. One area where this is most apparent is in the importing of *halal* and *kosher* meat to Europe.

The perceived inhumanness of non-stun slaughter motivates many European countries to ban non-stun religious slaughter outright, but those countries nonetheless allow for the import of *halal* and *kosher* meat. While this policy is a compromise between religious communities and animal welfarists, it still raises ethical issues.

The ability to import *kosher* and *halal* meat allows religious communities a minimal expression of religious liberty, but it simply shifts the harm to animals to another part of the globe.

For those animal welfare utilitarians who look to reduce the overall amount of suffering in the world, distance should make no moral difference when it comes to preventing harm to others without sacrificing something of equal moral significance (Singer 1972). Therefore, if a European country were to ban religious slaughter without pre-stunning on ethical grounds, then it ought as well to ban the import of *kosher* and *halal* meat. Yet this is not the case, and European countries simply attempt to shift the moral burden to other countries while failing to actually reduce the amount of global animal suffering.

I argue that from a utilitarian perspective it may be better for global animal welfare if European countries allow non-stun religious slaughter and instead ban the import of *kosher* and *halal* meat. Depending on which countries the meat is being exported from and imported to, there may be significant moral differences. There are many countries around the world which do not have the animal welfare standards found in Europe. Allowing imported *kosher* and *halal* meat while banning religious slaughter at home means that more animals are being raised in farms with significantly worse living conditions. Allowing religious slaughter at home offers the opportunity to keep track of animal welfare standards throughout the duration of the animals' lives. Ultimately, in some cases it may increase overall animal suffering to ban religious slaughter at home while allowing for *kosher* and *halal* meat to be imported from countries with poor animal welfare standards.

Alternatively, New Zealand, for example, only allows *halal* meat to be pre-stunned using an electrical stun. Many Muslims who reject penetrating bolt pre-stunning will nonetheless accept electrically pre-stunned meat as *halal* because the electric stun does not kill the animal, it instead renders the animal unconscious with the possibility of it regaining consciousness. However, this stunning equipment is prohibitively expensive for smaller farms to buy, so certain countries with a smaller Muslim population may not have enough demand for a local farm to be able to purchase such equipment. In this case, at least from a utilitarian perspective, it may be better to import *halal* meat (electrically pre-stunned) from New Zealand than other countries. In any case, the blanket ban on non-stunning slaughter while allowing the importing of *halal* and *kosher* meat may not achieve the ethical goals of improving global animal welfare. A more nuanced approach to importing meat could improve both animal welfare and religious freedom.

The proponents of animal rights are the second major strain of the animal movement. These advocates are not interested in simply improving animal welfare and reducing suffering, but actually giving rights to animals, which would include the right to not be killed altogether. While some animal welfare proponents are concerned with improving animal welfare incrementally, certain animal rights advocates, the 'animal abolitionists', believe that incremental welfare improvements only reinforce animal exploitation and reinforce the unjust status of animals as property.

One would initially assume that animal abolitionists ought to support a ban on non-stun religious slaughter because the ban might improve the welfare of animals, but this is surprisingly not the case. Animal abolitionist, Gary Francione, argues that the concept of 'humaneness' prohibits the application of the word 'humane' to animal slaughter. If the animal use is, in his view, considered unjustified (for example, slaughter for the purposes of consumption) then there must be a strong stance against that use altogether, even if a practice like stunning improves animal welfare incrementally. Francione compares the concept of 'humane' animal exploitation with "campaigns for 'humane' rape or 'humane' child molestation" (Unferth 2011). He believes that animal welfare campaigns do not ultimately create an overall better situation of protection for animals because the welfare reforms only help the exploiters become more efficient at exploitation. This simply leads to a reinforcing of the system of exploitation, not to the overall benefit of animal interests. Francione says:

> For example, the Humane Slaughter Act of 1958 [U.S.] was nothing more than a recognition that large animals who were not stunned at the moment of slaughter would cause injuries to workers and incur costly carcass damage. The current campaign for gassing chickens is based on the lower costs of gassing over current methods of slaughter; the campaign to eliminate the gestation crate is based on the increased production efficiency of alternative methods (Unferth 2011).

Banning non-stun religious slaughter because it is considered inhumane only reinforces the idea that stun slaughter is humane and more deeply entrenches the status of animals as property.

While the animal abolitionist and utilitarian positions take different approaches to improving the lives of animals, I argue both positions ought to reject the idea that banning non-stun religious slaughter in Europe as a more humane option. While I agree that it may be a morally better option to not slaughter animals at all, I do not believe the practical situation of animal lives in Europe is necessarily improved by banning non-stun religious slaughter.

23.3 Secularism, Moral Pluralism, and Justice

Today, western society is in what the philosopher, Charles Taylor, calls a "secular age" where religion no longer structures our social imaginaries but has instead become one option among many (Taylor 2007). Religion is compartmentalized in our lives among so many other commitments—it is one of many ways we give our lives meaning. In a liberal, secular society defined by a plurality of worldviews, freedom of conscience remains a highly valued protection. The secular state is supposed to be neutral to various conceptions of the good, allowing each person the freedom to set their own values and act on them to a reasonable extent. However, this extent also needs clarification. Should religious beliefs be given accommodation when it comes to religious practice? What about non-religious moral beliefs? How should the state practice moral egalitarianism when religious moral practices conflict with secular moral practices? This theoretical problem is at the heart of the practical debate over animal welfare and religious slaughter in a liberal democracy. I argue that members of a liberal democracy, including proponents of animal welfare and animal rights, ought to reject bans on non-stun religious slaughter in order to protect cultural pluralism and the fundamental rights of freedom of conscience and religion.

Some animal welfare and animal rights advocates are critical of the liberal commitment to moral pluralism. Animal 'Protectionist', Robert Garner, argues that traditional liberal conceptions of justice are anthropocentric, like John Rawls's 'political liberalism', and therefore fail to include the interests of animals as a matter of justice by relegating animal welfare concerns to one moral position among many (Rawls 1999, 2005; Garner 2005). This is why Garner rejects most animal rights theories. Animal rights theories only offer animals moral protections, that is to say, animals have standing as moral subjects, "subjects-of-a-life," but not as political subjects (Garner 2005). The liberal concept of moral pluralism, which is vigorously defended as a matter of justice, says that the state has an interest to protect human rights, like freedom of religion, speech, and assembly, but not an interest to legislate concepts of the good. In other words, liberal theories allow citizens to have pluralistic views of the moral status of animals and do not give animals protection under a theory of justice. According to Garner, this makes animal welfare "a matter for individual voluntary preferences rather than a set of compulsory obligations" (Garner 2002). Animals can gain some protections from the state, but only as a matter of protecting human moral interests, not as a result of animals having political protection under justice. So when humans change their moral positions, animal protections change as well (for better or worse). Animal protections are therefore at the mercy of voluntary moral preferences, rather than compulsory obligations to justice.

Religious slaughter is just one example of allowing certain groups (Jews and Muslims in this case) to practice their own conception of the good. Their moral view relegates animals to objects of slaughter. On the other hand, people who believe animals have moral standing which prohibits their exploitation are also free to practice their moral beliefs—practicing and advocating veganism, for example.

The animal protectionist point here is well taken. Animal welfare ought to be considered under a theory of justice and not simply as one moral position among many. However, in the case of banning religious slaughter, the problem is that animals are still not taken as subjects protected under justice; the state has simply favored one moral view over another. The moral view that animals are subjects who ought not to suffer the pain of religious slaughter without pre-stunning simply places the limits of animals' moral worth at suffering at death. This view does not include animals as subjects *who ought not to be killed at all*. The state has essentially decided, not that animals are subjects under justice, but that the moral view that animals should be killed with a bolt or electric stun is simply better than the moral view that animals should be killed by *shechita* or *dhabiha*. This is not a vindication of animal justice; it is a condemnation of religious freedom.

Animal protectionists are also rights incrementalists and may therefore support the bans on religious slaughter without pre-stunning because the ban makes a small step toward full animal rights. Then again, the ban on religious slaughter does not view various moral positions equally, since it rejects the moral claims of religious traditionalists, and so it favors one moral view over others. Animal protectionists may find this imbalance of moral interests to be in their favor in this situation, which initially looks like a positive incremental change for animal welfare, but there may be other situations where this imbalance of moral pluralism works against animal welfare.

There is one such situation in the French town, Chalon-sur-Saône. In 2015 the town banned pork-free lunches in public schools. The reason given by the Mayor, Gilles Platret, is that offering an alternative to the religious and vegetarian students is "discrimination" (Editorial 2015). This is an example of the French version of liberal secularism, '*laïcité*'. Platret's argument is that pork is a part of the French tradition and serving it to all students is therefore egalitarian. Exceptions to this practice would favor one moral view (religion or vegetarianism) over irreligion and moral neutrality, which violates the secular principle.

This is an example of how liberal secularism can be used to oppress moral viewpoints and freedom of conscience. While animal protectionists might want to reject the view that animal welfare is one moral position among many, in the meantime, proponents of this view ought to be cautious in supporting cases where moral pluralism is applied in an imbalanced way. In the case of mandatory pork lunches, the interests of religious traditionalists and vegetarians are aligned, and both are rejected by supposed demands of liberal secularism.

Some supporters of non-stun bans reject the religious traditionalists' claim of right to religious freedom when it comes to religious slaughter because eating meat is not a 'requirement' of Judaism or Islam, therefore banning religious slaughter it is not directly affecting the religious freedom of Jews and Muslims (Barry 2002; Singer 2012). However, there are two major issues with this view. First, it takes a mistaken view of religion as needing some sort of authoritative organizational hierarchy to lay out the 'official rules' of the religion. Neither Islam nor Judaism have total authoritative hierarchies, only a patchwork of religious tradition and scripture to draw upon. Second, this view assumes that prohibiting *halal* and *kosher* meat through banning religious slaughter and/or import bans—effectively prohibiting religious traditionalists from eating meat altogether—is not an unfair burden on the religious traditionalists or that this unfair burden is justified.

This second assumption is problematic because it views the moral positions of religious traditionalists as less valuable than the moral positions of non-Jews and Muslims. Even if we assume that there is no religious requirement to eat meat, thus negating non-stun religious slaughter as necessary at all, it nonetheless places an effective ban on meat eating as whole, but not for the general population, only for the religious traditionalists. This unfair burden is a particular problem in contemporary European society because it exacerbates the problem of limiting religious freedom for religious minorities. While it may be easy to look at the issue of non-stunning slaughter as a single issue, it is actually part of a larger set of issues that limit freedom of conscience for already marginalized groups.

While the utilitarian may view the ban on non-stun religious slaughter as a way to reduce overall suffering of animals, utilitarians are also committed to seeing animal and human interests improve. Religious freedom is an important human interest. Limiting religious freedom for marginalized groups has consequences that extend beyond this one particular issue. The rise in Islamophobic and anti-Semitic attacks and rhetoric is one of the most pressing issues facing Europe and the global community. The consequences of non-stun religious slaughter bans cannot be viewed in a vacuum. Limiting religious freedom for religious minorities in Europe has a larger effect on attitudes of religious minorities and their feeling of acceptance in a community, as well as the attitudes of non-Muslims and non-Jews towards these religious minorities. Banning religious slaughter as 'inhumane and cruel', while accepting stun slaughter as 'humane and ethical', only proliferates the ostracizing of religious traditionalists from the larger national community. This likely contributes to hate crimes against the religious traditionalists, and the radicalization of religious traditionalists who feel oppressed. From a utilitarian perspective, banning religious slaughter may have positive effects on animal interests, but may also have many widespread negative effects on human interests.

23.4   How to Improve Animal Welfare by Working with Religious Traditionalists

One assumption that seems to be taken for granted in animal rights and animal welfare literature is that by practicing non-stun slaughter, religious traditionalists are opposed to animal welfare. This is a mistake. Religious traditionalists ought to be considered animal welfarists. In both Judaism and Islam, the duty to religious slaughter is a duty to protect animal welfare and promote compassion.

The Jewish *Halacha* offers a number of specific rules on how to tend to the well-being of animals that includes, but also extends beyond, the slaughter and preparation of *kosher* food. An important concept in Judaism, referred to as the prohibition of *Tza'ar Ba'alei Chayim*, forbids causing any unnecessary suffering to an animal. This concept is drawn out of several passages in the *Tanahk* as well as through rabbinic tradition. Maimonides, the medieval Jewish theologian, philosopher, and legal scholar, described the purpose of *Tza'ar Ba'alei Chayim* as to "perfect us so that we should not acquire moral habits of cruelty, and should not inflict pain gratuitously; but we should rather act with gentleness and mercy to all living creatures except in situations of need" (Maimonides 2007, 288). Animal welfare has been a concern in the Jewish community for thousands of years.

Furthermore, there are several Jewish commentators who argue that eating meat is a 'compromise' because humans were not originally meant to eat animals. Rabbi and conservative Jewish theologian, Samuel H. Dresner, argues that that Adam is "clearly meant to be a vegetarian" (Dresner 1966). Professor Richard Swartz claims that "God's initial intention was that people be vegetarians" (Schwartz 2001). It is not until the story of Noah that humans are permitted to eat meat, and this is a result of "a divine concession to human weakness and human need" (Schwartz 2001). *Kashrut* is therefore a way to make this human weakness into something holy by teaching reverence for life (Schwartz 2001).

Islam accepts and incorporates all of the Abrahamic lineage, and many of the accepted theological positions of Judaism are also accepted by Muslims. Moreover, the prophet Mohammed is reported to have made numerous comments supporting animal welfare. In the *Hadith* 17, Mohammed calls for compassion in animal slaughter when he says, "If you kill, kill well, and if you slaughter, slaughter well. Let each of you sharpen his blade and let him spare suffering to the animal he slaughters." The *Sahih Bukhari*, one collection of *haidith*, details two stories which demonstrate the requirement for protecting animal welfare. In one story, a woman is doomed to hell because she mistreated a cat; in the other, a sinner gives water to a dog dying of thirst and is then saved by Allah (Foltz 2001, 44).

Contemporary Muslim commentators also interpret Islam as being favorable to animal welfare. Gamal alBanna, an Egyptian Islamic scholar, said in an interview with *the Guardian*: "We must not remain rigid in our understanding of faith to mean the blind acceptance of anything, killing living beings included. There is no obligation to kill" (Mayton 2010). Islamic theologian Al-Hafiz B. A. Masri comments that Muslims would refuse to eat flesh on religious grounds if they knew the horrors of contemporary factory farming (Masri 2007). Richard Foltz claims that in contemporary times, "for the vast majority of Muslims the eating of meat is not only unnecessary but is also directly responsible for causing grave ecological and social harm, as well as being less healthful than a balanced vegetarian regime" (Foltz 2001, 40). While Islam is anthropocentric, Muslims nonetheless have reasons to promote animal welfare.

Both Judaism and Islam offer theological reasons to improve animal welfare. Some of these reasons are anthropocentric but nonetheless have a positive effect on animal welfare. Rather than viewing religious traditionalists as adversaries, animal welfare advocates ought to consider them as co-empathizers. While the disagreement over pre-stunning may be difficult to overcome, there are plenty of other ways that religious traditionalists and animal welfarists can work together to make impactful change.

Animal welfarists, utilitarians, and protectionists have pragmatic reasons to denounce bans on religious slaughter and instead work with religious traditionalists. While animal advocates may still disagree with the practice of religious slaughter, they should work together with religious traditionalists to make impactful change in other areas of animal welfare. Regardless of the dispute over unnecessary suffering in the last few minutes of life, there are several other problems facing animal welfare that as a whole, may outweigh the possible downsides of legalized religious slaughter. For this reason, religious traditionalists ought to be viewed as co-empathizers who have an interest in improving animal welfare for the sake of compassion, or for the anthropocentric requirements of *kosher* and *halal* meat, cosmetics, and medicine. Admittedly, this partnership between animal advocates and religious traditionalists is a pragmatic solution and, as most pragmatic solutions, it may be a solution that only works for a period of time, but the value of the partnership today is nonetheless worthwhile.

In an excellent example of religious traditionalists and animal welfarists working together to improve animal interests, The DIALREL Project brought together stakeholders from around the globe to discuss religious slaughter. The project was funded by the European Commission, coordinated by Dr. Mara Miele at Cardiff University, and engaged with 11 partners in Europe, Turkey, Israel and Egypt. DIALREL is a multidisciplinary project that brought together veterinarians, food scientists, sociologists, NGOs, stakeholders in the *halal* and *kosher* supply chains, and religious jurists to discuss the issues relating to religious slaughter. The aims of the project were to "promote best practices of slaughter, both conventional and religious, and to establish a dialogue among religious authorities and market operators about *halal* and *kosher* certification" (Velarde et al. 2010). The project created a report on recommendations for best practices in religious slaughter, including ways animals can be better handled to reduce stress, suggestions on restraint methods, recommendations on post-cut management, and a list of areas for further dialogue, such as more research on animal pain at the time of the cut and development of standardized methods across abattoirs.

There are a number of ways that religious traditionalists and animal welfare advocates can work together towards improving animal interests both in regard to religious slaughter and in living conditions and treatment of animals throughout their lives. Animal welfarists neither need to support religious slaughter per se, nor do they need to support legal bans on religious slaughter. This may be a marriage of convenience, but for the time being it promotes all interested parties. There are several issues facing animal welfare outside of slaughter, for example, overuse of antibiotics, lack of pasture space and grazing, overcrowded feedlots, lack of natural behavior including breeding habits and diet, excessive stress, abusive handling, lack of proper veterinary access, environmental problems like air and water pollution, vivisection, and cosmetic testing. Furthermore, animal welfarists and religious traditionalists should work to improve oversight in concentrated animal feeding operations (CAFOs). Neither stun nor non-stun slaughter is ethically or legally acceptable when regulations are not followed. This requires impartial oversight and training.

Another point of contention in the debate over religious slaughter involves the labelling of *halal* and *kosher* meat. In many situations, of meat that is slaughtered through *shechita* is not labelled *kosher* because it is in fact not *kosher*. This is because certain parts of the animal are not considered *kosher* at all, such as parts of the hindquarters, including veins and lymphatic and sciatic nerve branches. This also includes animals who were slaughtered by *shechita* but not found to be *kosher* because upon inspection *treifa* was found. This is problematic for people who are morally opposed to *shechita* and *dhabiha*. These people prefer to be able to choose not to buy meat which has been slaughtered in this fashion and clear labelling may help alleviate this issue.

However, it can also be argued that labelling kosher meat is discriminatory because it assumes that *kosher* meat is not humanely killed (shechitauk.org). Required labelling for non-kosher and non-halal meat which is nonetheless slaughtered using *shechita* or *dhabiha* also may be unfair because other methods of stunning and slaughter, such as bolt or electrical stunning, are not required to be labelled as such. For the same reasons people are opposed to *shechita*, they may also be opposed to bolt, electrical, or gas stunning. It seems that all stunning and slaughter methods ought to be labelled if there is a need to label *shechita* or *dhabiha* slaughter. This is an area where religious traditionalists and animal welfarists ought to come together to find ways to label meat which is fair and equal and gives the consumer the most amount of information possible.

Religious slaughter without pre-stunning is a contentious debate. I have argued that both sides of the debate have reasons to view their opponents as co-empathizers in improving animal welfare rather than as political and moral adversaries. While it is clear that the animal movement will not be able to support religious slaughter forever, my claim is that they ought not to oppose religious slaughter because it impedes religious freedom, and in some cases, like banning non-stun slaughtering while allowing for imported *kosher* and *halal* meat, creates more suffering for animals globally. Also, animal welfarists have pragmatic reasons to work together with religious traditionalists, both to improve the methods of religious slaughter and to improve the welfare of animals in all other stages of life.

References

Anil, Haluk, Mara Miele, Karen Von Holleben, Florence Bergeaud-Blackler, and Antonio Velarde. 2010. Religious Rules and Requirements—*Halal* Slaughter. DIAREL Report. http://www.dialrel.eu/images/dialrel_report_halal.pdf. Accessed 9 June 2016.

Barry, Brian. 2002. *Culture and equality: An egalitarian critique of multiculturalism*. Cambridge: Harvard University Press.

Dresner, Samuel H. 1966. *The Jewish dietary laws*. New York: Burning Bush Press.

Editorial. 2015. French Secularism and School Lunch. *New York Times*. October 18. http://www.nytimes.com/2015/10/19/opinion/french-secularism-and-school-lunch.html. Accessed 9 June 2016.

European Food Safety Authority. 2004. Opinion of the scientific panel on animal health and welfare on a request from the commission related to welfare aspects of the main systems of stunning and killing the main commercial species of animals. *The EFSA Journal* 45: 1–29.

Foltz, Richard C. 2001. Is Vegetarianism Un-Islamic? *Studies in Contemporary Islam* 3: 39–54.

Garner, Robert. 2002. Animal rights, political theory and the liberal tradition. *Contemporary Politics* 8: 7–22.

Garner, Robert. 2005. *The political theory of animal rights*. Manchester: Manchester University Press.

Grandin, Temple. 2010. Discussion of research that shows that *Kosher* or *Halal* slaughter without stunning causes pain. http://www.grandin.com/ritual/slaughter.without.stunning.causes.pain.html. Accessed 9 June 2016.

Jewish Telegraphic Agency. 2014. Danish envoy blasts accusations of anti-semitism. *The Times of Israel*, February 18. http://www.timesofisrael.com/danish-envoy-blasts-accusations-of-anti-semitism. Accessed 9 June 2016.

Nelson, Sara C. 2014. *Halal & Kosher* Slaughter banned in denmark as minister insists 'animal rights come before religion'. *Huffington Post UK*, February 18. http://www.huffingtonpost.co.uk/2014/02/18/halal-kosher-slaughter-banned-denmark-animal-rights-before-religion_n_4807192.html?utm_hp_ref=uk. Accessed 9 June 2016.

Maimonides, Moses. 2007. *Guide for the perplexed*. New York: Cosimo Classics.

Masri, Al-Hafiz BasheerAhmad. 2007. *Animal welfare in islam*. Markfield: The Islamic Foundation.

Mayton, Joseph. 2010. Eating less meat is more islamic. *The Guardian*. August 26. http://www.theguardian.com/commentisfree/belief/2010/aug/26/meat-islam-vegetarianism-ramadan. Accessed 9 June 2016.

Rawls, John. 1999. *A theory of justice*. Cambridge: Harvard University Press.

Rawls, John. 2005. *Political liberalism*. New York: Columbia University Press.

Royal Society for the Prevention of Cruelty to Animals (RSPCA). n.d. Welfare at slaughter—Joint statement of principles by the British Veterinary Association (BVA), Humane Slaughter Association (HAS) and the RSPCA. http://www.rspca.org.uk/adviceandwelfare/farm/slaughter/religiousslaughter. Accessed 9 June 2016.

Schwartz, Richard. 2001. *Judaism and vegetarianism*. New York: Lantern Books.

Singer, Peter. 1972. Famine, affluence, and morality. *Philosophy & Public Affairs* 1: 229–243.

Singer, Peter. 2012. The use and abuse of religious freedom. *Project Syndicate*. June 11. https://www.project-syndicate.org/commentary/the-use-and-abuse-of-religious-freedom. Accessed 9 June 2016.

Taylor, Charles. 2007. *A secular age*. Cambridge: Harvard University Press.

Unferth, Deb Olin. 2011. Interview with Gary L. Francione. *The Believer*. February. http://www.believermag.com/issues/201102/?read=interview_francione. Accessed 9 June 2016.

Velarde, A, P. Rodriguez, C. Fuentes, P.Llonch, K.von Holleben, M von Wenzlawowicz, H. Anil, M. Miele, B.Cenci Gogo, B.Lambooij, A. Zivotofsky, N. Gregory, F. Bereaud-Blackler, and A. Dalmau. 2010. Improving Animal Welfare during Religious Slaughter, Recommendations for Good Practice. DIALREL Report. https://issuu.com/florencebergeaud-blackler/docs/dialrel-recommandations-final-edited. Accessed 9 June 2016.

Withnall, Adam. 2014. Denmark Bans *Kosher* and *Halal* Slaughter as Minister Says 'Animal Rights Come Before Religion'. *Independent*, February 18. http://www.independent.co.uk/news/world/europe/denmark-bans-halal-and-kosher-slaughter-as-minister-says-animal-rights-come-before-religion-9135580.html. Accessed 9 June 2016.

Zivotofsky, Ari. 2010. Religious rules and requirements—Judaism. DIALREL Report. http://www.dialrel.eu/images/dialrel-wp1-final.pdf. Accessed 9 June 2016.

Zivotofsky, Ari Z. 2012. Government regulations of Shechita (jewish religious slaughter) in the twenty-first century: Are they ethical? *Journal of Agricultural Environmental Ethics* 25: 747–763.

Chapter 24
An Ecophenomenological Approach to Hunting, Animal Studies, and Food Justice

Jonathan McConnell

Abstract A phenomenological description of hunting and eating animals provides insight into many timely questions about environmental justice, the ethical relationships of human and non-human animals, the politics of food production and consumption, and the even larger questions of alienation, authenticity, and mindfulness in our modern capitalist lifeworld. Despite this nexus of discourse, hunting has rarely been treated seriously as a site for food justice or ecological ethics. Informed by Iris Marion Young's concept of social justice and Val Plumwood's ecofeminist perspectives on predation, this essay proposes an ecophenomenological description of hunting whitetail deer in the Midwestern United States and other entanglements in the wild places of the Pacific Northwest where we ourselves may become prey. In particular, I offer examples from the decade-long public policy debate over 'canned hunting' in Indiana to show how our relationships with traditional game animals have ontological consequences that reverberate throughout broader environmental and food justice realities. Inspired by William Faulkner's phenomenological critique of private property in his novel *Go Down, Moses* and by my own evolution as an eater, I claim that our understandings of the arbitrary division of culture and nature play a key role in our ethical comportment towards nonhuman lifeworlds.

Keywords Ecophenomenology · Hunting · Predation · Deer

24.1 Ecophenomenology and Deer

Several years ago now, during a November deer-hunt with my Dad, I brought up the ongoing debate over high-fence deer hunting in Indiana, where I lived at the time. Also known as "canned-hunting," high-fence hunting is a business model where landowners breed and manage private herds of game animals in isolation

J. McConnell (✉)
Philosophy and Literature Program, Purdue University, West Lafayette, IN, USA
e-mail: jmcconn@purdue.edu

© Springer International Publishing AG 2017   299
I. Werkheiser and Z. Piso (eds.), *Food Justice in US and Global Contexts*,
The International Library of Environmental, Agricultural and Food Ethics 24,
DOI 10.1007/978-3-319-57174-4_24

from the general animal population, behind supposedly impervious fences, and charge customers to enter and kill those animals. Since 2005, the Indiana Department of Natural Resources has argued that canned hunting operations are illegal, because all wildlife in the state is the common property of the people of Indiana, held in a public trust, and all hunting of that wildlife is legally managed by the DNR (Sabalow 2014). While the decade-long legal and legislative dispute continues, at least four of these hunting preserves continue to operate. To my surprise, my Dad said that, as far as he was concerned, if those deer were bought and raised separately from the state game herds, then they were private property and the owners could do with them whatever they wanted. Even though the idea of trophy hunting in general and canned-hunting in particular goes against what my family values as the essence of hunting—hunting for food and with a respect for life—the most important principal in my Dad's reasoning was a doctrine of private property rights.

This discontinuity between values and rights seemed to point to an ontological problem we have regarding non-human life; namely that what we talk about when we talk about the relationships we value with animals is different from what we talk about when we talk about rights, whether property rights or animal rights. Furthermore, our own human being is of a different kind when considering relationships than when considering rights. I have been considering this discontinuity ever sense, both while in the academy and while in the woods. In this chapter, I detail how I approach the ethics of deer hunting in one specific time and place, and how attempting a similar analysis in a different ecosystem reveals an entirely different understanding. In a general sense, I find that certain kinds of comportment towards the life surrounding us reveals a deeper understanding of what it means to be an eater and, perhaps, the eaten.

Phenomenology has always been concerned with Nature, World, and Environment, in all the polyvalent ways those terms are deployed. Ecophenomenology takes the abstract concerns of classic phenomenology and explicitly turns towards the practical concerns of environmental activism (Brown and Toadvine 2003). In other words, rather than being merely broadly critical of the mind/world divide, or of naturalism, ecophenomenology adds the specific ethical task of motivating that critique towards solving 21st century environmental crises. It is within this larger ecological and environmental context that I want to think about animal rights and food justice, showing that the question of ethical eating is also a question of how necessarily different forms of life interact in the world.

In a basic sense, the ecophenomenological methodology I follow in this essay aligns with Erazim Kohak's statement that "the move from empirical ways of knowing to phenomenology is in part a shift from the *naivete* of approaching reality as a set of space-time objects in causal relations to approaching it as a system of interlocking roles" (Brown and Toadvine 2003, 24). Phenomenology insists that the philosophical investigator can examine those interlocking roles by bracketing (*epoche* to use Edmund Husserl's term) the naive ego-centric perspective of everyday life to briefly see the world as this matrix of co-constitutive phenomena. What we are, as humans and individuals is not the starting point of reality. Rather, the things in the world that we perceive and interact with give us our existence as humans and as individuals. This is not a static relationship, but one grounded in the experience of being in the world.

To approach the problem of hunting in an ecophenomenological way, I offer three brief ontological accounts of the human-deer relationship, or possible inter-locking roles. By ontological, I simply mean that the being of the things we encounter in the world, and therefore our own being, is revealed by the way in which we encounter them.

(1) Driving home from the pumpkin patch late one summer, my family passed a freshly killed deer mangled in a heap on the side of the road. A few dozen yards farther down the road was a car on the shoulder, shattered windshield, hood and fender bashed in. In the car (never leaving the car) sat a man talking on his cellphone and waving his arms around; I imagine his conversation: "I don't know what happened...it just jumped in front of me!" In the language of Paul Virilio, we would say that for this man a deer is "what crops up." According to Virilio, our perceptions and experiences are flattened by the Speed of modern life so that we encounter the world as if looking through a car windshield which, at highway speeds, becomes a flat screen in which "things crop up". Lost is all depth perception and peripheral vision. All we have is speed, the flat screen between self and world, and the accident (Virilio 2007). For many people in the 21st Century American Midwest—a world of medium-sized cities connected by highways traveling through fields of corn and soy—I suggest that this is exactly what a deer is: the potential automobile accident.[1] Additionally, the always ready to crop up accident is a human-animal relationship that reveals the deeper truth that modern society attempts to push the non-human world to the periphery only to have it come rushing back against that expulsion.

(2) Seeing that shattered animal, and that shattered car, and presumably the shat-tered perceptual field of that driver, I recall a recent experience I had deer hunting. Walking into the cold woods before dark, finding my spot against a downed tree at the edge of a clearing, sitting quietly while the woods slowly comes awake with the predawn light, thinking about how the slope of the land *should* funnel deer into the clearing just opposite of where I'm sitting. Every so often, reaching down slowly to pour another shot of coffee from my thermos, the movements and sounds of the woods jumping directly into my immediate and reflective consciousness almost simultaneously, and then, amazingly, a deer appears right where I expected her to. But also, in addition to this pastoral

---

[1]According to the Indiana DNR, there are more than 14,000 deer-vehicle collisions reported in Indiana annually. State Farm Insurance Company provides statistics on a driver's odds of being in a collision with an animal. Compared to the nation-leading West Virginia rate of 1 in 44, Indiana's 1 in 142 is actually fairly low. It is worth considering the amount of deer-accident studies put out by insurance companies as a way to sell higher-premium policies in the evolution of the meaning of deer.

peace, there is of course the violent outburst of my rifle that shattered the quiet morning, the warmth and smell of the viscera as I field-dress the dead animal, and later the messy and intricate work of butchering and wrapping the meat.

In one objective sense, both of these scenarios describe an identical outcome: the violent death of a deer. The second deer, however, became nourishment for my family that winter, and the eating reverberated with the experience of that slow cold morning in the woods. It was an experience in which I learned something of what it takes to live off the flesh of another animal, and presumably, the other non-humans in the woods that day had an experience which reinforced whatever existence humans bring forth in them. Hunters often point out that there was a more efficient use of the second animal, but that makes me uncomfortable for two reasons. First, because any such argument reintroduces the idea of domination and oppression by making the animal subordinate to economic relationships—not necessarily money economics, but the general balancing of ledgers that distributive justice pursues to the dehumanization of life. Secondly, as any hiker coming upon chewed bones or hair-filled scat knows, there is very little "waste" in a natural world filled with scavengers, not to mention the insects, microbes, and nematodes that depend upon corpses. Instead, I want to emphasize that the Being of the thing, in this case a deer, is determined by and changes with its relationships to other things in the world. And, of course, the Being of the humans involved is radically different based on those relationships. Not only is a hunter or eater a different kind of thing than a deer-striking motorist, but we can assume that for the deer, a relationship with a motorist-human and the hunter-human establish different modalities of being-deer.

(3) The third phenomenological account of a deer is from an Indianapolis Star newspaper report on canned hunting. This passage describes a video made at one of the deer farms that support the canned hunting industry. The video was available, for a time, on YouTube for the public to watch.

> The video opens with five men wrestling a white-tailed buck into a metal chute. The deer's eyeballs bulge as the animal thrashes in pure terror. One man crams a cuplike device wrapped in duct tape over the buck's mouth and nose. It's connected to a tube pumping tranquilizer gas. They wrap a pink blanket over the buck's eyes. Its thrashing slowly subsides.

> One of the men gets to work on the deer's hindquarters.

> He inserts a probe into the deer's anus and places a camouflage-covered bag with a funnel attachment over the deer's genitals. He gradually turns a knob on a yellow box. The deer's back legs shake as the probe's electric current causes the deer to ejaculate into the funnel.

> After collecting the semen, the men saw off the animal's antlers for safety in the pen.

Beyond the sadistic electro-sexual violation of this buck, which is a violent atrocity on the ontic level, there is the further violent instrumentalization of life itself, which is ontological violence. The buck with desirable characteristics is reduced to the use-value of those genetic traits, which are isolated then physically and metaphorically wrenched from his body to be re-assembled into a product fit to be sold in a one-to-one market exchange—the client at a canned hunting business may pick out a particular animal and pay a trophy fee to kill that animal. Some of the animals bred in this manner are simply grotesque, with racks many times larger than the average wild animal would produce. These monstrously endowed bucks stagger under the weight of their scientifically produced fetish.

Recognizing this genetic violence, and other sorts of ontological violence, leads to a kind of concern which goes beyond the focus on individual suffering that some rights-based animal philosophies prioritize (Singer 1975; Regan 1983). Anyone concerned with interspecies relationships must see that this 'institutionalized domination and repression' of life itself, by which I mean the essence of the deer has been instrumentalized into something to be re-arranged for no other possible ends than profit and pleasure, precludes the possibility of a relationship at all. Or, more precisely, it reverts to the naive anthropocentric relationship where humans have intentions and objects are merely means for fulfilling those intentions. Objectively speaking, we may recognize that we are still talking about a thing called deer, but our phenomenological examination reveals that the interspecies relationship constitutive of the concept of 'deer' has been corrupted into something with no similarity to either our hunting or our road hazard experiences.

Iris Marion Young's deceptively simple statement that social justice "means the elimination of institutionalized domination and oppression. Any aspect of social organization and practice relevant to domination and oppression is in principle subject to evaluation by ideals of justice" (Young 1990, 15). Although her work does not explicitly consider non-humans, her idea of institutionalized domination and oppression constantly guides my understanding of animal ethics. The captive buck being completely objectified, physically and genetically, with no possible future other than being breeding stock for trophy hunters represents a kind of domination and oppression that bores down to the very basic foundations of life as we understand it.

With these three examples, I've attempted to show how three things that are uncritically called deer are revealed through a phenomenological examination to be quite different based on the relationships in which they appear: (1) A road hazard revealed through technological speed as "what crops up" in the accidental, (2) a prey animal encountered through lived experience and is part of a larger food cycle, and (3) a fully over-coded object of scientific rationality tailored to the fetish market. Obviously, these three modalities barely begin to express the relationships involved with any living thing. Far from being an exhaustive list of ways we could interact with non-human animals, at this point I just want to point out that focusing on the co-constitutive relationships we have with other beings reframes a starting point of animal studies and food justice conversations.

As mentioned, I started thinking seriously about an ecophenomenological approach to interspecies relations because of my Dad's surprising acceptance of high-fence hunting based solely on an ideology of private property rights. A crucial insight to the threat I felt at that moment came when I was re-reading a passage from William Faulkner's novel *Go Down, Moses*. In this scene, Isaac McCaslin is renouncing his birthright to the McCaslin plantation and its violent history of slavery, rape, miscegenation and incest; a history that is metaphysically tied to the enclosure of wild lands by white colonists. The McCaslin history on the land begins with buying the plantation from the Chickesaw Chief, Ikkemotubbe:

> I can't repudiate it. It was never mine to repudiate. It was never Father's and Uncle Buddy's to bequeath me to repudiate because it was never Grandfather's to bequeath them to bequeath me to repudiate because it was never old Ikkemotubbe's to sell to Grandfather for bequeathment and repudiation. Because it was never Ikkemotubbe's fathers' fathers' to bequeath Ikkemotubbe to sell to Grandfather or any man because on the instant when Ikkemotubbe discovered, realized, that he could sell it for money, on that instant it ceased ever to have been his forever, father to father to father, and the man who bought it bought nothing (245).

The "it" in this passage that, defying logic, ceased to exist, is not the physical plot of land defined by the surveyor's measurements. It is instead a kind of meaning that is inconceivable in a worldview that considers the landscape, with its manifold systems of interlocking life, to be a simple object of ownership and exchange.

The temporal and ontological slipperiness of Faulkner's language illuminates what is at stake in shifting interspecies and ecosystem relationships. The land which would become the McCaslin plantation changes nature instantly across generations and temporalities by being captured in the world-exploding concept of exchange value. The land itself becomes a different kind thing, phenomenologically speaking, and with it the people on the land change. The land is an object of human perception but also an object that allows the human consciousness to spring forth in the reflection. If we allow the pay-to-shoot killing of genetically designed and dominated animals to be called hunting, then the nature of hunting, the nature of deer, and this particular possibility of engaging with the commons of earthly public trust through hunting and eating is lost not only for our own and future generations, but for past understandings of hunting as well. The ontological violence extends both directions in time.

Faulkner's Isaac McCaslin laments a lost worldview that he only knows through the teachings of his mentor, the half-Indian half-Slave hunter and son of Ikkemotubbe, Sam Fathers. Isaac, then, is driven to a certain extent by a nostalgia for a world that may never have existed. It is certainly true that, for all of Faulkner's genius with language, we can hardly take his idealization of a pre-Columbian native worldview uncritically. We can avoid a similar inertia towards essentialism by reminding ourselves that the relationships we have with the diverse parts of our food systems have always been shaped by human desires and techne. In fact, historian Charles Mann describes the Ohio River valley that the first European explorers encountered as "forests as open as English parks" as a result of the Indian's ecological exploitation of their environment through controlled burns. "Rather than domesticating animals for meat, Indians retooled whole ecosystems to grow bumper crops of elk, deer, and bison" (Mann, 10). Mann's natural history shows how problematic it is to considering the North American landscape separate from human culture, and provides a historical perspective on the modern North American hunter's conservation theories. The American hunting tradition stands out against European history because the wildlife is understood here as being a resource for all people to share in, not something owned by the heads of estates and the aristocracy. Our conservation tradition attempts to preserve this wilderness populism with a surprising continuity with Native American management.

24.2 Ecosystem Management and Being Prey

We should examine the false-dichotomy between Nature and Culture in both philosophical and everyday usage. In an everyday way, the idea that there is Nature as a space where we can point to and say 'leave that be' is false and seems to underwrite anti-hunting sentiments. As Peter Singer and Jim Mason write in their book about food choices, "unlike animals raised on farms, the animals killed by hunters would have existed quite independently of us, and their deaths mean that there are fewer animals enjoying their lives" (259). Most us would like to believe there are spaces where animals exist independently of humans. Maybe there are such places, but the Midwestern United States where thousands of hunters pursue deer every year is certainly not one of them, and hasn't been for thousands of years. This history alone, of course, is not enough to justify hunting if it otherwise proves to embody all of the characteristics of systematic domination and oppression. However, acknowledging that wild animals and human culture are historically interdependent works to soften the anthropocentric dualism exhibited in both the instrumental control of life and the rights-based animal liberation movement.

The mistaken animal rights activist belief that, if not hunted, wild animals would be enjoying their lives is a symptom of the paradox at the heart of trying to apply a rights-based ethics to the nonhuman world. The rights argument appeals to a supposed universal quality, usually the ability to suffer, and homogenizes the individuals within this category while pushing out anything that doesn't make the somewhat arbitrary qualifications. I say arbitrary because the line of sentience can be drawn in many places within the animal kingdom. Even drawing that line between the animal kingdom and other forms of life, including entire ecosystems can be problematic. The legacy of Humanism, in other words, is necessarily a legacy of exclusion through the process of framing. As Cary Wolfe uses the term, "framing decides what we recognize and what we don't, what counts and what doesn't; and it also determines the consequences of falling outside the frame" (Before the Law, 6). The Animal Rights movement extends the frame to include certain kinds of individual non-human life—i.e. those with sentience defined by sensations of suffering—but does not extend the frame to include species or ecosystems, which also exhibit intentionality, or to "lower" living things like insects, bacteria, and plants (Plumwood 1993; Haraway 2008). Condemning hunting with the broad stroke of animal rights talk, but then erasing the actual lives of animals with the naive belief that non-hunted animals will live happily in some cordoned-off 'nature' shows an alarming lack of responsibility, often referred to as the "Bambi Syndrome" (Cartmill 1993).

Plumwood points out that Tom Regan's own unsatisfactory answer of whether humans should intervene in "wolf and sheep" cases of predation, where it would seem that the concept of giving rights to individual nonhuman animals would require us to, shows that the concept of rights is not a very good strategy when dealing with natural environments (Plumwood 1993). The rights argument works better when applied to domesticated animals fully enclosed in human culture, where their connections with larger ecosystem demands are largely nonexistent. That would imply, however, that the domestic/wild demarcation is the new hard line which enframes moral beings from the outside other. Plumwood's critical feminist framework is suspicious of the rights-based approach to nature because the history of the dangers of *moral extensionism*, where "nonhumans are included just to the extent that they resemble humans, just as women are allowed in the institutional structure of the public sphere just to the extent that they can be seen as possessing masculine characteristics or analogues of them" (1993, 172). One way that animal rights activists have engaged with ecological issues precisely illustrates this stratification of rights: the call for forced sterilization of wildlife to deal with ecological imbalance. Here we see a strategy of giving animals just enough rights to bring them into the moral world but not the basic right of self-determination.

The great monoculture fields of corn and soy, combined with the lack of any large predators, make midwestern states like Indiana a relatively simple scenario for thinking about deer hunting.

In such a situation of overwhelming amounts of easy calories and lack of natural dangers (though coyotes are said to kill and eat young deer on occasion), the deer population, if left unchecked by hunters, would continue to grow. A well-researched 2013 article by The Nature Conservancy called the overpopulation of whitetail deer a bigger threat to Eastern forests than climate change (Pursell et al. 2013). In these overpopulation scenarios, deer over-browse young vegetation, destroying desirable native tree species and understory plants such as trillium while non-native exotic plants fill those niches. The destruction of this shrub or intermediate canopy layer also destroys the preferred nesting habitat for songbirds, linking increased deer populations with declining woodland bird populations continent-wide (deCalesta 1994). Indiana State Parks, which are off-limits to normal hunting seasons, have been hosting yearly deer-reduction hunts since 1993 to protect the woodland habitats from excessive browsing (Mycroft 2015). I participated in one of these deer reduction hunts at Prophetstown State Park in 2014. Prophetstown is a park designed around the cultural and ecological history of the prairie, including a working 1920s style farm and Native American village structures, as well as habitat restoration of wetlands, fens, open woodlands and high-grass prairie. Controlling the deer population through hunting allows those sensitive environments to thrive, and connects current area residents with the landscape in a way that evokes the wild food legacies of that area.

There may be non-hunting solutions to the overpopulation problem, such as controlling wildlife fertility, reintroducing natural predators, or even just deciding that ecosystem management is not something we should be engaged in. Fertility control is often mentioned in animal rights activist literature as a preferred strategy for non-lethal population control, but as a recent study points out, "all fertility control methods impose animal welfare impacts" including both immediate negative impacts such as physical and psychological distress, and the long-term denial of positive natural behaviors such as mating, play, and child rearing activities (Hampton et al. 2015). The reintroduction of major predator species to the Eastern United States is an interesting idea, but we may assume, politically impossible for several reasons; the lack of public land, the fear of human children being put in danger, and the resistance of livestock producers come immediately to mind. Finally, the idea of not managing wildlife and ecosystems, whether because of a naive urban-privileged belief that nature is "out there" somewhere beyond our human lives, or just a cynical ethics that says we will attempt to do no harm as individuals and then not concern ourselves with whatever ecological fallout results from our isolationism; both of these are unsatisfactory for a culture that values engagement with and care of nature (however problematically defined) in our public parks and forestlands.

If we agree that unlimited population growth is unhealthy for both individuals and ecosystems, there is no utilitarian calculus that would tell us whether hunting or sterilization would be a kinder method of control. However, with a relational ontology or phenomenology approach, we can examine how those two regimes of control shape the Being of the beings involved. Both approaches rely on professionals making decisions about life opportunities of wild animals, but the boundary zone between culture and nature shifts in position and in porousness with these two strategies. When wildlife control professionals use all of their technological ability to monitor, capture, and apply sterilization techniques, whether surgical or chemical, it exhibits a kind of instrumentalization that is the perverse negative image of the captured buck being milked for his semen. Additionally, by offloading our societal responsibility for ecological wellness to professionals, all of us become dangerously more isolated from the biopower management that frames our cultural freedom—invoking a selfish privilege to ignore our interdependence with the natural world. On the other hand, there are over 14 million Americans who hold hunting licenses annually, and far more who identify with hunting culture, even if they may not hunt during any particular year. Directed by state wildlife biologists who set hunting seasons and bag limits, this mass of hunters represents a participatory management plan engaging humans with the non-human world for the maintenance of ecological balance and the gathering of healthy and non-commodified food. This deep cultural engagement with ecological processes is not enough, on its own, to justify the killing of any particular living thing, but what I am trying to show is, precisely, that there is no single qualification for killing or letting live—not suffering, or ecological balance, or our physical sustenance. Instead, contemplating how our own lives fit into a world where being alive means limiting, in various ways, the lives of others requires the examination and weighing of manifold values and desires.

24.3 Both Predators and Prey

When I originally developed this phenomenology of hunting paper at the first Michigan State Workshop on Food Justice, I was living and hunting in Indiana, where the overabundance of whitetail deer combined with the cultural history of my family and the question of the terrifying instrumentalization and commodification of deer in high-fence hunting preserves led me to the ethical entanglements and conclusions I have outlined above. However, since then my family has relocated to Oregon, where very different ecological and historical entanglements reign. Like other Western states, Oregon largely avoided the massive monoculture agricultural transformation and privatization of the landscape that the Midwestern states faced. There is monoculture in Oregon, but it is in the great Douglas Fir plantations. Oregon's rugged mountains and dense forests do not provide near the amount of forage that calorie-rich corn and soybean fields do, and Oregon contains healthy predator populations of cougar, coyote, bobcat, bear, and more recently, wolf. The Oregon Department of Fish and Wildlife introduced a comprehensive Black-Tailed Deer management plan in 2008 to address declining populations of Black-Tailed Deer (Oregon Department of Fish and Wildlife 2008). The rest of this essay explores whether ecological ethics can support hunting for venison when hunting is not a strategy for controlling overpopulation of the animals. In fact, because legal deer hunting in Oregon is mostly limited to male animals and at rates that still allow healthy breeding numbers, hunting is not shown to be a factor affecting annual deer populations. On the contrary, one goal of deer management in Oregon is to provide viable numbers of animals for what the Department calls recreational hunting—a designation that I find problematic.

I did purchase a deer license my first fall in Oregon and spent many hours exploring our local coniferous rain forests looking for prey, although I never saw an animal to take. Columbian black-tailed deer are known as an edge-adapted species because they prefer the deep cover of mature forests to shelter in during the day, while their primary food source is in early successional habitats created by logging or forest fire. To hunt such an animal means to move back and forth between the nearly impenetrable dense vegetation of fern, vine maple, and red alders below the great Douglas Fir canopy, where one can easily imagine large mammals, including humans, untraceably disappearing, and the huge slash-piles of recent timber operations where human presence is brutally apparent. Just months after moving across the country to a radically different cultural and ecological landscape, I was able to immerse myself in this rich natural history legacy of logging and fire suppression by forcing myself to imagine how animals navigate that landscape and consider what ways our recent human history impacts those animals. Within our competing cultural values of mobility and localness, the experience of beginning to connect with the landscape in such a immersive way can relieve some of our uprootedness.

One of those early mornings, after watching the sunrise from a ridge above the clearcut that on clear days opened the view to the gleaming white pyramids, one of them truncated, of Mt St Helens to the North and Mt Hood far away to the East, I picked my way over the slash down to the deep riparian gully where the saws never reached. This particular morning the mountains were hidden behind the clouds that released a persistent light rain and, returning home, I wrote the following passage:

> Sitting in a slow rain, or heavy precipitating mist, far from the mechanical noises of civilization, hearing becomes one's primary sense. The many layers of slow rhythmic sound, as each surface reacts to the primary (from the sky) and secondary (dripping from other ground-tethered objects) droplets in its own way. Hunting for a specific animal, sitting and listening becomes multilayered flow of changing perceptions in which consciousness must swim, giving more attention to this plop plop plop or this bip bip bip and listening for any crunch or rustle that disrupts that rhythmic overall fuzzy hmmm of background wetness. The aural tapestry is so lush that one may experience deep doubt about one's own ability to recognize a change—is that lower drumming a new sound or something constant that I wasn't able to focus on until it came to the foreground of reflection?

> A simple rain-hood magnifies both the signal and the noise as dripping water echoes on the covering material but the opening cone focuses and magnifies the earthly sounds. This directional receptor captures and focuses sound to such an extent that any change in the listener's head position radically changes the received aural sensations. The hunter now understands somewhat the deer, standing stock still but rotating her ears this way and that trying to decide which part of this enveloping sensual world represents mortal danger.

My brief sojourn with being-deer brings to fore another reality about the Western wilderness: I am not the only predator in these mountains and, in fact, am also possible prey. James Hatley refers to this feeling in the title of his essay "The Uncanny Goodness of Being Edible to Bears" (2004). When thinking about the interpenetration of all life through the vector of consumption, the existence of large predators in our environments forcibly destroys the false notion that humans transcend those relationships as being only eaters and never eaten. I agree with Hatley that just as seeing an animal as *merely* something to be consumed, as a means for our own lives and nothing more, as they are in most industrial agricultural models, is "an arrogant and presumptuous forgetfulness of our irretrievable involvement in all other flesh…the converse is true as well—to be human and to expect not to be eaten by other animals would be an act of *hubris,* a failure to acknowledge the goodness of the order of living flesh, in which we are intractably enmeshed" (Hatley 2004, 19–20). Although the likelihood of ending up food for a bear, wolf pack, or cougar is remote, the possibility shows the experience of Oregon wilderness to be that much more of an antidote to everyday anthropocentric metaphysics. As I have said already, there is nowhere in North America where ecosystems are untouched or unmanaged by human intentions, and encouraging those apex predators that could, conceivably, prey upon us is an important consideration when proposing management plans.

Hatley finds it notable that in a survey of humans who have been victims of grizzly bears and lived, or had loved ones preyed upon, there was no sense of vengeance towards the predator as individual or species (2004). On the contrary, the human victims often became or remained advocates for protecting the rights of those animals to remain in the wild where further violent interactions are possible. Similarly, after Australian philosopher Val Plumwood barely survived a crocodile

attack in 1985, she credited her fear that the attack would lead to a massive crocodile slaughter as part of the reason she didn't share her account of the attempted
predation publicly for over a decade (1996). When she did write about the event,
she described the way being prey disrupted her subject-centered view of the world
(the everyday way of seeing the world that we all presumably share) and forced her
into another reality, a reality where she saw herself from outside for the first time,
because of the very likely possibility of being food for another creature. Plumwood
makes explicit the link between never thinking of ourselves as food and treating
other animals so inhumanely:

> This is one reason why we now treat so inhumanely the animals we make our food, for we
> cannot imagine ourselves similarly positioned as food. We act as if we live in a separate
> realm of culture in which we are never food, while other animals inhabit a different world of
> nature in which they are no more than food, and their lives can be utterly distorted in the
> service of this end (1995, 34).

I strikes me as absolutely true that the inhumane treatment of livestock is directly
tied to our metaphysical separation of culture and nature. Whether or not people
who encounter natural spaces that include potentially man-eating predators are
more attuned to the ontological violence against food species is more difficult, if not
impossible, to prove. My own evolution about food systems is continually informed
by the ecological worlds I encounter through hunting. After being raised mostly on
wild game, meat from my uncle's small herd of cattle and sheep, and garden
vegetables, I became a strict vegetarian for roughly 8 years after I left my parents'
home. This decision was largely based on the environmental impact of factory
farming and, to a lesser extent, on the inhumane conditions of the animals in those
farms. My fairly dogmatic rejection of meat seems, in retrospect, as an important
step to re-orienting my own maturing relationship to food systems. When I decided
to eat meat again it was initially only venison, fish, or chickens that my parents
provided. These were all animals whose stories I was familiar with. And after
traveling and living through much of the mountain west and encountering healthy
ecosystems full of large carnivores, it *could be* that my criticisms of factory-farmed
meat was subtly reaffirmed.

The closest that I came to being prey was while fly fishing a small mountain river
in British Columbia several years ago. I should say it's the closest that I know
about—there could be any number of times when I was shadowed by a prowling
cougar or separated from a grizzly by nothing but a hedge of salmon berry bushes
without knowing it. Even this time it would have just been a vague feeling of
unease if Erin had not have been watching the grizzly over my shoulder from the
bluff on the other side of the river. It had been a nice afternoon of hiking and
wading but not catching any fish when, just as dusk was approaching I had two
back-to-back strikes on my tiny feather-fashioned insect as clouds of the real things
were hatching all around me. The irrational part is that I heard heavy footsteps in
the bushes behind me and my first thought was "there is a bear right behind me" but
just then I hooked into a trout and never turned around, focusing instead on landing
my fish as quickly and gently as possible. The two modalities of predator and prey

came crashing together within me in that instance and, whether because of evolutionary conditioning or the anthropocentric arrogance of the unthinkability of being food, I ignored my being-prey and chose instead being-predator. When it became too dark to fish and I returned to camp, Erin told me she saw the bear right behind me and then had followed the big grizzly downstream a ways, with the river and the bluff between them. Whenever I stand beside a roaring mountain stream, I remember what may have been a very bad decision on my part and wonder what creatures are going about their lives in the bushes behind me.

Even in my own narrative, however, I cannot say with any certainty that the possibility of being prey is directly linked to empathy for other animals. It is true that that was one of the last fishing trips I took to catch-and-release waters. BC manages much of their trout populations for truly recreational access instead of a food source. This makes sense for their goal of providing fishing experiences to the greatest number of people, but since then I've decided that whatever trauma we inflict on fish, and that is a matter of debate by biologists, there is something I find creepy about going through those actions for pure recreation. Did my pantomime of predation while a true predator was lurking just behind me affect that decision? I still don't know for sure.

What I am sure of, however, is that consciously engaging with the reality of our fleshy existence helps to orient the role I choose to play as an earthly being. As Donna Haraway notes, "outside Eden, eating means also killing, directly or indirectly, and killing well is an obligation akin to eating well. This applies to a vegan as much as to a human carnivore" (2008, 296). Additionally, eating well and killing well, to my mind, requires a conscious engagement with one's place in the cycles of living and dying that constitute our existence as earthly creatures. There are no easy answers to the question of how to do this, and I hope that the conversations we're having as seekers of Food Justice continue to inform our decisions. I will go into the Oregon woods this fall to hunt black-tailed deer and elk and hope to successfully feed my family with the meat from animals whose lives are as free from institutional domination and oppression as possible. Part of that commitment includes doing what I can, politically, to support the vibrant populations of large predators that actually make it more difficult to find prey of my own. The same value system that informs my hunting also informs my participation in a vegetable CSA farm share, and the meat I cook at home comes from chickens whose farm I have visited or from a butcher who sources his animals from animal welfare-inspected farms. My hope is to continue to treat food, this immensely complex and politically divergent topic, as constitutive of what it means to be human and nonhuman at the same time.

The biggest challenge that I see to an truly mindful understanding of our place in a world of life and death is thinking that there are easy answers to these questions. Whether it is the instrumentalization of life through genetic oppression and commodification of deer for high-fence hunting preserves, or the belief that hunting or being carnivorous is somehow natural and therefore beyond reproach, or the dogmatic belief that we can choose to do no harm by adopting the self-assured logic of mainstream animal rights activists, I suggest that we avoid any theory that claims authority over our ethical choices. We may resent the fact that being alive means to eat and that none of our food choices place us beyond feelings of guilt and the recognition of injustice. However, being open to the relational systems of edibility and the constant devouring of others by ourselves and of ourselves by others also engenders some deep feelings of goodness and understanding with the human and the nonhuman parts of our lifeworld.

References

Brown, Charles S., and Ted Toadvine. 2003. *Eco-phenomenology: Back to the Earth Itself.* SUNY Series in Environmental Philosophy and Ethics. Albany, NY: State University of New York Press.

Cartmill, Matt. 1993. The Bambi syndrome. *Natural History.* 102 (6): 6–10.

deCalesta, D.S. 1994. Effect of white-tailed deer on songbirds within managed forests in Pennsylvania. *Journal of Wildlife Management* 58 (4): 711–718.

Hampton, Jordan O., Timothy H. Hyndman, Anne Barnes, and Teresa Collins. 2015. Is wildlife fertility control always humane? *Animals* 5 (4): 1047–1071. doi:10.3390/ani5040398.

Haraway, Donna Jeanne. 2008. *When Species Meet.* Posthumanities. Minneapolis: University of Minnesota Press.

Hatley, James. 2004. The uncanny goodness of being edible to bears. In *Rethinking Nature: Essays in Environmental Philosophy*, ed. Bruce Foltz, and Robert Frodeman, 13–31. Indianapolis: Indiana University Press.

Mycroft, Mike. 2015 State Park Deer Reduction Results. Indiana State Parks. https://www.in.gov/dnr/parklake/files/sp-DeerRMRR.pdf. Accessed 1 June 2016.

Oregon Department of Fish and Wildlife. 2008. Oregon Black-Tailed Deer Management Plan. Retrieved from www.dfw.state.or.us/wildlife/docs/oregon_black-tailed_deer_management_plan.pdf. Accessed 1 June 2016.

Plumwood, Val. 1993. *Feminism and the Mastery of Nature.* London: Routledge.

Plumwood, Val. 1995. Human vulnerability and the experience of being prey. *Quadrant* 29 (3): 29–34 (March).

Plumwood, Val. 1996. Being prey. *Terra Nova* 1 (3): 33–44.

Pursell, Allen, Troy Weldy, and Mark White. 2013. Too Many Deer: A Bigger Threat to Eastern Forests than Climate Change? The Nature Conservancy. blog.nature.org/science/2013/08/22/too-many-deer/. Accessed 1 June 2016.

Regan, Tom. 1983. *The Case for Animal Rights.* Berkeley: University of California Press.

Sabalow, R. 2014. Preserves set rules. Indianapolis Star. Retrieved from http://search.proquest.com/docview/1512315216?accountid=13360 (April 2).

Singer, Peter. 1975. *Animal Liberation: A New Ethics for Our Treatment of Animals.* A New York Review Book. New York: New York Review: Distributed by Random House.

Virilio, Paul. 2007. *The Original Accident.* Cambridge; Malden, MA: Polity.

Young, Iris Marion. 1990. *Justice and the Politics of Difference.* Princeton, NJ: Princeton University Press.

Conclusion
Fish, Fowl, and Food Justice

Neither Fish Nor Fowl

This is an unusual volume, in part because it flaunts the conventions of two different genres of writing. It is not a comprehensive introduction to scholarly theories of food justice, though authors do deftly introduce theories, theorists, and traditions as resources for thinking through practical and ethical food justice challenges. Nor is it a systematic handbook for practitioners of food justice, though authors do make recommendations for on-the-ground food justice projects based on their experience working in specific, and carefully circumscribed, contexts. As one reviewer remarked, the volume is neither fish nor fowl. In this conclusion, we want to take a moment to grapple with what the book is not, to build toward a cohesive articulation of what it is, and to discuss the unique challenges and success conditions for kindred volumes and scholarly projects. Our view, of course, is that there is a need for food justice projects that are neither fish nor fowl, but are another sort of creature altogether, one whose home is much more grounded. The problem is that such beasts can be a bit chimerical—each chapter has its own lineage and its own function, with these different lineages and functions giving rise to different questions, methods, and voices. We think it worth the time to unpack these lineages and functions to reveal a unity that might not be apparent on first sight.

Lineage can be important in a couple different senses to the more theoretical chapters of the volume, where authors share perspectives that are commonly marginalized in the discourse of food justice. On first sight, this might suggest that these perspectives are highly idiosyncratic; where, one might ask, are the mainstream perspectives that have defined the discourse? Here we want to stress that the volume, and the workshops that accompany it, are intended as a space to work

Z. Piso (✉)
Department of Philosophy, Michigan State University, East Lansing, USA
e-mail: pisozach@msu.edu

I. Werkheiser
Department of Philosophy, University of Texas—Rio Grande Valley, Edinburg, USA

© Springer International Publishing AG 2017
I. Werkheiser and Z. Piso (eds.), *Food Justice in US and Global Contexts*,
The International Library of Environmental, Agricultural and Food Ethics 24,
DOI 10.1007/978-3-319-57174-4

through recalcitrant problems at the intersection of theory and practice. This is part of the immediate history in which they were conceived. Authors had to look to marginal perspectives *because* the mainstream perspectives were inadequate to the problems that these mainstream perspectives could not address. As noted in the introduction, it is all too easy to stay trapped in prevailing frameworks, even when these frameworks struggle to navigate the rough ground of practice. Not infrequently, the inadequacy of mainstream perspectives is due, at least in part, to their own lineage, and the fact that the marginal perspectives were born out of histories that remain unrecognized by the mainstream. Chapters like Anne Portman's "Agriculture, Equality, and the Problem of Incorporation" draws on Val Plumhood's anti-dualist philosophy as a necessary antidote to prevailing theories of justice, while Emily Holmes' and Chris Peterson's "Race, Religion, and Justice: From Privilege to Solidarity in the Mid-South Food Movement" draws from critical race theory and liberation theology to locate virtues and concrete practices that grapple with the challenges they face as white practitioners engaging a majority-black community. It is important to read these projects, and many of the more theoretical moments in this book, as a critical (and often times political) effort to decenter the discourses surrounding food justice. Each offers a response to Christopher Long's call for public philosophy, providing models for "collaborative activity in which philosophers engage dialogically with activists, professionals, scientists, policy-makers, and affected parties whose work and lives are bound up with issues of public concern" (2013). That is an important part of their function. It comes at the expense of a more comprehensive survey of theories of food justice, and the absence of particularly influential theorists may seem conspicuous. But we wouldn't expect a beastly volume on food justice to be especially inconspicuous. To continue with the fish nor fowl metaphor, we suggest that sometimes, fowl theories can be a bit too "in the clouds," whereas beastly theories prefer to stay closer to the ground.

On the other hand, we want to stress that the volume does not aspire to be a systematic handbook for practitioners, one which provides a clear roadmap for how to initiate food justice projects and carry them out successfully. In part, this is because we, the chapter authors, and the participants of past workshops are fairly sure that we don't yet know what would mark a food justice project as a "success" (we come back around to the success conditions for this volume in the next, and final, section of this conclusion). Indeed, this is exactly what we would expect when engaging a "wicked problem"; if we cannot agree on a definition for food justice, we will certainly struggle to agree on success conditions for food justice work (Whyte and Thompson 2012). In part, we find it premature to provide a clear roadmap because there is, as far as we can tell, too few venues for sharing experiences and sustaining a conversation about particular food justice projects and their challenges and accomplishments. At the time that this volume was written, food justice scholars were in the process of building multiple networks to connect food justice organizations and create space for dialogue (New Ethics of Food 2017; Food Tank 2016). Our view is that the theories that inform food justice work, even the marginalized theories just celebrated, will always confront obstacles when brought to bear in concrete situations. The relationship between theory and practice is not simply one of applying theory to practice, and responsible theory and practice require that we reflect on what works and doesn't work as we reconstruct our practices in line with our most cherished ideals. Again the fish nor fowl metaphor can be instructive—if a handbook for practitioners represents the way that a scholarly project can be a fish, then the streamlined current through which fish swim is decidedly not the habitat in which food justice activists and practitioners find themselves. Food justice activism and practice encounter much more friction. Sarah Stapleton's "Views from the Classroom: Teachers on Food in a Low-Income Urban School District" grapples with the structural and institutional constraints that impinge on the pursuit of justice at every turn, while Seven Mattes' "Save the Whale? Ecological Memory and the Human-Whale Bond in Japan's Small Coastal Villages" reflects on the very real cultural losses that can accompany overzealous food justice activism. Chapters like these share experiences from carefully positioned practical projects, and then testify to these experiences in problematizing the prevailing discourses dominating the academy. These authors model a sort of amphibiousness, a sense in which they spend some of their lives navigating the complexities of particular communities and part of their lives navigating the complexities of academic scholarship.

Food Justice: Bridging Theory and Practice Together aspires to be something of a beast, to remain grounded whenever possible and to model the sort of amphibiousness that characterizes public philosophy and related engaged scholarship. If you're reading a volume with these aspirations, it's likely that your work too is neither fish nor fowl, or if it is, you see the value in occasional flights of fancy into theory or dipping your toes into practice. It's also likely that your work is still evolving—perhaps you are a student preparing to enter a workforce of diverse food justice non-profit organizations and civic service opportunities, or you are a long-time practitioner bringing your experience to bear on the philosophy and science of food justice. In these cases we hope that the diversity of projects featured in this volume gives some sense of the different ways that someone can work in the liminal space between theory and practice. Marisela Chavez's "It's Not Just About Us: Food as a Mechanism for Environmental and Social Justice in Mato Grosso, Brazil" engages in ethnographic research to render visible the struggle to build solidarity across diverse coalitions; Joseph A. Tuminello's "Farm Forward: A Pragmatist Approach to Advocacy in Agriculture" demonstrates the ways that a theorist can benefit from outreach and engagement, which renews their commitment to ideals such as pluralism and meliorism that remain underemphasized in the scholarly discourse. Other chapters model the kind of participation to which this volume is dedicated; Justine William's "Building Community Capacity for Food and Agricultural Justice: Lessons from the Cuban Permaculture Movement" attests to the value of critical social science when elaborating theoretical notions such as "community capacity" and "skillful disclosure" in a way more sensitive to these notions' construction in particular communities; Melanie Bowman's "Institutions and Solidarity: Wild Rice Research, Relationships, and the Commodification of Knowledge" cautioned against conceptions of knowledge that stand in the way of collaboration, or worse, allow conversation to act as colonization. All of these chapters model engaged scholarship with the underlying commitment to renegotiate epistemic authority and power differentials in a way that promotes social justice (Healy 2001). In the world of beastly food justice scholarship and practice, all of the chapters in this volume reflect different ways of existing between theory and practice. Though neither fish nor fowl, such beasts are a necessary part of the ecology of food justice.

Success Conditions for Beastly Food Justice Projects

Part of the trouble with being a beast, especially a rather chimerical beast, is that the standards for evaluating success are still embryonic. Such standards are an invaluable part of growth, for they provide goals to which to aspire and benchmarks to measure progress. Fish and fowl benefit from clear procedures for assessing success; a good handbook is one that clearly lays out a procedure for conducting practical endeavors, while a good theoretical or scientific manuscript is one that builds incrementally on the work of one's peers, to the satisfaction of one's peers. Both sorts of standards, though, assume a context different from the habitat of food justice work—practitioners and theorists of food justice lament, respectively, the diverse goals that complicate clear practical guidelines and the rough ground that complicates the applicability of theory. This unique context is the motivation for this volume, and as the introduction details, it is a context that informs the structure and composition of the volume and the workshops that precede it. Here we want to take a moment to articulate success conditions for such beastly projects. Our method and conclusions are still nascent, but in the eyes of most participants, the volume and workshops have been a success, and we think it important to locate some reasons for this sentiment. We focus here on three reasons why we think of this effort as a success: the volume and workshops have produced lively conversation, built community, and supported the individual achievement of participants.

Lively conversation: Much has been made of how disciplinary specialization creates silos, wherein researchers with different interests pursue their passions in relative isolation (Norton 2005; Klein 1996). Then there is the equally pronounced rift between academic work and practical work (Frodeman 2014). These rifts challenge lively conversation, in part because we rarely know enough about the goings-on of other fields to recognize the significance of their work for our own. Indeed, good scholarly manners usually demands that we evaluate others' work with respect to the norms of the others' community, and while we support such etiquette, it has its downside. In editing this volume, and in curating workshops on food justice, we often found ourselves posing methodological questions rather than inviting scholars and practitioners to expound on the broader meaning of their work.

The design of this volume, and sessions at the workshop, made a point to bring together philosophical, scientific, and practical projects that share common themes or address common problems. Convening such interdisciplinary assemblages is successful when conversation coalesces around the space between fields. Philosophers can draw our attention to the ways that concepts work in a particular social scientific investigation, pointing out how disciplinary jargon may have deviated from ordinary language. Scientists can uncover mechanisms that complicate practice, helping practitioners develop projects that can cope with flexible and adaptive social systems. Activists and practitioners provide invaluable expertise about how we realize our ideals in practice, reporting back to the unintended consequences of our theories or the pernicious problems that our theories leave unaddressed. In our experience, beastly projects fail when theorists are willing to "bite the bullet" and abide by their theories despite unattractive consequences, or when practitioners wave off theoretical objections because they would never work "in the real world." They succeed when individuals learn from one another, taking the time to understand exactly why a practical problem reveals a weakness in a theory or how an ideal like food sovereignty can be worked into local projects and policies.

Community building: In order to engage in lively conversation, you really need to know a little bit about your interlocutor's field of study, and it also helps to know a little bit about your interlocutor's specific interests and assumptions. This takes time that is rarely available when attending a brief workshop. We were fortunate that many of the participants at the workshop returned year after year, learned a bit about a wide range of issues, and modeled the sort of critical engagement constitutive of lively conversation. Not incidentally, their interactions with one another also informed their own work, and many of the chapters of this volume were authored by workshop participants who returned year after year. Building a community pays dividends, but it is also worth emphasizing that community building is a success in its own right. Projects like this don't just introduce future collaborators to one another, but they create collaborators out of one another, creating spaces for individuals to appreciate the significance of their work for others', and others' work for their own.

Building a community is difficult in the first place, but building the right sort of community is a further mark of success. We were careful in compiling this volume to discourage polemics, and in convening the workshops, we encouraged participants to present work-in-progress that they were sincerely open to revising, perhaps dramatically, in light of the feedback that they received. Diversity, both in terms of discipline and form of engagement, and in terms of social categories such as race and gender, was another ideal that a successful project would live up to. Not surprisingly, the community we built skewed academic, and at least initially, skewed philosophical, though we did succeed to spotlight philosophy, science, and practice in this volume and in recent workshops. Perhaps also of little surprise, the volume and workshops struggled to promote diversity with respect to multiple social categories (progress has been made on this front, but hardly success). We continue to work toward a more diverse and inclusive community, as it is a clear success condition for bridging the theory and practice of food justice.

Individual achievement: Though lively conversation and community building accord with the general ethos of this volume, we would be remiss if we didn't also emphasize success conditions that are more individualistic. Our experience editing this volume and organizing workshops brings into high relief the specific material constraints faced by both theorists and practitioners. In order to sustain community goals like lively conversation and mutual understanding, participants do need to benefit individually. It was often the workshop participants who had found success publishing articles on which they'd received feedback who returned the next year. Volumes like this one, conference proceedings, or special issues of journals represent alternative venues to promote individual success.

Such tangible compensation is difficult to secure for practitioners. Though we were thrilled to feature the stories of four participating organizations through the vignettes at the opening to each section, we acknowledge that publications don't carry the same currency for practitioners and activists. Frankly, edited volumes and interdisciplinary workshops are not the most suitable spaces to support activists' and practitioners' individual achievement. Many such individuals reported an interest in attending the workshop, but simply could not justify attendance to their offices. Our tack for navigating these sorts of constraints was to identify barriers to attendance and to soften these barriers whenever possible. Dates were selected to accommodate local organizations' availability, registration fees were waived for non-academics, and site visits were arranged so that local organizations could participate in dialogue even when they couldn't get away from their workspaces or farms. However, the elimination of barriers is not the same as providing compensation. The most tangible compensation for many non-profit organizations and other practitioners is the financial support necessary to sustain their work. While our particular food justice beast falls short with respect to such a standard, kindred projects should aspire to coordinate grant writing sessions or related collaborations between theory and practice.

We would have a hard time imagining a successful project that does not facilitate lively conversation, build a diverse and supportive community, and support the individual success of participants facing very different material constraints. But there are other reasons to think the project a success that are too frequently neglected. Workshops should be fun—there should be time to socialize with old and new friends, and to deliberate over the novel ideas discovered over the course of the day. It helps to have good food to nourish these deliberations; after all, it is a workshop about food. In a similar vein, the most rewarding aspects of editing this volume has been the discussions about the chapters through which they were revised and polished. Academic writing can be a slog, and often the only feedback you receive is the dry reviews dished out by overworked referees. Thoughtful and constructive criticism is not just a way to certify the quality of scholarly work; it is a way to grow as individuals, and to learn how one's ideas are being received by those whose judgment one trusts. Beastly food justice projects succeed because they produce lively conversation, community, and individual achievement, and this success is nourished by fun and by thoughtful criticism.

References

Food Tank. 2016. https://foodtank.com/. Accessed 17 Jan 2017.

Frodeman, Robert. 2014. *Sustainable Knowledge: A Theory of Interdisciplinarity*. London: Palgrave MacMillan.

Healy, Karen. 2001. Participatory action research and social work: A critical appraisal. *International Social Work* 44 (1): 93–105.

Klein, Julie Thompson. 1996. *Crossing Boundaries: Knowledge, Disciplinarities, and Interdisciplinarities*. Charlottesville: University Press of Virgina.

Long, Christopher. 2013. What is Public Philosophy? cplong.org.

New Ethics of Food. 2017. http://foodethics.publicphilosophyjournal.org. Accessed 17 Jan 2017.

Norton, Bryan G. 2005. *Sustainability: A Philosophy of Adaptive Ecosystem Management*. Chicago: University of Chicago Press.

Whyte, Kyle, and Paul Thompson. 2012. Ideas for how to take wicked problems seriously. *Journal of Agricultural and Environmental Ethics* 25 (4): 441–445.

© Springer International Publishing AG 2017

I. Werkheiser and Z. Piso (eds.), *Food Justice in US and Global Contexts*,
The International Library of Environmental, Agricultural and Food Ethics 24,
DOI 10.1007/978-3-319-57174-4

CPSIA information can be obtained
at www.ICGtesting.com
Printed in the USA
BVOW06*0815070817
491357BV00001B/1/P